DESIGN MANAGEMENT

CREATE, DEVELOP, AND LEAD EFFECTIVE DESIGN TEAMS

Andrea Picchi

**Foreword by Professor Clive Grinyer,
Head of Service Design, Royal College of Art, London**

Apress®

Design Management: Create, Develop, and Lead Effective Design Teams

Andrea Picchi
London, UK

ISBN-13 (pbk): 978-1-4842-6953-4   ISBN-13 (electronic): 978-1-4842-6954-1
https://doi.org/10.1007/978-1-4842-6954-1

Managing Director, Apress Media LLC: Welmoed Spahr
Acquisitions Editor: Shiva Ramachandran
Development Editor: James Markham
Coordinating Editor: Jessica Vakili

Distributed to the book trade worldwide by Springer Science+Business Media New York, 1 New York Plaza, New York, NY 100043. Phone 1-800-SPRINGER, fax (201) 348-4505, e-mail orders-ny@springer-sbm.com, or visit www.springeronline.com. Apress Media, LLC is a California LLC and the sole member (owner) is Springer Science + Business Media Finance Inc (SSBM Finance Inc). SSBM Finance Inc is a **Delaware** corporation.

For information on translations, please e-mail booktranslations@springernature.com; for reprint, paperback, or audio rights, please e-mail bookpermissions@springernature.com.

Apress titles may be purchased in bulk for academic, corporate, or promotional use. eBook versions and licenses are also available for most titles. For more information, reference our Print and eBook Bulk Sales web page at http://www.apress.com/bulk-sales.

Any source code or other supplementary material referenced by the author in this book is available to readers on GitHub via the book's product page, located at https://github.com/Apress/Design-Management. For more detailed information, please visit http://www.apress.com/source-code.

Printed on acid-free paper

To Simona.
I only know one way to measure time,
with you and without you.

Contents

About the Author

Andrea Picchi is Global Head of Design at Kroo. With a background in cognitive psychology and computer science, he subsequently learned design thinking at the Hasso Plattner d.school at Stanford University and studied its business applications at the MIT Sloan School of Management.

Andrea developed his leadership and management knowledge working with and for companies such as Apple, Google, Samsung, Sony, and Nokia.

His research in human-computer interaction has been used to develop patents for products such as Google Glass, and in the past decade, he has been building, leading, and managing successful teams that created products and services used by millions of people worldwide.

Preface

There has always been a mismatch between what science knows and what business does. This book aims to repair that breach concerning some of the core concepts that underpin design management and leadership practices. This knowledge gap is twofold: first, there is still not adequate comprehension of how design management and leadership interplay, and, second, there are no books available on design leadership and management that go sufficiently deep on these concepts from a neuroscientific, psychological, and social perspective.

This book intends to present these concepts with an adequate degree of formalism without sacrificing a practical outlook that can enrich design managers and leaders at any level in an actionable manner. The Introduction and Part 1 of this book that encompasses the first two chapters primarily elaborate on this interplay, while Parts 2, 3, and 4, with their seven chapters, predominantly explore these essential concepts.

Intended Audience

I wrote this book for design managers and leaders at any level and every designer interested in considering this type of role.

While this book is concentrated on design, all the core concepts presented can be applied to every managerial role irrespective of its operational context. Management and leadership principles and their neuroscientific, psychological, and social aspects are not a function of a specific discipline and do not change across departments.

The content presented in this book also aims to raise the current business maturity in the design community and facilitate the development of designers from junior roles to Chief Design Officer positions.

On the Shoulders of Giants

As you will learn in this book, we do not see the world as it is, but we only perceive it through our reflective assumptions. As human beings, we are merely a function of our past experiences, and the content of this book is no exception.

I have developed my understanding of Neuroscience and Cognitive Psychology at the University of Padova and my affinity with Computer Science at the University of Pisa, where I also taught as an adjunct professor.

After these initial academic formations, my human-centered mindset has been shaped by my experience at the Design School at Stanford University and its business application by the one at MIT Sloan, the business school of the Massachusetts Institute of Technology.

In this book, I condensed two decades of knowledge matured working with and for organizations such as Apple, Google, Samsung, Sony, Nokia, and studios like IDEO.

That practical experience is presented in conjunction with neuroscientific, psychological, and social research findings based on the invaluable work of innumerable researchers. I tried to demonstrate my respect for their contribution by providing continuous references to their work that you can follow to continue your journey into these topics.

The Leadership Tent model and all the interrelated concepts presented in this book are adapted from the original work of John H. Zenger and Joseph R. Folkman, which is based on what is probably the most extensive, reliable, and insightful research on management and leadership available. The core concepts of this research are a function of a comprehensive initiative based on 25,000+ successful managers and leaders and 200.000+ correlated stakeholders composed of their direct reports, peers, and bosses across different industries, including many established organizations in the Silicon Valley.

How the Content of This Book Is Organized

The book is comprised of an introduction and nine chapters divided into four main parts. Each chapter presents specific learning outcomes, following the central ones.

Introduction: Is Management Needed? At the end of this section, you will be able to

- Describe the evolution of management as a practice
- Recognize the interdependence between management and leadership
- Distinguish the attitudinal and behavioral differences between these two mindsets
- Analyze the critical aspects of managing a leader

Part 1: Manage like a Leader

Chapter 1 – The Behavioral Elements of a Design Manager. At the end of this chapter, you will be able to

- Describe the social cognitive neuroscientific elements of leadership and management and analyze the consequences of their imbalance

- Distinguish the behavioral abilities of the design manager, analyze the interrelation between hard skills and soft skills, and assess your perceived managerial inclination

- Discuss the five pillars of high-performing teams and analyze and assess their underpinning social needs

- Identify the behavioral requirements of your working environment and architect an adequate spatial configuration based on the specific needs of your team

Chapter 2 – The Building Blocks of a Design Leader. At the end of this chapter, you will be able to

- Describe the core concepts of design leadership

- Discuss the differentiating design leadership abilities and their underpinning intertwined relationship

- Recognize the substrate of design leadership by describing the four domains of emotional intelligence and its core competencies

- Assess your levels of emotional intelligence and examine how it interlinks with leadership and management abilities

Part 2: Managing Yourself

Chapter 3 – Create Your Developmental Program. At the end of this chapter, you will be able to

- Describe the Four Stages of Contribution model and indicate how the design focus and contribution evolve along the managerial career track

- Explain the context dependency of leadership and discuss how the brain extracts meanings from the physical world, creating the reflecting assumptions used to drive creative and critical thinking

- Create a competency model and assess your current capability levels based on the specific need of the statistical population of your organization

- Assess your current perceptual gap, define a development strategy, and create an actionable developmental plan

Chapter 4 – Establish Your Core Practices. At the end of this chapter, you will be able to

- Describe the executive functions enabled by the brain's prefrontal cortex and discuss how these functionalities support creative and critical thinking

- Connect with your professional purpose and create short-term and long-term objectives

- Align your behavioral configuration with your intentions, strategize your weekly workload accordingly, and maintain a balance between leadership and managerial activities

- Identify and engage with mentors and sponsors based on your professional aspirations, invite reflection, promote experiential learning, and cope with adverse events when necessary

Chapter 5 – Build and Project Influence. At the end of this chapter, you will be able to

- Describe the dimensions and domains of influence, analyze the political abilities necessary to project it, and assess your current level of proficiency in this area

- Analyze the types of social power and identify the social repercussions of their use

- Create the intention to perform a behavior, convert the willingness to act into action, and manage counterarguments

- Build your social system, develop healthy relationships, and maintain healthy connections recognizing and resolving moments of irritation

- Assess and manage your social capital inside and outside your organization and leverage different types of initiatives to expand it

Part 3: Managing Designers

Chapter 6 – Create the Team. At the end of this chapter, you will be able to

- Build and assess trust, develop productive relationships across cultures, and project confidence inside and outside your team

- Create and assess psychological safety, describe the social cognitive neuroscientific elements of fear, and establish the preconditions for continuous learning and intelligent failure

- Describe the principles of effective hiring, create a statistically effective hiring process, and prioritize your approach based on the four possible outcomes of an interview

- Define behavioral standards, write compelling job descriptions, and assess candidates using a statistical approach

- Develop a career framework, onboard new hires, and maximize new team members' engagement

Chapter 7 – Develop the Team. At the end of this chapter, you will be able to

- Create the preconditions for experiential learning and describe the correlation between development effectiveness and discretionary effort

- Distinguish and assess distinct approaches to development and employ them under the different modalities of creative leadership

- Connect with your team, identify their superpowers, and analyze their potential contribution

- Communicate about performance, offer appreciative and constructive feedback, promote development, and minimize social pain

- Coach the team for performance and development, consider the different conative mentalities present within the group, instill a growth mindset, and investigate behavioral and emotional attitudes across cultures

- Leverage nonreactive empathy, employ active listening, and define clear social contracts with the team members

- Delegate work, prevent reverse delegation, promote learning, and manage task complexity

Part 4: Managing Design Teams and Workgroups

Chapter 8 – Cultivate Creative Collaboration. At the end of this chapter, you will be able to

- Describe the social cognitive neuroscientific elements of creativity, analyze the phases of the creative process, and describe the contribution of the prefrontal and parietal cortex of the brain

- Create the precondition necessary to unlock creativity, analyze the behaviors that promote creative problem-solving, and identify the paths of least resistance

- Leverage the three core forms of design leadership, manage creative tension, support experimentation, and craft a learning-oriented design culture

- Frame challenges into actionable problems, optimize the creative outcome, and guide experiments

- Craft compelling stories, lead with a vision, engage stakeholders, and implement meaningful design directions

- Establish a design culture, create rituals, manage creative tension, and assess the emotional journey of the team

Chapter 9 – Optimize Design Operations. At the end of this chapter, you will be able to

- Build the team infrastructure, describe the different types of design value, find and activate the team purpose, and institute social norms

- Increase group cohesiveness and create a healthy, sustainable, and productive environment for diversity

- Establish individual and collective objectives for the team members, set goals, and increase the zone of proximal development

- Support the team activities, create a design strategy, communicate about design, conduct design critiques, and manage creative meetings

- Report to your line manager, nurture collaboration with individual accountability, and integrate intuition into the decisional process

- Scale design management and leadership in your organization

Foreword

Design is often seen as an individual pursuit. Observing problems, analyzing causes, creating solutions, and the craft of design to render, shape, and deliver new ideas are associated as the personal journey of a creative individual. The cult of the hero designer is a difficult one to challenge, and many like to search for the individual star who might be at the center of innovation and creativity.

Whereas the 20th century was indeed the century of hero designers, from Raymond Loewy and Dieter Rams to Jonny Ive, the 21st century might be described as the century of collaboration with individual designers working in teams. As the quantity and scale of challenges that designers are required to solve increases across every sector from digital transformation, engineering innovation, and the design of services, it is the team we now acknowledge and celebrate. Whether Government Digital Services (GDS) in the UK or in-house design teams in financial services companies, healthcare, and technology platforms, we are less likely to associate creativity and transformation with a single individual. Change now comes from the collaboration of many designers who are experts in research, capable of designing the full experience through time and with the ability to navigate through the politics of organizations.

Designers have tools, methods, skills, and craft at their center, but increasingly it is their soft power, their behavior, guile, resilience, diplomacy, and inspiration that are required to achieve truly transformational and effective outcomes. We look for designers who have curiosity, beyond normal reason, to comprehend how we interact and experience a service or product. Designers who have the courage to articulate the reality of that human experience and identify the opportunities for improvement and more successful solutions. Designers who have the spirit of collaboration to facilitate collective endeavor and lead from both within and at the top, to effect positive change.

For this reason, the qualities of management and leadership are increasingly at the forefront of what we require to create effective designers. As design has become an essential resource for organizations to move into the age of digital and service design, the management of creative teams and creative leadership have become essential to success.

But what is special about design management and leadership? Why is it different to mainstream business management?

When I was a young student, I led a rock band. The band consisted of several brilliant musicians skilled in their instruments, brimming with creative ideas and ego, who had to collaborate to make music. As the self-appointed leader, I tried to marshal a single direction for these talented individuals so that our music would please our audience. I learned a lot that was to become useful to me when I stopped being a musician and became a designer. The need for individual passion combined with the ability to collaborate was vital to the design teams I was part of and eventually led. Respect, empathy, a definition of the direction we were traveling on together, and the need to complete and deliver were all just as vital to a design team as they had been to my fledgling band of musicians. I had to learn the hard way, unsuccessfully with music and more successfully in design. The soft and the hard nature of decision-making and inspiring people to their creative heights are both equally essential to make progress. I wish I had had a book that explained to me the way our brains work and how to understand the drivers and leadership required to create a great team.

In reading Andrea Picchi's words, I now have that book. From his experience of being a designer inside teams and then leading them, Andrea captures and brilliantly explains the underlying reasons behind how people think, leaders lead, and, most importantly, how we can combine the hard and the soft elements of leadership to build creative teams that thrive and deliver positive impact.

Our decisions, whether we are designers or not, shape the world we live in. Design is not an option; we are doing it all the time. But we don't always do it consciously or with thought of the impact on our user and customer or those who deliver or are impacted by the products and services we make. We need designers to define that interface along with the exchange of value between any organization and human users. So, in this age of design collaboration, developing and sustaining creative design teams is of such value that we need to think carefully how we use and manage this important resource.

The lessons and wisdom of this book lay out how we can do that. But more than that, the advice and explanation in this book is for every leader. As we move through the age of empathy in our management styles, the essence of management and leadership described here becomes relevant and essential to all people who lead. Our perception of leaders is often based on historical and political examples that, in times of new challenges, serve us less and less well. Now is the time for leadership that embraces our inner strengths and our external responsibilities to shape the world in a human fashion and maintain a livable planet. These are huge challenges for designers and all leaders. This book is a thoughtful, practical, tangible blueprint to design leadership and beyond. I wish I had this book when I was running my band.

—Professor Clive Grinyer

Head of Service Design, Royal College of Art, London

April 2022

Praises for the Book

Design Management: Create, Develop, and Lead Effective Design Teams meticulously references and synthesizes research findings from neuroscience, psychology, sociology, and management and offers valuable insights, frameworks, and practical approaches that both new and experienced leaders can apply to their everyday challenges.

—Andy Polaine

Co-author of *Service Design*

Former Group Director of Design at Fjord

In *Design Management: Create, Develop, and Lead Effective Design Teams*, Andrea leads us through a comprehensive and detailed range of the issues encountered by any design leader, coupling those challenges with proven strategies and best practices for tackling them. Taken together, he's given us a set of frameworks, teachings, and ideas that are a valuable and welcome resource for any creative leader working to build a more thoughtfully designed world.

—Bob Baxley

Senior Vice President of Design, Thoughtspot

Former Head of Design at Pinterest and Director of Design at Apple and Yahoo

In times of rapidly changing realities, complexities increase, and design managers will be critical to the success of their design teams, who are expected to drive value and progress. *Design Management: Create, Develop, and Lead Effective Design Teams* will share insights and frameworks to shape the appropriate capabilities, behavior, and mindset to blend design management and leadership into a symbiotic role that is both inspirational while delivering design excellence. A must-read for designers with an ambition to advance their careers.

—Eric Quint

Author of *Design Leadership Ignited*

Former Chief Brand and Design Officer at 3M

Design Management: Create, Develop, and Lead Effective Design Teams is full of tangible and actionable advice for design leaders wanting to steer their teams toward success. Andrea's thinking will be shaping design teams for years to come.

—Grace Francis

Global Chief Creative and Design Officer, WONGDOODY

Former Chief Experience Officer at Karmarama and Droga5

Design Management: Create, Develop, and Lead Effective Design Teams provides academic theories and practical recommendations for design leaders, and the readers can get both the Why and the How. It can benefit both people who'd like to get into design management and mature design managers with excellent guidance to establish and elevate their leadership in the real world.

—Joann Wu

Head of Design, Uber

Former Head of Design at LinkedIn

Whether you are any kind of design practitioner or are currently at any stage of design career development, *Design Management: Create, Develop, and Lead Effective Design Teams* covers in detail how to develop the psychological qualities and practical experience of being a good design manager. It is an indispensable guidebook for future design leaders who are willing to forge strong leadership.

—Shane Lee

General Manager, Design Innovation Center of TCL Technology Group

Former Head of Design at Samsung and Design Director at Google and Motorola

In today's working environment, managing like a leader by combining managerial and leadership abilities is essential to unlocking high levels of team performance. *Design Management: Create, Develop, and Lead Effective Design Teams* accurately discusses the neuroscientific, psychological, and social aspects of those two disciplines within a creative environment and offers tools, frameworks, and practical examples to develop these competencies following a systematic program. I wish I had it the moment I started developing my career into a managerial role; it's an essential read for design leaders at any level.

—Silke Bochat

Head of Design, Colgate

Former Head of Design at PepsiCo

A must-read for the aspiring or existing manager. Andrea's book, *Design Management: Create, Develop, and Lead Effective Design Teams*, breaks down foundational concepts and skills necessary to "manage like a leader" and provides helpful tips and tactics for transitioning to and performing effectively in a design management role.

—Willy Lai

Chief Design Officer, Haggleland

Former VP of Design at Macy's and Director of UX at Apple and Samsung

Acknowledgments

I spent circa two years writing this book, and I would like to thank all the individuals who dedicated some of their time to providing early feedback that promoted crucial moments of reflection during the development of the value proposition of this book; these persons are listed in the following in alphabetical order.

Andy Dahley, Head of Design at Facebook; Chris Smith, Design Director at Netflix; Erez Kikin, Design Director at Microsoft; Erico Fileno, Head of Innovation at Visa; Greg Storey, Design Director at InVision; Gülay Birand, Head of Platform Design at Miro; Jamin Hegeman, VP of Design at Capital One; Lane Kuhlman, Staff Designer at Google; Lee Sun Ooi, Design Manager at Motorola; Mark Vitazko, Senior Designer at Facebook; Moni Wolf, Principal Design Director at Microsoft; Paridhi Verma, Design Manager at Google; Paul Parson, Senior Design Manager at Microsoft; Thom Franey, Head of Design at Paypal; Tom Broxton, Design Director at Google; Umberto Abate, Senior Design Director at Citrix.

A special thanks to Clive Grinyer for writing the foreword of this book and being a fantastic role model for myself and many other designers across the globe: since the early 1990s, when he co-founded Tangerine with Jonathan Ive and Martin Darbyshire, envisioning the future of portable computers for Apple, through the times at the Design Council, where he contributed to articulate the design process model "double diamond," to today, at the Royal College of Art, where he created and developed the service design program to an unprecedented level.

Let Me Know What You Think

Design management and leadership is a life-long journey. Writing a book on such a nuanced subject within a predefined number of pages represents a challenge that requires compromises in terms of content.

The content presented in this manuscript reflects my firm conviction of what a design manager and leader should be: a person with integrity, strong technical abilities, but also a deep comprehension of pivotal aspects of neuroscience, psychology, and sociology.

If you feel that something relevant to this book's vision has not been included, and you want to see that topic included in a potential second edition, don't hesitate to get in contact. I will also appreciate any feedback from you on the current content that can help me reflect on my work and refine my thinking.

You can contact me at

andrea@andreapicchi.it

www.linkedin.com/in/andreapicchi/

Introduction: Is Management Needed?

In 2002, Google experimented with a completely flat organization. They eliminated engineering managers to break down barriers to promote rapid idea development and replicate the collegial environment that the two co-founders, Larry Page and Sergey Brin, enjoyed in graduate school.[i] Google at the time was a company built by engineers for engineers, and the shared sentiment on management was apparent and transparent: managers don't understand our work, they hold us down, and they're power-crazed bureaucrats.[ii] That experiment lasted only a few months. They relented when an impressive number of Googlers went directly to Page with questions about interpersonal conflicts and other nontechnical issues. The intangible value provided by managers became suddenly visible.

Despite the disastrous result of that attempt, that antimanager sentiment was still going strong among some of Google's leaders and engineers in 2008. As a result of that belief, the company launched *Project Oxygen* to use data to formally prove that managers didn't have any part in the company's success or, according to the project hypothesis, that the quality of the manager does not impact the performance of the team.[iii]

As social animals, the quality of the work that we collaboratively generate is not solely a function of our technical abilities and, more broadly, results-oriented skills, as you will learn in the following chapters. In consonance with this foundational fact of our existence, the result of that research validated the "uncontrolled experiment" effectuated six years prior, corroborating what psychology and sociology literature articulated from many different perspectives for decades: groups can improve their performance under *healthy and inspiring guidance*.[iv] Healthy and inspiring guidance can be achieved by *managing like a leader*, leveraging managerial and leadership capabilities; this is another foundational concept that you will explore in this book. When a designer manages a group like a leader, that person acts as a catalyzer to the team's performance.[v]

In the following two sections, you will explore the principal preconditions that originated the necessity to develop and combine these specific sets of competencies and how this need reshaped the foundation of our profession as designers.

A New Working Landscape

Today's ecosystem is pervaded by systemic challenges that require a change in perspective. Approaching these types of challenges, known as *wicked*,[1] demands an in-depth and comprehensive knowledge domain.[vi] In this scenario, a multitude of expertise and perspectives are necessary to navigate the ambiguity and uncertainty that characterize the problems that originated from those challenges.[vii]

Multidisciplinary teams are now an accepted requirement in modern working environments. This form of social, cultural, and technical multifariousness requires the fulfillment of specific social and cognitive needs to establish healthy and productive dynamics. Psychological and behavioral synchrony is necessary to leverage diversity. Trust and psychological safety are essential to promote creative collaboration. In this book, you will explore these and all the other conditions indispensable to unlock high levels of performance in a design team.

A New Management Age

Management and working classes evolved significantly during the past centuries. From an original idea of Ian MacMillan, Emeritus Professor of Innovation and Entrepreneurship at the Wharton Business School, Rita Gunther McGrath, Professor of Management at the Columbia Business School, proposed that we have seen three "ages" of management since the industrial revolution, with each emphasizing a different theme: execution, expertise, and empathy.[viii] In Figure 1, you can see the timeline underpinning these three periods. Before the industrial revolution, outside the military environment, the concept of what today we call "management" was not professionally present.

---

[1] The term "wicked problem" was first coined by Horst Rittel, design theorist and Professor of Design Methodology at the Ulm School of Design, Germany.

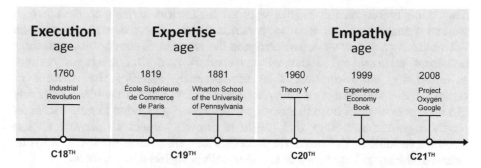

Figure 1. The three ages of management

In the 18th century, with the rise of the industrial revolution and the need to coordinate large organizations, owners needed to depend on others to function at scale; the manager role was born. During that age, the focus was on "execution," competition with other companies was not a factor, and optimization became the primary objective of management.

In the 19th century, management competence began accumulating, and in 1819 the world's first business school was founded in Europe, Paris: the ESCP (École Supérieure de Commerce de Paris). In the following years, other schools started to offer business classes, and in 1881 the Wharton School of the University of Pennsylvania became the first business school on American soil. These chains of events initiated the transition to the second age of management, the "expertise," an era where companies streamlined their operations, and *command-and-control* became the default managerial mindset.

In the 20th century, a new category of workers began to create value by leveraging personal information instead of performing physical tasks. Peter Drucker coined the term *knowledge workers* for this new class, and one year later, in 1960, Douglas McGregor developed the "Theory Y" that shifted the focus away from the command-and-control paradigm, emphasizing the importance of the motivation and engagement of workers. More organizational theorists began to explore these new ideas, and the managerial shift gained further momentum. At the end of that century, in 1999, B. Joseph Pine and James H. Gilmore introduced the concept of *Experience Economy*, completing a shift toward an environment where connecting the company to the customers and the managers to their teams at a human level was a primary intent.

Today, in the 21st century, we operate in the age of "empathy," where the capacity to understand what other individuals are experiencing from within their frame of reference applies to teams and extends to customers.[ix] This age is the era where every significant problem has systemic ramifications and where design managers with leadership capabilities use questions as the atomic unit of exploration.

This book embraces this change without hesitation, striving to illuminate a way to manage like a leader using a mindset that knows when to *lead ahead* with a strong point of view, *lead alongside* the team in direct contact with their activities, or *lead behind* through culture.[2] A role that leverages results-oriented and social-oriented skills concurrently, blending the concept of design leader and design manager into something novel that transcends any old interpretation of both functions. An approach grounded in neuroscience, psychology, and sociology that, while it doesn't expect to provide all the answers, intends to delineate the preconditions that can propel you and your team's thinking to high levels of creative collaboration and performance.

The Interdependence of Leadership and Management

The debate around the interdependence of leadership and management initiated in 1977 when Harvard Business School Professor Abraham Zaleznik published a controversial, groundbreaking article[x] arguing that the theoreticians of scientific management at the time were missing an essential factor defined by the vision, inspiration, and the full spectrum of human drives and desires.[xi] The seminal work of Abraham Zaleznik contributed to surface the attitudinal and behavioral differences between leadership and management, embodied by his famous provoking statement, "leaders have much more in common with artists than they do with managers."[xii] Table 1 captures the foundational concepts of Zaleznik's work.

---

[2] You will explore these three leadership modalities in Chapter 8.

Table 1. Attitudinal and behavioral differences between leadership and management

	Managers	Leaders
Attitudes toward goals	Prefer an impersonal, reactive outlook.	Prefer a personal, active outlook.
	Administer objectives emerged out of necessity.	Articulate strategies out of new visions.
Conceptions of work	Negotiate and balance opposing viewpoints.	Reframe and develop new viewpoints.
	Limit choices, define compromises, and avoid risks.	Increase choices, explore opportunities, and take risks.
Relations with others	Establish unemotional connections.	Establish emotional connections.
	Concentrate primarily on the process of the decisions.	Concentrate primarily on the outcome of the decision.
Sense of self	Come from perpetuating human and economic dynamics.	Come from altering existing human and economic dynamics.
	Support the status quo.	Challenge the status quo.

In 1990, John Kotter, Emeritus Professor at the Harvard Business School, deepened and extended the insights from that publication with an article[xiii] that introduced for the first time the concept that management and leadership mindsets[3] are different but complementary, and that in the management age of empathy, one cannot function without the other. Kotter argued that management is about *coping with complexity*; it brings order and predictability to a situation. Leadership, in contrast, is about *coping with change*. Managers promote stability while leaders press for change, and only organizations that embrace both sides of that dichotomy can thrive in this new era.[xiv] Each mentality involves deciding a course of action, leveraging a network of persons and relationships that can accomplish an agenda, and ensuring that those individuals achieve the desired result. Despite these similarities, each mindset accomplishes these objectives in different ways.[xv] Table 2 juxtaposes the differences between these two attitudinal and behavioral systems.

---

[3] A mindset is a state of mind that influences how individuals conceive and enact their goal-directed activities.

Table 2. Attitudinal and behavioral differences between leadership and management mindsets

Mindset	Management	Leadership
Core activity	Coping with complexity.	Coping with change.
Course of actions	1. Establish an objective.	1. Articulate a vision.
	2. Implement a plan.	2. Create a strategy.
	3. Accomplish a plan.	3. Achieve a strategy.

Establish an objective vs. articulate a vision. A manager mindset focuses on planning short-term intentions allocating resources to support the implementation of those plans. Instead, a leader mentality focuses on setting long-term intentions setting directions to support the creation of those strategies.

Implement a plan vs. create a strategy. A manager mentality focuses on organizing individuals to develop an organizational structure that can execute, monitor, and ultimately accomplish a specific and measurable objective. Instead, a leader mindset focuses on aligning individuals to create a coalition that can understand a vision and commit to the achievement of a strategy.

Accomplish a plan vs. achieve a strategy. A manager mindset focuses on controlling and adjusting by rationally monitoring the results, coping with complexity identifying and resolving deviations. Instead, a leader mentality focuses on motivating and inspiring by emotionally stimulating the coalition, coping with change pushing individuals to overcome barriers.

Manage like a Leader

In order to be a manager that also demonstrates leadership capabilities, you need to embrace both mindsets and understand how to leverage them concurrently. In the age of empathy, your chances of success are connected directly with your propensity to develop both managerial and leadership capabilities. A leader without managerial abilities is solely a dreamer who cannot promote change and materialize visions. A manager without leadership abilities is only a blind executor who cannot navigate complexity and define meaningful objectives.

Should I Be a Manager?

There is no single general answer that can help you to decide if you should pursue a managerial career. Designing your professional life represents a fascinating wicked problem that can only be solved using insights, exploration, and reflection. In that regard, Chapter 4 will help you to find, activate, and reflect on your professional purpose.

If you are contemplating a managerial career instead of continuing on an individual contributor trajectory, the following questions can help you reflect on this decision. If you are already in a manager role, instead, you can jump directly to Chapter 1 and start to learn how to manage like a leader.

The first question can help you explore the "why" aspect behind a job as a design manager and leader, while the subsequent three can help investigate their "what" component. If your answers to these questions are not strongly correlated, you should consider it a red flag. In that case, there may be a misalignment between your unconscious values and your conscious desires.

Why am I considering a role as a manager? Some examples of answers that may suggest the possibility that you may need to revisit your idea of pursuing or accepting a role as a design manager could be

- I want to be the decision-maker. You may be the designated decider under certain circumstances, but your focus must be on other areas of the problem-solving process.

- I want to progress in my career. You may advance in the hierarchy, but that progress is pointless if it does not fulfill your professional purpose.

- I was asked to be a manager. You may have received an offer, but you don't have to necessarily accept it if it does not fulfill your professional purpose.

Do I feel fulfilled in shaping the conditions that enable the work of other persons? As you will explore in different contexts reading this book, when you are in charge of a team, you are evaluated primarily on the success of that given group. For the most part, you contribute via others, and you cannot be perceived as a successful person if your team fails to achieve its objectives. Consequently, your focus must be on optimizing the effectiveness of the *collective intelligence*. If the team struggles with solving problems, you must spend your time nurturing creative collaboration. If the group lack capabilities, you must invest your time coaching and hiring. If the team develops a toxic dynamic, you must use your time investigating the root causes.

Do I feel fulfilled in supporting the conditions that enable the fulfillment of other persons? You are accountable for your team's results, but you are also indirectly responsible for the happiness of its team members. In this scenario, you have to know the designers in your group personally, connect with their professional aspirations, and, when possible, assign them challenges that can allow each individual to feel fulfilled.

Do I feel fulfilled in creating the conditions that enable the connection of other persons? If your team doesn't develop healthy relationships, they cannot collaborate efficiently. In this context, you need to regulate interpersonal relationships and orient those interactions toward the achievement of collective objectives.

If answering these questions did not dissipate all the doubts, the following chapters will help you to acquire enough knowledge to develop an informed opinion on the subject.

If envisioning the answer to these questions instead resonated with your professional purpose, this content presented in this book will provide you with the knowledge and tools necessary to fulfill your aspiration.

Endnotes

i. Garvin, David A., et al. Google's Project Oxygen: Do Managers Matter? Harvard Business School, 2013.

ii. Casserly, Meghan. "Google's Failed Quest To Prove Managers Are Evil – And Why You Should Care." Forbes, Forbes Magazine, 18 July 2013, www.forbes.com/sites/meghancasserly/2013/07/17/google-management-is-evil-harvard-study-startups/.

iii. Garvin, David A., et al. Google's Project Oxygen: Do Managers Matter? Harvard Business School, 2013.

iv. Kotter, John P. John P. Kotter on What Leaders Really Do. Harvard Business School Press, 2004.

v. Kotter, John P. John P. Kotter on What Leaders Really Do. Harvard Business School Press, 2004.

vi. Buchanan, Richard. "Wicked Problems in Design Thinking." Design Issues, vol. 8, no. 2, 1992, pp. 5–21, doi:10.2307/1511637.

vii. Buchanan, Richard. "Wicked Problems in Design Thinking." Design Issues, vol. 8, no. 2, 1992, pp. 5–21, doi:10.2307/1511637.

viii. McGrath, Rita Gunther. "Management's Three Eras: A Brief History." Harvard Business Review, 2 Nov. 2014, hbr.org/2014/07/managements-three-eras-a-brief-history.

ix. Bellet, Paul S. "The Importance of Empathy as an Interviewing Skill in Medicine." JAMA, American Medical Association, 2 Oct. 1991, jamanetwork.com/journals/jama/article-abstract/392335.

x. Zaleznik, Abraham. "Managers and Leaders: Are They Different?" Harvard Business Review, 22 May 2015, hbr.org/2004/01/managers-and-leaders-are-they-different.

xi. Zaleznik, Abraham. "Managers and Leaders: Are They Different?" Harvard Business Review, 22 May 2015, hbr.org/2004/01/managers-and-leaders-are-they-different.

xii. Kotter, John P. John P. Kotter on What Leaders Really Do. Harvard Business School Press, 2004.

xiii. Kotter, John P. John P. Kotter on What Leaders Really Do. Harvard Business School Press, 2004.

xiv. Kotter, John P. John P. Kotter on What Leaders Really Do. Harvard Business School Press, 2004.

xv. Kotter, John P. John P. Kotter on What Leaders Really Do. Harvard Business School Press, 2004.

Manage like a Leader

The Behavioral Elements of a Design Manager

In 2009, Jack Zenger and Joe Folkman published a fascinating study of over 60,000 leaders ideated to capture what identifies a manager as "excellent."[i] The findings showed that if a manager displays strong results-oriented skills, the chance of being seen as "excellent" is 14%.[ii] If a manager demonstrates strong social-oriented skills instead, the possibility of being seen as "excellent" is 12%.[iii] However, if a manager possesses strong results-oriented and social-oriented skills, the likelihood of being seen as "excellent" significantly increases to 72%[iv] (Figure 1-1). These findings must be considered gender-agnostic, because as research literature in this field abundantly demonstrates, male and female leaders generally do not genetically differ in their potential levels of results-oriented or social-oriented abilities.[v]

A. Picchi, *Design Management*, https://doi.org/10.1007/978-1-4842-6954-1_1

Figure 1-1. Correlation between results-oriented and social-oriented skills and the likelihood of being perceived as an "excellent" manager

If we analyze these research results[vi] in association with the four possible behavioral permutations of individuals characterized by results-oriented skills (Ros) and social-oriented skills (Sos) (Figure 1-2), it becomes evident that in the management age of empathy, a person in charge of a team must focus on tenaciously achieving results and purposely managing relationships equally.

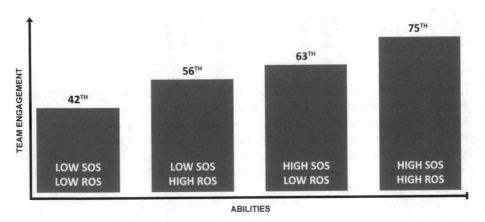

Figure 1-2. Correlation between the results-oriented skills (Ros) and social-oriented skills (Sos) of a manager and team engagement

Subsequently, the study analyzed their 360-degree assessment data and revealed that only 13% of the statistical population ranked high in both results-oriented and social-oriented skills, while 87% had an imbalance[vii] (Figure 1-3).

The data illustrated that 78% of that 87% of leaders with an imbalance of these skills were predominantly results-oriented.[viii] This condition arises from a few interconnected factors. In this chapter, you will explore the behavioral elements of a design manager. You will learn how the neural networks that underlie the human component in design extend directly to the way we deploy leadership and management abilities. You will analyze the management and leadership preconceptions, the consequences of an imbalance between these two competencies, and how to assess your behavioral inclination. You will also examine the social needs of a team and the functional requirements of the working environment learning how to architect an optimal physical and digital configuration based on the needs of a specific group.

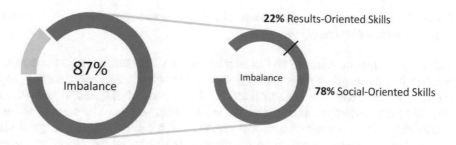

Figure 1-3. Relationship between results-oriented and social-oriented imbalances

The Human Element

Design is the practice of generating value through problem-solving.[ix] This audacious and empowering way to intend this discipline reminds us of the centrality of the human element as the discriminator between "good" design and "bad" design. The neural networks that underlie the human component in design extend directly to the way we deploy leadership and management abilities.

Analytical and Social Networks

Findings from a relatively young field called *social cognitive neuroscience*[x] illuminated that evolution has built our brains with different networks to handle analytical thinking and social thinking.[xi] Brain regions that are reliably associated with analytical, results-oriented tasks, and related cognitive abilities, such as working memory and reasoning, tend to be the outer, lateral surface of the brain. In contrast (Figure 1-4), the regions associated with social-oriented tasks, including oneself, primarily utilize the medial regions of the brain.[xii]

Analytical Network

Social Network

Figure 1-4. Brain activation of the analytical and social network measured via functional magnetic resonance imaging (fMRI)

The pivotal characteristic of this scenario is that the network used for social-oriented tasks tends to be quieted down by other kinds of thinking and relative system, like the analytical one.[xiii] Moreover, these neural networks that support analytical and social thinking often work at cross-purposes: resembling the two ends of a *neural seesaw*.[xiv] Typically, the more engaged is the analytical network, the more disengaged is the social network, and vice versa.[xv] This inversely related interplay between the analytical and social networks impedes our ability to leverage results-oriented and social-oriented skills simultaneously.[xvi] Utilizing both systems concurrently, while possible, does not come naturally, and it requires active training.[xvii]

Management and Leadership Preconceptions

Robert Lord, Professor of Psychology at the University of Akron, studied perceptions of leaders for decades and surfaced the difference in general opinion between analytical and social intelligence. His research demonstrated how the current understanding of a leader is still considerably biased by dated mental models that do not include social skills and are pronominally developed around attributes like "intelligence," "dominance," and "masculinity."[xviii]

In this context, social skills are considered "soft skills" in the pejorative sense of the term and unequally important. Furthermore, as a corollary to the previous point, most organizations tend to favor a rational approach to business at the expense of other essential ways of thinking, such as social and often even creative mindsets.[xix] The 13% of managers who exhibit results-oriented and social-oriented behaviors overcome that tendency using the latter to magnify the former: using social skills to amplify the performance of the collective intelligence and, indirectly, the happiness and fulfillment of the team.[xx]

This book openly aims to uplift that 13% by indicating a way to *manage like a leader*: combining leadership and managerial abilities with the intent to drive results and relationships simultaneously, an approach that, ultimately, can make you successful in the management age of empathy. This new breed of design manager with leadership capabilities encapsulates the definition of effectiveness tacitly adopted in this book.

Imbalance Consequences

When a design manager and leader demonstrates an imbalance between results-oriented and social-oriented abilities, that person typically exhibits specific profiles delineated by peculiar behavioral tendencies. In the following list, you can explore these behavioral tendencies and compare them with the characteristics of a balanced attitude.

A person who exhibits a predominantly results-focused behavior

- Does not provide adequate vision and strategic direction for the team

- Spends most of the time on creating deliverables

- Shows little regard for the team's personal and professional needs

- Avoids personal conversations

- Neglects the team's development and focuses on expediency

- Dispenses solutions and commands

- Defends personal points of view obstructing cooperation

- Makes decisions considering primarily results-oriented implications

- Views leadership responsibilities as a distraction

- Does not retain designers and fails to achieve team objectives

A person who exhibits a predominantly social-focused behavior

- Does not provide adequate vision and strategic direction for the organization

- Spends most of the time on developing connections

- Shows little concern for the team's road map and deadlines

- Avoids performance conversations

- Tolerates a low level of performance and focuses on harmony

- Asks questions and opinions
- Invites different points of view and struggles to achieve alignment
- Makes decisions considering primarily social-oriented implications
- Views management responsibilities as an incumbency
- Does not deliver results and fails to achieve company objectives

In contrast, a person who exhibits a predominantly balanced behavior

- Creates a vision for the company and the team
- Balances the time allocated to deliverable creation and team development
- Demonstrates equal interest in projects and designers' needs
- Engages in personal and performance conversation
- Develops collective abilities and pushes individual performance
- Leads with questions and engender inclusivity
- Supports a multitude of perspectives and achieves collective alignment and cooperation
- Makes decisions seeing and considering social- and results-oriented implications
- Believes that leadership and management are equally essential to deliver results
- Delivers results and retains designers achieving company and team objectives

In Figure 1-5, you can see the *Leadership Tent*,[xxi] a model developed by Jack Zenger and Joe Folkman that illustrates the interconnection between results-oriented and social-oriented skills. Chapter 2 will introduce the remaining components of this framework that identify the building blocks of a design manager who also possesses leadership capabilities.

As Tim Brown, former CEO and now Chair of IDEO, reminds us, an idea has value only if implemented.[xxii] As a design manager and leader, you must leverage creative collaboration inside and outside the team to bring concepts to life, and neglecting the development of social-oriented skills can significantly decrease your effectiveness. In the management age of empathy, developing both results-oriented and social-oriented competencies and combining leadership and managerial abilities is the only way to generate sustainable success: for yourself, your team, and your organization.

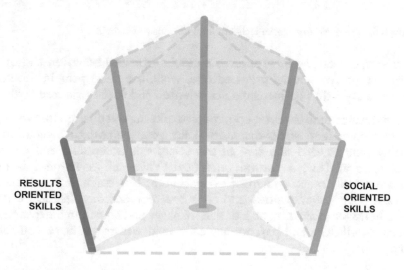

Figure 1-5. The Leadership Tent model

In today's human-centered business landscape, the ultimate aspiration of a person in charge of a group is to achieve the organization's results while retaining the team members. Ultimately, this performance indicator must be the measure of success of every design manager and leader.

Assess Your Behavioral Inclination

A necessary prerequisite for personal development is self-awareness. Before moving forward in your journey, it is beneficial for you to acquire an understanding of your current behavioral inclination. The questionnaire presented in Table 1-1 can help you to self-assess your perceived managerial propensity. The questions span across ten domains and are evaluated using a seven-point scale with three descriptors (Figure 1-6).

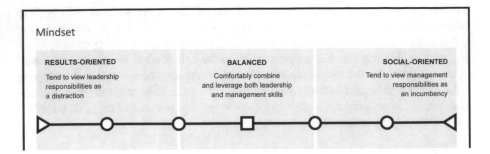

Figure 1-6. Perceived managerial inclination questionnaire example

The descriptor on the left captures a results-oriented behavior, the one on the right captures a social-oriented one, while the descriptor in the center identifies a style that balances results-oriented and social-oriented abilities.

If this is the first time that you commit to exploring the dualism between your results-oriented and social-oriented skills, your behavioral inclination will probably tend toward one side of the scale; it is acceptable and expected because we all have a natural tendency. While self-assessment is not a psychometrically reliable form of evaluation, at this stage, it is sufficient to stimulate your self-awareness and an initial moment of reflection on your natural behavior inclination as a design manager and leader. In Chapter 3, you will use an efficient and psychometrically valid way to evaluate your ability profile.

Table 1-1. Perceived managerial inclination questionnaire

Mindset
  • Tend to view leadership responsibilities as a distraction.
  • Tend to view management responsibilities as an incumbency.
  • Comfortably combine and leverage both leadership and management skills.

Vision and Strategy
  • Tend to struggle to create an adequate vision and strategic direction for the team.
  • Tend to struggle to create an adequate vision and strategic direction for the organization.
  • Comfortably create an adequate vision and strategic direction for the team and the organization.

Time Allocation
  • Tend to spend most of the time on creating deliverables.
  • Tend to spend most of the time on developing connections.
  • Comfortably balance the time allocated to deliverable creation and team development.

(continued)

Table 1-1. (*continued*)

Decision-Making

- Tend to make decisions considering primarily results-oriented implications.
- Tend to make decisions considering primarily social-oriented implications.
- Comfortably make decisions seeing and considering social- and results-oriented implications.

Collaboration

- Tend to defend personal points of view obstructing cooperation.
- Tend to invite different points of view and struggle to achieve alignment.
- Comfortably support a multitude of perspectives and achieve collective alignment and cooperation.

Inclusion

- Tend to dispense solutions and commands.
- Tend to ask questions and opinions.
- Comfortably lead with questions and engender inclusivity.

Communication

- Tend to avoid personal conversations.
- Tend to avoid performance conversations.
- Comfortably engage in personal and performance conversations.

Engagement

- Tend to show little regard for the team's personal and professional needs.
- Tend to show little concern for the team's road map and deadlines.
- Comfortably demonstrate an equal interest in projects and designers' needs.

Coaching

- Tend to neglect the team's development and focus on expediency.
- Tend to tolerate a low level of performance and focus on harmony.
- Comfortably develop collective abilities and push individual performance.

Achievements

- Tend not to retain designers and fail to achieve team objectives.
- Tend not to deliver results and fail to achieve company objectives.
- Comfortably deliver results and retain designers.

7-Point scale with extreme descriptors: Results-oriented, Social-oriented, Balanced.

In the last part of this chapter, on the foundation of another pivotal study, you will continue to explore the concept of connecting results-oriented and social-oriented skills analyzing and scrutinizing what characterizes an effective design team.

The Social Needs of the Team

A social need represents an observable activity motivated by internal stimuli that negatively affect the individual's health when it is not manifested.[xxiii] As a design manager and leader, your responsibility is to establish a social environment for the team that is able to develop a healthy and sustainable level of creative collaboration while optimizing the efficiency of the design operations. Establishing this environment requires the fulfillment of five specific social needs, listed as follows in order of significance:[xxiv]

1. **Psychological safety:** Do team members take risks without feeling insecure or embarrassed?

2. **Dependability:** Do team members count on each other to do excellent work on time?

3. **Structure and clarity:** Do team members clearly understand roles, responsibilities, and decisional processes?

4. **Meaning of work:** Do team members perceive their work as personally valuable?

5. **Impact of work:** Do team members perceive their work as socially significant?

The order of significance emphasizes the necessity to understand the social interdependence between these five social needs. Irrespective of the fact that you are creating a new team or joining an existing one, your plan of action to build the social environment of your group must account for this interdependency; Figure 1-7 depicts that relationship.

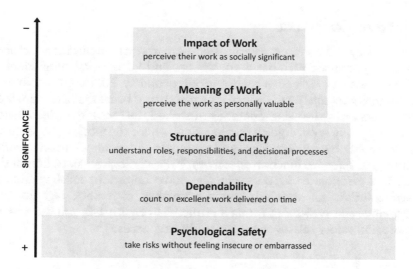

Figure 1-7. The five dynamics of an effective team in order of significance

While the interdependent nature of these five behavioral requirements demands you to leverage concurrent initiatives to maximize the impact of your plan of action, you must always manage them following that specific order. Eliciting the impact and meaning of work requires structure and clarity, which necessitates dependability, which, similarly to many other aspects of creative collaboration, demands psychological safety.

Psychological Safety

In past decades, different researches and studies approached from varied perspectives the challenge of understanding the dynamics of an effective team, and they all came to similar conclusions: before anything else, the members of a group need to feel *socially safe*. The following are some of the research and studies that punctuated the evolution of this fundamental construct.

Social Safeness

In 1965, Edgar H. Schein and Warren G. Bennis initially explored the need for social safety within a team as a critical requirement to drive organizational change.[xxv] Schein and Bennis proposed that fulfilling this need reduces barriers to change within a group and creates a context that encourages provisional tries and tolerates failure without retaliation, renunciation, or guilt.[xxvi]

In 1990, William A. Kahn suggested that feeling safe within a team represents a necessary precondition to develop and sustain engagement with a given group.[xxvii] Kahn defined this condition as the feeling of being able to manifest the true self without fear of negative consequences to social status or professional career.[xxviii]

Collective Construct

In 1999, Amy Edmondson, during her research on medication errors in hospitals, in contrast with her predecessors and for the first time, identified social safeness as a collective need, a group property. Edmondson coined the term *psychological safety* and defined it as the shared belief that the team is safe for interpersonal risk-taking. Edmondson shifted the focus from the individual level to the group level, arguing that in a time where the working environment is permeated by uncertainty and interdependence, a team requires psychological safety to function effectively.[xxix] Psychological safety, Edmondson explains, is essential to support effective communication, collaboration, and experimentation, all social dynamics necessary to unlock high levels of performance and well-being within a group.[xxx] In Figure 1-8, you can see how psychological safety relates to performance standards.[xxxi]

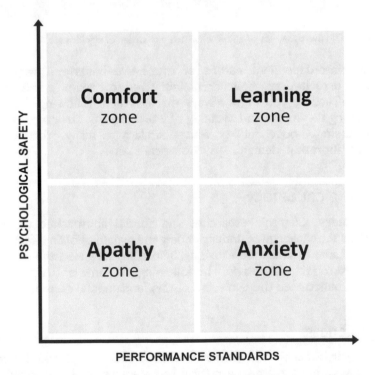

Figure 1-8. Correlation between psychological safety and performance standards

In the lower-left quadrant, when psychological safety and standards are low, the team experiences "apathy." When the team experiences apathy, individuals tend to be disengaged and favor personal interests over collective objectives.[xxxii]

In the upper-left quadrant, when psychological safety is high and standards are low, the team experiences "comfort." When the team experiences comfort, individuals tend to enjoy the work, but the absence of challenges and, therefore, learning decreases their engagement and contribution.[xxxiii]

In the lower-right quadrant, when psychological safety is low and standards are high, the team experiences "anxiety." When the team experiences anxiety, individuals tend to prioritize self-protection and minimize their social interaction to reduce risk.[xxxiv] In Simon Sinek's words, when we don't feel safe within a team, we are forced to expend our time and energy to protect ourselves from each other; and that inherently weakens the organization.[xxxv]

In the upper-right quadrant, when psychological safety is high and standards are high, the team experiences "learning" and consequently high performance. When the team experiences learning, individuals tend to prioritize problem-solving and maximize their social interactions to increase collaboration.[xxxvi]

In Chapter 9, you will learn how to establish objectives and goals, calibrating the inevitable cognitive stress generated by this process with the intent to push the group into the "learning zone."

Psychological Safety and Trust

Framing psychological safety as a collective construct also allowed Edmondson to differentiate it from other individual constructs like trust and, as you will learn in Chapter 6, its interconnected elements: respect and confidence. Edmondson proposed that while both constructs involve a willingness to be selectively vulnerable to others' efforts, they are conceptually and theoretically distinct (Figure 1-9).

Psychological Safety　　　　Trust

Figure 1-9. The contextual distinction between psychological safety and trust

Trust is defined as the expectation that others' future actions will be favorable to one's interests.[xxxvii] Instead, psychological safety goes beyond interpersonal trust and refers to a climate in which individuals are comfortable being and expressing themselves.[xxxviii] In particular, psychological safety is centrally tied to a learning-oriented behavior, while trust is oriented to lower transaction costs and reduce the need to monitor behavior.[xxxix]

Project Aristotle

In 2012, Julia Rozovsky and her team at Google ran Project Aristotle, named after the Greek philosopher and his famous quotation, "the whole is greater than the sum of its parts."[xl] The research team analyzed 180 teams and arrived at the same conclusion, demonstrating that psychological safety is the most critical element of a team, and it underpins its existence; without it, everything else is insufficient.[xli]

Project Aristotle and all the other research before it tell us that a team's "average intellect" is not as essential as the team's ability to understand internal moods, manage emotions, and nurture a psychologically safe collaboration. We have all encountered a group composed of designers without an elitarian education or a prestigious company on their curriculum vitae that generated unmatched value and achieved stellar performances against any formal prediction. In the management age of empathy, "soft" skills have a "hard" impact on the economy of a team.

Sport is a perfect exemplification of this fascinating human dynamic; the best team does not always win, and that is precisely what makes every event exciting: the Florida Marlins winning the World Series of Baseball over the New York Yankees in 2003, Italy winning the World Cup of Football beating Germany in the semi-finals and France in the final in 2006. These are only two memorable examples of underdogs resulting victorious over more accredited adversaries.

Typical signs that your team needs to improve psychological safety are[xlii]

- Fear of communicating about performance and reluctance to offer and receive feedback
- Fear of expressing contrasting ideas and hesitance in asking conflicting questions

The following are some of the things that you will learn in the book that you can use to improve psychological safety:

- Communicate about performance minimizing social pain.
- Support candid communication.
- Clarify the need for voice.

- Frame challenges as learning journeys.

- Demonstrate selective vulnerability.

- Demonstrate situational humility.

The primary responsibility of a person in charge of a team is establishing a psychologically safe environment for the group: creating what Simon Sinek calls a *circle of safety* where the team members feel safe and sense they belong.[xliii] Subsequently, on the foundation of a psychologically safe environment, to drive both results and relationships, you must fulfill four supplementary needs: dependability, structure and clarity, personal meaning of work, and impact of work.

Dependability

The team solves problems as a collective entity; this means that every team member needs to know that they can rely on other members to deliver excellent work on time. Everyone in the team needs to know *who* does *what* by *when* and to rely on that. Creating dependability means establishing the belief that everyone in the group will contribute to the collective effort and the confidence that everyone is accountable and responsible for a specific part of the work.

Typical signs that your team needs to improve dependability are[xliv]

- Insufficient visibility on priorities, objectives, goals, and progress

- Avoidance of ownership in favor of diffusion of accountability

The following are some of the things that you will learn in the book that you can use to establish dependability:

- Establish clear priorities, objectives, and goals for the team.

- Define individual and collective accountability.

- Lead alongside the team to provide support to the problem-solving activities.

Structure and Clarity

The team comprises different roles; this means that every team member needs to understand how their specific part of the work contributes to the collective effort. Every team member needs to have a clear idea of the behavioral expectation associated with their role. The team also needs to know how to make decisions and clearly understand the decision-preparation and decision-making process.

Dependability and structure and clarity are intimately connected but with a significant difference. Dependability is built on the confidence that every team member is professionally capable of providing their contribution. It is a social contract with the team members. Instead, structure and clarity are built on the confidence that every team member is able to provide their contribution. It is a social contract with oneself via the organization. Similarly to the relationship between psychological safety and trust, one is a collective construct, while the other is an individual one.

Typical signs that your team needs to improve structure and clarity:[xlv]

- Absence of clarity around roles and responsibilities
- Inadequate understanding of the decisional process and its rationale

The following are some of the things that you will learn in the book that you can use to establish structure and clarity:

- Define individual and collective roles and responsibilities.
- Frame decision connecting the purpose, vision, and problem.
- Identify and activate the team superpowers.

Meaning of Work

The team has a purpose; this means that every team member needs to believe that the reason to exist beyond making profits of the group, and by extension their part in it, is aligned with their professional purpose. In the case of this social need, irrespective of your effort, individuals with low self-awareness tend to have a limited perception of the meaning of work because they cannot clearly articulate their underlying purpose.[xlvi]

Typical signs that your team needs to improve the meaning of work:[xlvii]

- Purposeless assignment of work driven solely by expertise and workload requirements
- Scarceness of recognition for individual contributions and collective achievements

The following are some of the things that you will learn in the book that you can use to improve the perceived meaning of work:

- Assign work considering individual developmental needs and interests.

- Recognize individual contributions without depreciating collective achievements.

- Create stories that symbolize the purpose of the team.

Impact of Work

The team attacks challenges and solves problems; this means that every team member needs to believe that the solution produced by the group, and by extension their contribution to it, is aligned with their societal purpose. Even in the case of this social need, irrespective of your effort, individuals with low self-awareness tend to have a limited perception of the impact of work because they cannot clearly articulate their underlying purpose.[xlviii]

Typical signs that your team needs to improve the impact of work:[xlix]

- Inadequate visibility on the connection between team purpose and company vision

- Inadequate focus and inability to make significant progress

The following are some of the things that you will learn in the book that you can use to improve the perceived impact of work:

- Create narratives that connect the purpose, vision, and challenges.

- Create strategies that provide focus and orientation to the team.

- Adopt human-centered metrics to evaluate contributions.

Assess the Fulfillment of the Team

Fulfilling the social needs of your team represents a ceaseless endeavor that requires careful consideration and awareness. The questionnaire[l,li] presented in Table 1-2 can help you assess these needs and develop a statistical database that can inform your actions.

Table 1-2. Team social needs questionnaire

Advocacy$^{\alpha}$

    1. I would recommend working in this team to others.

Psychological Safety$^{\beta}$

    1. In this team, if you make a mistake, it is often held against you.^{R}

    2. In this team, everyone can bring up problems and severe issues.

    3. In this team, sometimes members reject others for being different.^{R}

    4. In this team, it is safe to take a risk.

    5. In this team, it is difficult to ask other members for help.^{R}

    6. In this team, no one would deliberately act in a way that undermines my efforts.

    7. In this team, my unique skills and talents are valued and utilized.

Dependability$^{\beta}$

    1. This team has a clear idea of the organizational purpose.

    2. This team has a clear idea of the collective objectives.

    3. This team has a clear idea of the individual goals.

Structure and Clarity$^{\beta}$

    1. This team receives all the information required to work and plan our schedule.

    2. This team obtains expert assistance when something comes up that we don't know how to handle.

    3. This team has visibility on current developments and plans that may affect its work.

Meaning of Work$^{\beta}$

    1. My manager provides the autonomy I need to do my job.

    2. My manager assigns stretch opportunities to help me develop in my career.

    3. My manager gives me actionable feedback regularly.

    4. My manager shows appreciation for my contribution to the team.

    5. My manager shows consideration for me as a person.

Impact of Work$^{\beta}$

    1. This team creates something that has value for the organization.

    2. This team creates something that has value for the people who receive or use it.

    3. This team discusses the impact of the work produced.

Recommendations$^{\chi}$

    • What would you recommend the team keep doing?

    • What would you have the team change?

R Reverse score: Value 1 is converted to a value of 7, 2 to a 6, and 3 to a 5.

α Binary answer with "Yes" and "No."

β 5-Point scale: Never, Rarely, Sometimes, Usually, Always.

χ Open response within a text field.

The questionnaire is evaluated using a seven-point scale where three questions aimed to investigate psychological safety use a reverse score. You can survey your team every 12 months or after significant structural changes. Fulfilling the needs of your team's social environment represents an important milestone on your journey as a design manager and leader, but there is one additional requirement that you must fulfill to unlock your team's full creative potential: the working environment.

The Functional Requirements of the Working Environment

The correlation between human beings and the surrounding environment is profound, especially in the workplace. From a social cognitive standpoint,[lii] it can promote specific social interaction that, among other dynamics, facilitates the development of psychological safety and creative collaboration.

The digital and physical spaces can consciously and unconsciously direct normative behavior,[1] because, in our mind, situational norms[2] are represented as associations between a given environment and an associated set of behaviors.[liii] The mental model associated with a given expected behavior is activated automatically when the achievement of a specific goal drives the act of visiting the associated environment.[liv]

Imagine walking into a library full of persons concentrated with their heads down reading at their desks; your behavioral configuration immediately switches to reduce any possible form of noise.

In these situations, the strongest is the cognitive association between the expected behavior and the environment, and the highest is the likelihood that the person will manifest the behavior.[lv] In Chapter 9, you will learn how to institute social norms to establish a human-centered culture within the team.

Can a team express its full potential in a nonsupportive environment? The short answer is no. Can you run using the wrong shoes? Probably yes. Can you beat your personal best wearing them? Probably not.

The environment delineates a behavioral path of least resistance that establishes and reinforces sustainability in a given conduct.[lvi] Perceptual decisions are always biased by the cost to act, and desired behaviors can be triggered architecting a working environment that makes these decisions less

---

[1] A normative behavior refers to observable and measurable actions aligned with a given set of social norms, standards, and conventions.

[2] A situational norm represents a socially determined consensual standard that indicates how to behave in a particular context (descriptive) and what behaviors are considered appropriate in that given circumstance (prescriptive).

cognitive intense.[lvii] The team uses the digital and physical working environment as a tool to do their work. Like a post-it or a laptop, if it is not the correct tool for the challenge at hand, the outcome suffers; in some cases, so much that it neutralizes every other type of effort.

Identify Your Optimal Spatial Configuration

The working environment of a team is a function of its operational requirements. Different groups have different necessities, and as the person in charge of the team, your responsibility is to identify an optimal spatial structure that can nurture creative collaboration and supports the team's expected behaviors. Based on your organizational context, there are two additional determinants that you may have to consider: access to talents and real estate costs. These are all critical factors in theory, but it is impossible to optimize toward one determinant without penalizing the other two in practice. There are three main spatial configurations to consider, where each one tends to optimize in a specific direction.

Co-located: The team is resident in the same physical location.

Distributed: The team is disseminated across different locations.

Composite: The team is both co-located and distributed.

In Table 1-3, you can see the juxtaposition of these three spatial configurations in relation to creative collaboration, access to talents, and real estate costs. Each structure is rated using a three-point Likert scale.

Table I-3. The characteristics of the three main team structures

Team Structure	Creative Collaboration	Access to Talents	Real Estate Costs
Co-located	High	Low	High
Distributed	Low	High	Null
Composite	Medium	Medium	Low

"Co-located" teams prioritize creative collaboration at the cost of access to talents and real estate costs. In this scenario, if your office is in an area with a large design community like San Francisco or London, you may be able to mitigate this downside, but the real estate costs remain significant.

"Distributed" teams prioritize access to teams leveraging a worldwide sourcing strategy at the cost of creative collaboration. In this scenario, an optimal digital space can partially mitigate this downside, but depending on the nature of your design team, this option may not be possible.

"Composite" teams prioritize a balance between creative collaboration and access to talents while reducing real estate costs. In this scenario, if the team does not operate in a domain that requires exceptional creativity levels or the group expresses the desire to achieve a less demanding life-work balance, this can represent an effective configuration to adopt.

In this hybrid configuration, there are two necessary conditions to consider. The physical space needs to be reframed as a collaborative environment and architected accordingly. Under this configuration, individual tasks are completed remotely, while intensive creative collaboration sessions are conducted in the office. The team's weekly calendar needs to be organized around these two different types of activities. The time spent in the physical space needs to be socially regulated to prevent frictions and conflicts and preserve a climate of trust and psychological safety. Access to the office needs to be punctuated by the requirements of the problem-solving process, not the personal need of an individual. The calendar is the perfect tool to manage the team's remote and in-office presence.

When you need to identify the optimal spatial configuration for your team, you have to reflect on your team and your organization's priority, considering two primary factors that affect culture and performance: the type of work that needs to be produced and the environment required to support that effort. The following two questions can help you to investigate these requirements:

What is the output of the design team?

If you manage and lead an industrial design team, for instance, the group needs to be physically present in the studio to access CNC machines and create physical prototypes, while if you are in charge of a digital design team, they can rely on their laptop to achieve the same objective irrespective of their location.

What is the degree of creative collaboration required?

If you solve uncommonly complex problems, for example, the group needs intensive sessions where they can also leverage social proximity to maximize their creative potential. If you operate in a less ambiguous environment, you are more likely to achieve high performances in any spatial configuration. In Chapter 8, you will explore the social component of creativity in detail.

Identifying the optimal spatial configuration for your team is a challenge that requires experimentation because every organization is unique and diverse groups have different needs. Architecting an appropriate physical and digital space represents another design problem requiring a prototyping mindset to explore possibilities and determine what is functioning as expected and what needs to be reconsidered to improve the outcome.

Architect the Physical Space

Human beings are able to read the physical environment like they read human faces: the form, functionality, and finish of space reflect the culture, behaviors, and priorities of the individuals inhabiting it.[lviii] The physical space is, ultimately, the "body language" of an organization, and it tells a story that reflects its values and group norms and, consequently, engenders specific congruent behaviors.[lix] This peculiarity embodies the reason why manipulating an organization's physical space is often politically challenging because it directly attacks the status quo and everything it represents.

The team processes this environmental information to construct a coherent narrative that serves as the foundation of their experience with the company.[lx] This cognitive appraisal of the working environment also represents a crucial determinant that enables or hinders personal engagement, an antecedent of discretionary effort.[lxi]

Figure 1-10. Hasso Plattner Institute of Design at Stanford University[lxii]

The d.school at Stanford University (Figure 1-10) is an excellent example of a space that engenders psychological safety and, among other attitudes, a bias toward action. The raw materials, such as wood, steel, concrete, glass, and leather, remind us that design is an explorative journey, that the problem-solving process can be convoluted, and invite us to investigate solutions tentatively and share unpolished ideas. The separation between areas dedicated to idea generation and selection reminds us that we need to spend enough time exploring and diverging before focusing on downselecting and converging. We need to protect creative ideas allowing them to flourish before we start to challenge and test them. The adaptability of its spaces and the mobility of its objects like T-Walls, Z-Rack whiteboards, and a variety of different tables and foam cubes remind us that design is a mindset, not a rigid process, that every challenge is different, and that we need to be flexible on our approach.

Every space is unique, and we cannot predict its specific requirements without knowing the exact details of the teams and organizational needs. Despite that, there are some guidelines to consider when you architect the physical environment of the group:

- Allocate an entire macro area to a single team.

- Allocate separate spaces for divergent and convergent thinking to protect creativity.

- Allocate an open area to support collective thinking and encourage collective contribution.

- Allocate a private area to support individual thinking and increase individual contribution.

- Allow space saturation to support visual thinking and optimize the use of working memory.

- Allow space personalization to promote ownership and self-expression.

- Utilize movable elements to allow the team to adapt the spaces to the requirements of the challenge at hand.

- Expose raw materials to promote bias toward actions, exploration, and selective vulnerability.

- Allocate a breakout area between macro areas to create a collision space and promote lateral thinking.

- Create signs to label the different areas to frame the demanded mindset and convey the emotional state and mindset of the environment.

In the next section, you will explore a design template that you can use to facilitate the design of the space, optimizing the interplay between your team's needs and the characteristics of a given area.

The Design Template

Adam Royalty and Dave Baggeroer at the d.school at Stanford University developed a *design template* that deciphers the needs and opportunities presented by an existing area.[lxiii] The template articulates four principles: *Place* and *Properties* to guide the division of the space and *Actions* and *Attitudes* to define the focus of the area.

Places: Broad spatial types that share an overall purpose, such as doorways or openings, including thresholds and transitions, such as hallways.

Properties: Specific aspects of individuals or space that can be enhanced or altered to impact behavior, such as seating modifications to adjust designers' posture and light modifications to affect their mood.

Actions: Specific behaviors and tasks performed by the individuals inhabiting the space; for instance, designers tend to saturate workspaces visually with project inspiration and artifacts.

Attitudes: Specific cultural values and habits that can be perceived by the individuals inhabiting the space; for instance, bias toward Action or radical collaboration.

Space Characteristics

The characteristics of the space must meet the requirements of the challenge at hand. Failing to fulfill this need impedes your team and forces them to waste cognitive resources to overcome an uncollaborative environment, manifesting a specific behavior that research identifies as *adaptation*. Figure 1-11 shows the characteristics of areas generated by the permutations of different space and seating configurations.[lxiv]

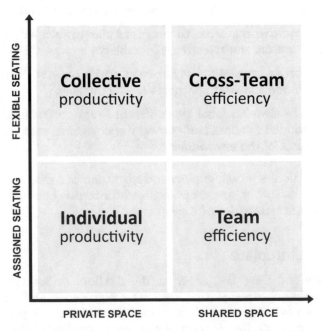

Figure 1-11. Characteristics of space openness and seating flexibility

In the lower-left quadrant, a private space with assigned seats is indicated to support individual productivity; this configuration allows designers to think deep individually.

In the upper-left quadrant, a private space with flexible seatings is indicated to support collective productivity; this configuration allows designers to think deep collectively.

In the lower-right quadrant, a shared space with assigned seats is indicated to support team efficiency; this configuration allows designers to think wide individually.

In the upper-right quadrant, a shared space with flexible seatings is indicated to support cross-team efficiency; this configuration allows designers to think wide collectively.

Moreover, you also have to consider that designers need to switch between individual and collective activities continuously. This requirement implies the necessity to architect a physical environment that facilitates an effortless transition between diverse areas.

Architect the Digital Space

The way the digital space of a team is architected can significantly affect the productivity of a team. Its spatial structure can be particularly decisive if it operates in a distributed or composite spatial configuration. Every space is unique, and it is impossible to predict its requirements without knowing the specific needs of a given team or organization. Despite that, there are three criteria to consider when you architect the digital environment of the group.

Communication: This is how the team exchanges information. The group needs tools that allow everyone to initiate a conversation, assign urgency and reminders to tasks, and complete any interaction that maintains the team continuously connected.

Accessibility: This is how the team retrieves information. The group needs tools that allow everyone to open files such as research insights, whiteboard sessions, and prototypes and consult any data contributing to the problem-solving process.

Collaboration: This is how the team manipulates information. The group needs tools that allow everyone to frame and attack problems, present and discuss concepts, manage and share documents, and initiate any activity that contributes to engaging team members and external stakeholders during the problem-solving process.

Considering these three criteria, in combination with the expected design output and the degree of creative collaboration required, will allow you to select the appropriate software platforms to architect the digital environment of your group.

Endnotes

  i. Zenger, John H., and Joseph R. Folkman. The Extraordinary Leader. McGraw-Hill, 2009.

  ii. Zenger, John H., and Joseph R. Folkman. The Extraordinary Leader. McGraw-Hill, 2009.

  iii. Zenger, John H., and Joseph R. Folkman. The Extraordinary Leader. McGraw-Hill, 2009.

  iv. Zenger, John H., and Joseph R. Folkman. The Extraordinary Leader. McGraw-Hill, 2009.

  v. Van Engen, Marloes L., and Tineke M. Willemsen. "Sex and Leadership Styles: A Meta-Analysis of Research Published." Psychological Reports, vol. 94, no. 1, 2004, pp. 3–18., doi:10.2466/pr0.94.1.3-18.

  vi. Zenger, John H., and Joseph R. Folkman. The Extraordinary Leader. McGraw-Hill, 2009.

  vii. Zenger, Jack, and Joe Folkman. "Extraordinary Leader 360-Degree Survey." Zenger | Folkman, 7 Apr. 2019, zengerfolkman.com/white-papers/el360/.

  viii. Zenger, John H., and Joseph R. Folkman. The Extraordinary Leader. McGraw-Hill, 2009.

  ix. Picchi, Andrea. "The 3 Dimensions of Design: A Model to Scale the Human-Centered Problem-Solving Practice across the Organization." ResearchGate, Jan. 2017, www.researchgate.net/publication/330634631_The_3_Dimensions_of_Design_A_Model_to_Scale_the_Human-Centered_Problem-Solving_practice_across_the_Organization.

  x. Lieberman, Matthew D. "Social Cognitive Neuroscience: A Review of Core Processes." Annual Review of Psychology, vol. 58, no. 1, 2007, pp. 259–289., doi:10.1146/annurev.psych.58.110405.085654.

  xi. Lieberman, Matthew D. Social: Why Our Brains Are Wired to Connect. Oxford University Press, 2015.

  xii. Lieberman, Matthew D. Social: Why Our Brains Are Wired to Connect. Oxford University Press, 2015.

  xiii. Lieberman, Matthew D. Social: Why Our Brains Are Wired to Connect. Oxford University Press, 2015.

xiv. Lieberman, Matthew D. Social: Why Our Brains Are Wired to Connect. Oxford University Press, 2015.

xv. Lieberman, Matthew D. Social: Why Our Brains Are Wired to Connect. Oxford University Press, 2015.

xvi. Lieberman, Matthew D. Social: Why Our Brains Are Wired to Connect. Oxford University Press, 2015.

xvii. Lieberman, Matthew. "Should Leaders Focus on Results, or on People?" Harvard Business Review, 6 Aug. 2015, hbr.org/2013/12/should-leaders-focus-on-results-or-on-people.

xviii. Lord, Robert G., et al. "A Meta-Analysis of the Relation between Personality Traits and Leadership Perceptions: An Application of Validity Generalization Procedures." Journal of Applied Psychology, vol. 71, no. 3, 1986, pp. 402–410., doi:10.1037//0021-9010.71.3.402.

xix. Lord, Robert G., et al. "A Meta-Analysis of the Relation between Personality Traits and Leadership Perceptions: An Application of Validity Generalization Procedures." Journal of Applied Psychology, vol. 71, no. 3, 1986, pp. 402–410., doi:10.1037//0021-9010.71.3.402.

xx. Zenger, Jack, and Joe Folkman. "Extraordinary Leader 360-Degree Survey." Zenger | Folkman, 7 Apr. 2019, zengerfolkman.com/white-papers/el360/.

xxi. Zenger, John H., and Joseph R. Folkman. The Extraordinary Leader. McGraw-Hill, 2009.

xxii. Howard, Suzanne Gibbs, and Tim Brown. "What Leaders Need to Thrive in Uncertainty." Creative Confidence Podcast, 11 Oct. 2019, www.ideou.com/blogs/inspiration/what-leaders-need-to-thrive-in-uncertainty.

xxiii. Greifeneder, Rainer, et al. Social Cognition: How Individuals Construct Social Reality. Routledge, 2018.

xxiv. Rozovsky, Julia. "Re:Work - The Five Keys to a Successful Google Team." Re:Work, Google, 17 Nov. 2015, rework.withgoogle.com/blog/five-keys-to-a-successful-google-team/.

xxv. Schein, Edgar H., and Warren G. Bennis. Personal and Organizational Change through Group Methods: The Laboratory Approach. Wiley, 1965.

xxvi.   Schein, Edgar H., and Warren G. Bennis. Personal and Organizational Change through Group Methods: The Laboratory Approach. Wiley, 1965.

xxvii.   Kahn, William A. "Psychological Conditions of Personal Engagement and Disengagement at Work." Academy of Management Journal, vol. 33, no. 4, 1990, pp. 692–724., doi:10.5465/256287.

xxviii. Kahn, William A. "Psychological Conditions of Personal Engagement and Disengagement at Work." Academy of Management Journal, vol. 33, no. 4, 1990, pp. 692–724., doi:10.5465/256287.

xxix.   Edmondson, Amy. "Psychological Safety and Learning Behavior in Work Teams." Administrative Science Quarterly, vol. 44, no. 2, 1999, p. 350., doi:10.2307/2666999.

xxx.   Edmondson, Amy C. The Fearless Organization: Creating Psychological Safety in the Workplace for Learning, Innovation, and Growth. John Wiley & Sons, 2019.

xxxi.   Edmondson, Amy. Teaming: How Organizations Learn, Innovate, and Compete in the Knowledge Economy. Wiley, J., 2012.

xxxii.   Edmondson, Amy C. The Fearless Organization: Creating Psychological Safety in the Workplace for Learning, Innovation, and Growth. John Wiley & Sons, 2019.

xxxiii. Edmondson, Amy C. The Fearless Organization: Creating Psychological Safety in the Workplace for Learning, Innovation, and Growth. John Wiley & Sons, 2019.

xxxiv.   Edmondson, Amy C. The Fearless Organization: Creating Psychological Safety in the Workplace for Learning, Innovation, and Growth. John Wiley & Sons, 2019.

xxxv.   "Why Good Leaders Make You Feel Safe." Performance by Simon Sinek, TED, YouTube, 19 May 2014, www.youtube.com/watch?v=lmyZMtPVodo.

xxxvi.   Edmondson, Amy C. The Fearless Organization: Creating Psychological Safety in the Workplace for Learning, Innovation, and Growth. John Wiley & Sons, 2019.

xxxvii. Robinson, Sandra L. "Trust and Breach of the Psychological Contract." Administrative Science Quarterly, vol. 41, no. 4, 1996, p. 574., doi:10.2307/2393868.

Design Management

31

xxxviii. Edmondson, Amy C. "Psychological Safety, Trust, and Learning in Organizations: A Group-Level Lens." Trust and Distrust in Organizations: Dilemmas and Approaches, by Roderick M. Kramer, Russell Sage Foundation, 2004, pp. 239–272.

xxxix. Edmondson, Amy C. "Psychological Safety, Trust, and Learning in Organizations: A Group-Level Lens." Trust and Distrust in Organizations: Dilemmas and Approaches, by Roderick M. Kramer, Russell Sage Foundation, 2004, pp. 239–272.

xl. Rozovsky, Julia. "Re:Work - Google's Project Aristotle." Google, Google, Sept. 2012, rework.withgoogle.com/print/guides/5721312655835136/.

xli. Rozovsky, Julia. "Re:Work - The Five Keys to a Successful Google Team." Re:Work, Google, 17 Nov. 2015, rework.withgoogle.com/blog/five-keys-to-a-successful-google-team/.

xlii. Rozovsky, Julia. "Re:Work - The Five Keys to a Successful Google Team." Re:Work, Google, 17 Nov. 2015, rework.withgoogle.com/blog/five-keys-to-a-successful-google-team/.

xliii. Sinek, Simon. Leaders Eat Last. Portfolio Penguin, 2017.

xliv. Rozovsky, Julia. "Re:Work - The Five Keys to a Successful Google Team." Re:Work, Google, 17 Nov. 2015, rework.withgoogle.com/blog/five-keys-to-a-successful-google-team/.

xlv. Rozovsky, Julia. "Re:Work - The Five Keys to a Successful Google Team." Re:Work, Google, 17 Nov. 2015, rework.withgoogle.com/blog/five-keys-to-a-successful-google-team/.

xlvi. Duval, Shelley, and Robert A. Wicklund. A Theory of Objective Self-Awareness. Academic Press, 1972.

xlvii. Rozovsky, Julia. "Re:Work - The Five Keys to a Successful Google Team." Re:Work, Google, 17 Nov. 2015, rework.withgoogle.com/blog/five-keys-to-a-successful-google-team/.

xlviii. Duval, Shelley, and Robert A. Wicklund. A Theory of Objective Self-Awareness. Academic Press, 1972.

xlix. Rozovsky, Julia. "Re:Work - The Five Keys to a Successful Google Team." Re:Work, Google, 17 Nov. 2015, rework.withgoogle.com/blog/five-keys-to-a-successful-google-team/.

l. Rozovsky, Julia. "Re:Work - Guide: Give Feedback to Managers." Google, Google, Sept. 2017, rework.withgoogle.com/guides/managers-give-feedback-to-managers/steps/try-googles-manager-feedback-survey/.

li. Edmondson, Amy. "Psychological Safety and Learning Behavior in Work Teams." Administrative Science Quarterly, vol. 44, no. 2, 1999, p. 350., doi:10.2307/2666999.

lii. Lieberman, Matthew D. "Social Cognitive Neuroscience: A Review of Core Processes." Annual Review of Psychology, vol. 58, no. 1, 2007, pp. 259–289., doi:10.1146/annurev.psych.58.110405.085654.

liii. Aarts, Henk, and Ap Dijksterhuis. "The Silence of the Library: Environment, Situational Norm, and Social Behavior." Journal of Personality and Social Psychology, vol. 84, no. 1, 2003, pp. 18–28., doi:10.1037/0022-3514.84.1.18.

liv. Martin, Garry, and Joseph Pear. Behavior Modification: What It Is and How to Do It. Routledge, 2019.

lv. Aarts, Henk, and Ap Dijksterhuis. "The Silence of the Library: Environment, Situational Norm, and Social Behavior." Journal of Personality and Social Psychology, vol. 84, no. 1, 2003, pp. 18–28., doi:10.1037/0022-3514.84.1.18.

lvi. Picchi, Andrea. "The 3 Dimensions of Design: A Model to Scale the Human-Centered Problem-Solving Practice across the Organization." ResearchGate, Jan. 2017, www.researchgate.net/publication/330634631_The_3_Dimensions_of_Design_A_Model_to_Scale_the_Human-Centered_Problem-Solving_practice_across_the_Organization.

lvii. Hagura, Nobuhiro, et al. "Correction & Perceptual Decisions Are Biased by the Cost to Act." ELife, vol. 6, 2017, doi:10.7554/elife.26902.

lviii. Doorley, Scott, et al. Make Space: How to Set the Stage for Creative Collaboration. John Wiley & Sons, 2012.

lix. Aarts, Henk, and Ap Dijksterhuis. "The Silence of the Library: Environment, Situational Norm, and Social Behavior." Journal of Personality and Social Psychology, vol. 84, no. 1, 2003, pp. 18–28., doi:10.1037/0022-3514.84.1.18.

lx. Bruner, Jerome. "The Narrative Construction of Reality." Critical Inquiry, vol. 18, no. 1, 1991, pp. 1–21., doi:10.1086/448619.

lxi. Zigarmi, Drea, et al. "Beyond Engagement: Toward a Framework and Operational Definition for Employee Work Passion." Human Resource Development Review, vol. 8, no. 3, 2009, pp. 300–326., doi:10.1177/1534484309338171.

lxii. Herman, Jeff. "Stanford D.school | DTH_3125, DTH_3124." Flickr, 11 May 2010, www.flickr.com/photos/jefftopia/4599231453/in/photostream/. Reproduced with permission.

lxiii. Doorley, Scott, et al. Make Space: How to Set the Stage for Creative Collaboration. John Wiley & Sons, 2012.

lxiv. Waber, Ben, et al. "Workspaces That Move People." Harvard Business Review, 31 Oct. 2014, hbr.org/2014/10/workspaces-that-move-people.

The Building Blocks of a Design Leader

More than 10,000 articles have been published on the subject of leadership in the past century, and more than 100,000 books are available on Amazon alone. Whereas some publications on leadership are based on research, the vast majority reflect the personal opinion or the passive observation of practicing managers with leadership capabilities. Despite the extensive literature available, the essence of leadership remains a challenging subject to comprehend. Primarily, discerning the nature of leadership represents a wicked problem that cannot be framed and solved without deliberate research and exploration. Secondarily, the absence of a structured, evidence-based, and contextualized approach limits the opportunity to discern this fascinating human characteristic. Like Gary Yukl, Professor Emeritus of Management at the University at Albany, reminds us, the results of many studies on leadership are contradictory or lack any clear conclusion.[i] In this chapter, you will explore the building blocks of a design leader. You will learn the differentiating design leadership competencies and the interrelation between these abilities. You will examine the business impact of the design leadership capabilities and analyze the social cognitive neuroscientific origin of its emotional substrate. You will also learn how to assess your level of emotional intelligence.

© Andrea Picchi 2022
A. Picchi, *Design Management*, https://doi.org/10.1007/978-1-4842-6954-1_2

Demystifying Design Leadership

In 2009, Jack Zenger and Joe Folkman published another study to complement and advance the work introduced in Chapter 1 and capture what identifies an "excellent" manager.[ii] The research initiative analyzed over 25,000 managers across different industries and identified the top 10% performers through the eyes of over 200,000 stakeholders composed of their direct reports, peers, and line managers.[iii] Subsequently, the study compared the top 10%, the high-performing group with the highest aggregate score, to the bottom 10%, the low-performing group with the lowest aggregate score, with the intent to isolate the capabilities and attributes that separated these two groups.[iv] The findings surfaced that being a successful design manager and leader requires possessing five intertwined *building blocks* (Figure 2-1) that correspond to five clusters of capabilities.[v]

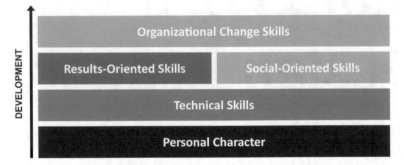

Figure 2-1. The developmental relationship between the five building blocks of leadership

While intertwined in practice, these five clusters of capabilities must be developed following a precise direction.[vi] Your "personal character" represents the foundational layer. On top of that stratum, there are your "technical skills" that define your design area of expertise. From that point, results-oriented and social-oriented skills are leveraged to achieve different types of objectives via your team or workgroup. With a balanced development of these three layers, the organizational change skills allow you to scale change to the entire organization. Before investigating the details of these five clusters of capabilities, and their sixteen differentiating design leadership abilities, we need to introduce four essential concepts about leadership.

Leadership is contextual: It is manifested under the needs and requirements of a specific circumstance.[vii] Being successful in an organization does not guarantee the same level of success in another company.

Leadership is a holistic construct: It does not equal any of its constituent talents in isolation.[viii] Its effectiveness is a function of all the capabilities necessary to achieve a given outcome in a particular context.

Leadership requires followers: It is manifested when other individuals voluntarily participate in the achievement of a given objective. Its effectiveness is a function of other individuals' perceptions.

Leadership is connected to adversities: It is required during difficult moments that necessitate a form of change. The more profound is the challenge, the more distinguished is the opportunity to display leadership abilities. Rudolph Giuliani affirmed his status after the attacks on the World Trade Center, Winston Churchill established his position during World War II, and Steve Jobs consolidated his role returning to Apple in 1996 after 11 years of exile and saving the company from insolvency.

The Differentiating Design Leadership Abilities

The Zenger Folkman study, investigating the five clusters of capabilities, revealed that managers with one strong ability at the 90th percentile would not be rated at the 90th percentile in the overall leadership effectiveness analysis.[ix] Leadership requires a relatively homogenous development of its five clusters of capabilities; otherwise, it's statistically improbable to be perceived as an effective design manager and leader.[x] In Figure 2-2, you can see how managers with one strength are in the 60th percentile, with three strengths are in the 80th percentile, but with five strengths, they are propelled in the 90th percentile.[xi]

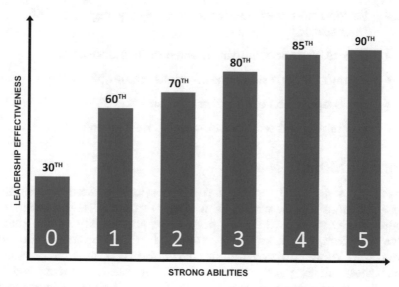

Figure 2-2. Relationship between strong leadership abilities and leadership effectiveness

The following sections will introduce a decomposition of the five clusters of leadership capabilities into their 16 differentiating abilities adapted to the design context.

Personal Character

The personal character (Pch) represents the foundational cluster of abilities that project your essence as a design manager and leader. These competencies influence the atmosphere of the working environment and are so crucial to be considered by many authors, researchers, and successful entrepreneurs the de facto equivalent of leadership. While you can leverage "role power" to accomplish an agenda, without these skills, your chances of building trust, respect, and being accepted as a leader are statistically insignificant. The following are some of the things that promote the display of your character:[xii]

Ability to lead by example demonstrating high integrity and credibility:

- Be approachable by anyone.
- Be adherent to ethical values.
- Be consistent with moral principles.
- Be trustworthy and keep your commitments.
- Be kind and treat everyone with dignity regardless of their status.
- Be inclusive and transparent when making decisions.
- Be emotionally resilient and tenacious during difficult times.
- Be receptive to constructive criticism.
- Be humble and practice self-development openly.

Technical Skills

The technical skills (Tsk) represent the professional cluster of abilities that define your area of expertise as a design manager and leader. These competencies constitute the core portion of your ability profile that also enables the comprehensive development of results-oriented and social-oriented skills. While you provide your contribution via teams and workgroups, without these skills, your chances of building trust, respect, and being perceived as an effective leader are statistically low. The following are some of the things that promote the display of your technical skills:[xiii]

Ability to understand persons, business, and technological needs and implications:

- Understand human-centered implications of business and technology.

- Create an overarching vision for the organization.

- Define a strategic direction for the organization.

- Translate the strategic direction into personal objectives for the team.

- Concurrently use a holistic and operational mindset to navigate challenges.

- Allocate resources to multiple teams adequately.

Ability to solve problems and analyze issues:

- Understand and contextualize problems and issues holistically.

- Promote group cohesiveness.

- Promote data-informed, insight-driven explorations.

- Protect and nurture diverse points of view.

- Protect and encourage divergent ideas.

- Make decisions leveraging analysis, experience, and judgment.

Ability to innovate:

- Scan the horizon for change, monitoring persons, business, and technological trends.

- Protect the creative nature and approach of problem-solving.

- Continuously refine and improve the way of working and its environment.

- Empower persons to find efficient new ways to overcome challenges.

- Establish a culture based on continuous learning.

- Create a healthy, sustainable, and productive environment for diversity.

Ability to seek, process, and implement feedback:

- Reserve time to meet individuals and request feedback.
- Reserve time to analyze and process feedback.
- Reserve time to work on your developmental strategy and plan.
- Reserve time to work on your skills and abilities.
- Reserve time to process adverse events.

Results-Oriented Skills

Results-oriented skills (Ros) represent the cluster of abilities that allows you to undertake the initiative necessary to achieve objectives via teams or workgroups. In collaboration with social-oriented skills, these competencies define your capacity to produce a tangible contribution via teams or workgroups. The following are some of the things that promote the display of your results-oriented skills:[xvi]

Ability to produce outcomes:

- Represent customer's needs within the organization.
- Define and pursue high standards of excellence.
- Connect the design effort to the business impact.
- Connect organizational targets to departmental objectives and goals.
- Set collective objectives and goals for the team.
- Set individual objectives and goals for the team members.
- Define objectives and goals balancing long-term and short-term targets.
- Pursue projects and tasks until completion.
- Reserve time to develop relationship power.
- Reserve time for reflection.

Ability to establish challenging objectives:

- Promote a mindset of continuous improvement.
- Define measurable standards and metrics for yourself.
- Define measurable standards and metrics for the team.

- Understand the individual and collective potential of the team.
- Articulate the meaning and impact of objectives and goals.
- Encourage individuals and groups to go beyond their comfort zone.

Ability to take initiative:

- Align your behavior with your priorities.
- Develop comfort with ambiguity.
- Strategize your time and find your focus.
- Initiate programs, projects, and processes.
- Personally sponsor initiatives and campaigns.
- Take accountability for personal outcomes.
- Take accountability for the outcomes of the team.
- Push yourself beyond your comfort zone to exceed expectations.

Social-Oriented Skills

Social-oriented skills (Sos) represent the cluster of abilities that allows you to engage with the social capital necessary to achieve objectives via teams or workgroups. In collaboration with results-oriented skills, these competencies define your capacity to produce a tangible contribution via teams or workgroups. The following are some of the things that promote the display of your social-oriented skills:[xv]

Ability to communicate powerfully and prolifically:

- Listen actively.
- Articulate new insights.
- Articulate strategic directions.
- Illuminate connections between individual work and collective objectives.
- Exercise influence at every level of the organization.

Ability to inspire and motivate others to high performance:

- Find and activate the team purpose.
- Champion others to grow personally and professionally.

- Reserve time to coach for performance.
- Understand individual abilities, drivers, and ambitions.
- Understand collective abilities, drivers, and ambitions.
- Drive individuals to stretch and exceed their expectations healthily.
- Infuse energy and curiosity into daily activity.

Ability to build relationships and networks:

- Reserve time to know individuals personally and professionally.
- Handle stressful situations constructively and tactfully.
- Balance concern for productivity with sensitivity for personal needs.
- Recognize and reward individual and collective contributions.
- Remain approachable, friendly, and constructive during complaints.

Ability to develop individual and collective skills:

- Make time in your schedule to coach for development.
- Show genuine concern for the work and success of individuals and the group.
- Allocate responsibility considering individual abilities, needs, and passions.
- Promote candidate conversations.
- Provide continuous feedback to individuals and groups.
- Provide an appropriate balance of positive and negative feedback.
- Identify and activate the team members' superpowers.
- Integrate the team members' shadow side.

Ability to foster collaboration and teamwork:

- Build trust and respect.
- Create a psychologically safe climate.
- Institute appropriate social norms.

- Promote a mindset of inclusiveness and cooperation.

- Nurture collaboration with individual accountability.

- Adopt a transparent decision-preparation and decision-making process.

- Adapt your decisional approach to tactical and strategic challenges.

- Develop cooperative working relationships with other groups.

- Ensure healthy collaboration dynamics between groups.

- Create rituals and establish behavioral patterns.

- Frame challenges to face uncertainty with curiosity.

- Guide experiments and support experiential learning.

- Choreograph meetings and conversations.

- Manage creative tension and frictions.

- Manage and resolve conflicts and crises.

Organizational Change Skills

Organizational-oriented skills (Ocs) represent the cluster of abilities that allows you to produce and scale change inside and outside the company. These competencies leverage all the expertise characterizing the underlying layers and can be considered the highest expression of leadership. The following are some of the things that promote the display of your capability in leading organizational change:[xvi]

Ability to champion change:

- Build your social system.

- Develop and project influence.

- Promote a change mindset within the organization.

- Lead and present initiatives to engage collaborative support in the organization.

- Promote the work and achievements of the group in the organization.

- Represent the work and achievements of the group outside the organization.

- Hire and develop the appropriate talents.

Ability to articulate visions and strategies:

- Develop a strategic perspective on the future of the organization.

- Develop a strategy balancing short- and long-term organizational needs.

- Connect the contribution of your group to the organization's strategy.

- Connect organizational needs to departmental priorities.

Ability to connect the group to the outside world:

- Reserve time to discuss persons, business, and technological trends.

- Reserve time to discuss external event engagements.

- Organize internal events inviting external contributors.

- Organize internal events extending the invite to the outside community.

Figure 2-3 represents the *Leadership Tent*, the model introduced in the Zenger Folkman study adapted to the design context.[xvii] This structure represents the intertwined nature of the five clusters of leadership capabilities and their 16 differentiating abilities that separated the top 10% of managers in the study.[xviii]

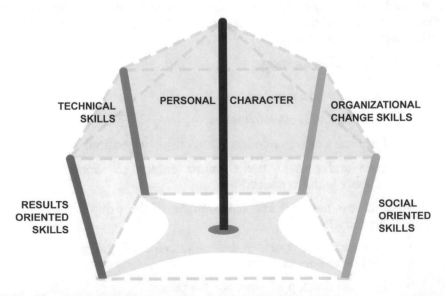

Figure 2-3. The Leadership Tent model adapted to the design context

These clusters of capabilities rarely operate in isolation, and each skill leverages other abilities to function proficiently.[xix] The ability to innovate demands the ability to champion change. The ability to develop others necessitates the ability to build relationships. The ability to produce outcomes requires the ability to inspire and motivate others to high performance. In the next section, you will explore this intertwined nature learning how they interrelate with each other.

The Interrelation of Design Leadership Capabilities

The whole is greater than the sum of its parts; we already saw this Aristotelian concept applied to the behavioral needs of the social environment of a team in Chapter 1. The origin of this effect resides in the diverse and intertwined nature of the elements involved. Leadership expresses itself under the same paradigm. The following are the most relevant interrelations between the five clusters of leadership capabilities identified by the study.[xx]

Personal character is connected to social-oriented skills.[xxi] Personal character (Figure 2-4) is the central behavior of a successful design manager and leader. It is palpable during face-to-face conversations and pivotal in forming healthy relationships and, as a consequence of that, in establishing psychological safety, which in turn impacts results-oriented skills. We naturally do not enjoy engaging with spurious, dishonest, and arrogant personalities, and if we have to do it, we proceed with caution refraining from expressing our true selves. Demonstrating high integrity and credibility relies on the interrelation between personal character and social-oriented skills; this condition elevates this foundational partnership above any other defined by the Leadership Tent model.

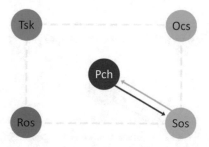

Figure 2-4. Relationship between personal character and social-oriented skills

Personal character is connected to organizational change skills.[xxii] Personal character (Figure 2-5) is pivotal in forming healthy relationships and, in turn, in mobilizing the social capital necessary to achieve organizational change. The stronger the individuals inhabiting your social system in your organization, the more likely these persons will support the changes you propose.

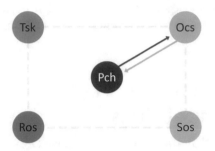

Figure 2-5. Relationship between personal character and organizational change skills

Personal character is connected to results-oriented skills.[xxiii] Personal character (Figure 2-6) is crucial in driving performances and achieving results. Human beings are social animals and the members of a social system and continuously look up to assess the leader's behavior and ability to lead continuously; it is a natural and inevitable behavior. If you exhibit an admirable conduct aligned with the values and beliefs shared within the group, the individuals around you tend to be inspired and motivated to follow the same behavioral patterns. Furthermore, like the Harvard psychologist David McClelland demonstrated with his research,[xxiv] achievement motivation can be developed. As a corollary to the previous point, if the group models the leader's behavior and achieves a positive result experiencing success, their attitude and character will begin to change, increasing the individual and collective achievement motivation. This phenomenon impacts the individual and collective ability of the group to achieve high levels of performance and results.

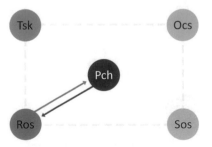

Figure 2-6. Relationship between personal character and results-oriented skills

Technical skills are connected to social-oriented skills.[xxv] Technical skills (Figure 2-7) operate via social-oriented abilities. When you are in charge of a group, you don't deploy your technical skills directly; you actualize and validate them interacting with other individuals. You leverage your social skills to communicate, collaborate, and indirectly contribute via your team or workgroup. Furthermore, combining technical skills and social abilities magnifies the group's perception of the respect and esteem associated with you.

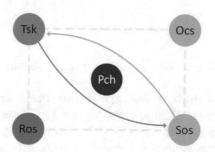

Figure 2-7. Relationship between technical skills and social-oriented skills

Technical skills are connected to results-oriented skills.[xxvi] Technical skills (Figure 2-8) are validated via results-oriented abilities. When you are in charge of a group, you don't manifest your technical skills directly; you demonstrate them leading by example, enabling the success of your team. When you lead by example, displaying efficiency and productivity, you become a role model that enhances the intrinsic motivation of the group and, consequently, its ability to achieve results. Furthermore, leading by example magnifies the group's perception of your technical expertise.

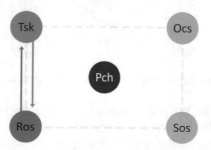

Figure 2-8. Relationship between technical skills and results-oriented skills

Organizational change skills are connected to social-oriented skills.[xxvii] Organizational change skills (Figure 2-9) operate via social-oriented abilities. Leadership culminates in championing change, and profound strategic improvements cannot be delivered without social expertise; radical collaboration and social-oriented skills are necessary. Leveraging your social-oriented skills, you can build the trust and influence required to create the preconditions to promote and establish the behavioral adaptations needed to actualize your vision.

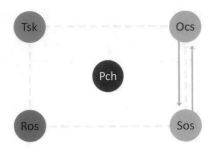

Figure 2-9. Relationship between organizational change skills and social-oriented skills

Organizational change skills are connected to results-oriented skills.[xxviii] Organizational change skills (Figure 2-10) are validated via results-oriented abilities. Leadership culminates in championing change, and accurate tactical improvements cannot be delivered without operational expertise; calibrated achievements and results-oriented skills are necessary. Leveraging your results-oriented skills, you can generate transformational momentum and build a positive reputation for you and the team inside and outside the organization. Furthermore, this widespread belief will help you to build trust and, consequently, influence within the organization.

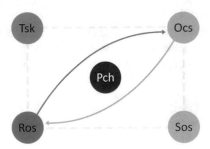

Figure 2-10. Relationship between organizational change skills and results-oriented skills

Results-oriented skills are connected to social-oriented skills.[xxix] Results-oriented skills (Figure 2-11) operate via social-oriented abilities and vice versa. These two clusters of capabilities are equally required to drive results and relationships simultaneously and achieve sustainable success for yourself and your organization. This correlation presents mutual implications in practice. On one side, driving results setting challenging objectives requires healthy relationships to obtain a genuine commitment to those intentions. Healthy relationships are created by establishing bonds leveraging your social-oriented skills. On the other side, driving relationships supporting others requires professional esteem to obtain a genuine commitment to those interactions. Professional esteem is generated by achieving results leveraging your results-oriented skills.

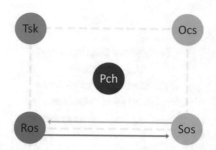

Figure 2-11. Relationship between results-oriented skills and social-oriented skills

One of the critical steps of your developmental journey as a design manager and leader is to learn how to balance results-oriented and social-oriented skills. The following section will introduce the specific behaviors that allow you to develop a balanced behavioral configuration that simultaneously drives results and relationships.

Connect Results-Oriented and Social-Oriented Skills

As a continuation of their research initiative, Jack Zenger and Joe Folkman analyzed the statistical population that ranked high in both results-oriented and social-oriented skills by asking the question:[xxx]

> *What are the observed behaviors of that 13% of managers rated in the top quartile of both results-oriented and social-oriented skills and are 72% more likely to be considered "excellent"?*

The study analyzed 40 behaviors performing a statistical T-Test to contrast both groups' results and subsequently a factor analysis to identify potential commonalities between these conducts interconnected to the five clusters of capabilities.[xxxi] The outcome identified six macro-behaviors that have the capacity to connect results-oriented and social-oriented skills bridging the gap between these two groups of abilities; the study categorized these six conducts as *behavioral bridges* (Figure 2-12).

Results-Oriented   **Balanced**   **Social**-Oriented

Professional Development
Inspiring Stories
High-Performing Environment
Continuous Learning Environment
Human-Centered Environment

Psychological Safety
(Trust and Respect)

Figure 2-12. The behavioral bridges adapted to the design context

During Parts 2, 3, and 4 of the book, you will explore these six macro-behaviors in different contexts using three perspectives: managing yourself, designers, and design teams and workgroups. The following are the "behavioral bridges" identified at the end of the study reframed within a design context.[xxxii] Each conduct reports an example of, among other dynamics, their primary contribution to the achievements of results and the development of relationships:

1. **Create a psychologically safe environment.**

    - **Contribution to results:** A climate that embraces ambiguity and risk

    - **Contribution to relationships:** A climate that embraces candor and creative tension

2. **Create a human-centered environment.**

    - **Contribution to results:** A condition that nurtures curiosity and bias toward action

    - **Contribution to relationships:** A condition that nurtures empathy and cooperation

3. **Create a continuous learning environment.**

    - **Contribution to results:** A condition that supports experimentation and reframes failure

    - **Contribution to relationships:** A condition that supports situational humility and encourages feedback

4. **Create a high-performing environment.**

- **Contribution to results:** A condition that redefines the comfort zone continuously

- **Contribution to relationships:** A condition that engenders a sense of individual pride and collective belonging

5. **Narrate inspiring stories.**

- **Contribution to results:** A situation that provides a sense of purpose and direction

- **Contribution to relationships:** A situation that provides a sense of ownership and camaraderie

6. **Support professional development.**

- **Contribution to results:** A situation that evolves hard and soft skills

- **Contribution to relationships:** A situation that demonstrates altruistic care

Managers that exhibited these macro-behaviors positioned themselves in the 91st percentile in their overall effectiveness.[xxxiii] Developing and nurturing these six conducts and leveraging them to drive results and relationships simultaneously create a *powerful combination* that dramatically enhances your impact within the organization.

The Business Impact of Design Leadership Capabilities

We all experienced the differences between environments shaped by managers with excellent and poor leadership capabilities. The impact of effective leadership on an organization can be profound. Leveraging our social network in the brain, we can recognize exceptional leadership when we encounter it, but can that difference in leadership effectiveness be quantified? As we previously introduced, leadership culminates in championing change and always happens within a given context; therefore, the most reliable way to measure its effectiveness is to capture a snapshot of the environment that it had contributed to develop.

The Zenger Folkman study demonstrated that, when we look at individuals in senior roles with a relatively efficient distribution of hard and soft skills, they are 127% more productive than the average person and infinitely more productive than the 100th person in that curve.[xxxiv] Furthermore, the study revealed a strong statistically significant relationship between leadership effectiveness and a range of desirable business outcomes such as profitability, low turnover, employee commitment, leaving intention, and customer satisfaction[xxxv] (Figure 2-13).

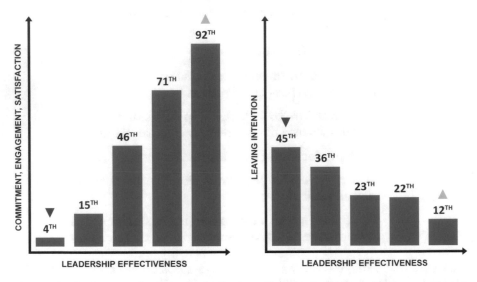

Figure 2-13. Relationship between leadership effectiveness and employee commitment, engagement, and satisfaction (left) and leaving intention (right)

Regrettably, the vast majority of companies implicitly assume a direct connection between hard skills and organizational performances, when in reality, as we learned, hard skills and soft skills continuously leverage one another and never act in isolation. This mindset is one of the prevalent limiting factors that impede many organizations from developing new managers with leadership capabilities and diminish the potential impact of the ones already in the company. In the management age of empathy, the reality is that an organization's success is inherently connected to their managers' leadership proficiency and their inclination to develop and purposefully balance hard and soft skills.

The Substrate of Design Leadership

In the management age of empathy, comprehending the emotional and behavioral substrate of your conduct is essential to lead and manage successfully. Your emotional intelligence (EI) represents the substratum of your leadership and management abilities (Figure 2-14).

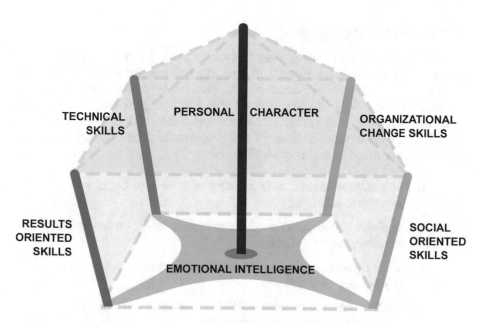

Figure 2-14. Emotional intelligence in relation to the Leadership Tent model

For decades, researchers have studied the reasons why a high intelligence quotient (IQ) does not necessarily guarantee success. In the following section, you can explore the main events that punctuated the evolution of this foundational construct.

Unified Concept

The contextual construction of emotional intelligence started at the end of the 19th century. At that time, psychologists, biologists, and philosophers like Henry Noble Day used the context of emotional intelligence to examine the contemplative and practical unification of feelings and intelligence.[xxxvi] Day argued that the mental process was dominated by a cognitive interconnection between feelings, intelligence, and endeavor.[xxxvii] In the middle of the 20th century, the rise of a psychological perspective known as *humanistic psychology* saw thinkers like Abraham Maslow concentrating their attention on the different contexts within which an individual can develop emotional strength. Subsequently, Howard Gardner introduced the concept of *multiple intelligences*, a notion based on the idea that intrapersonal and interpersonal intelligence is as relevant as the intelligence quotient (IQ).[xxxviii]

Emotional Intelligence

In 1990, psychologists Peter Salovey and John Mayer published their landmark article, "Emotional Intelligence," in the journal *Imagination, Cognition, and Personality*,[xxxix] and five years later, in 1995, inspired by that article, the science journalist Daniel Goleman popularized the concept with the homonymous book collaborating with Richard Boyatzis and Annie McKee to streamline a model composed of four domains.[xl] The four domains of emotional intelligence are as follows, listed in order of significance:

1. **Self-awareness:** Understanding of your emotional and behavioral substrate

2. **Self-management:** Active control of one's emotional and behavioral substrate

3. **Social awareness:** Understanding of others' emotional and behavioral substrate

4. **Social management:** Passive control of others' emotional and behavioral substrate

Each of the four domains is associated with a set of interconnected competencies (Figure 2-15) that, with different degrees of involvement, always collaborate to accomplish a given objective.

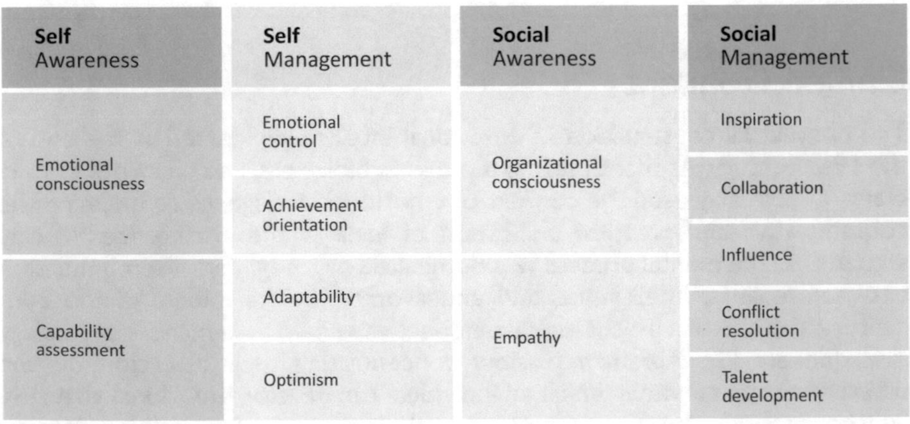

Self Awareness	Self Management	Social Awareness	Social Management
Emotional consciousness	Emotional control	Organizational consciousness	Inspiration
	Achievement orientation		Collaboration
			Influence
Capability assessment	Adaptability	Empathy	Conflict resolution
	Optimism		Talent development

Figure 2-15. The four domains and competencies of emotional intelligence

As you learned in the case of the five dynamics of a team, the order of significance emphasizes a vital concept: your self-development needs to follow a precise order. While the objective is to develop a relatively homogenous range of competencies across the four domains of emotional intelligence, an effective self-development plan conceived to achieve emotional intelligence maturity requires following a precise direction (Figure 2-16), particularly if it needs to recover severe deficiencies.

Figure 2-16. The four domains of emotional intelligence in order of significance

Similarly to psychological safety and personal character in their respective contexts, self-awareness nurtures the other elements of the model and, therefore, possesses more significance.[xli] It is improbable for you to become emotionally intelligent without first developing self-awareness of your emotional and behavioral substrate.

Social Brain

In 2000, Matthew Lieberman and Kevin Ochsner coined the term social cognitive neuroscience[xlii] to explore what they called *social brain* or *social network* (Figure 2-17). This part of the brain is dedicated to social cognition, supporting functions like, but not limited to, thinking about oneself, other individuals, and the relation of oneself to those individuals.[xliii] The findings from social cognitive neuroscience scientifically validated and intellectually consolidated the role of emotional intelligence as a foundational ability.[xliv]

Social Network

Figure 2-17. The part of the brain that is activated during social cognition

Emotional intelligence has the ability to support or hinder every one of the 16 differentiating design leadership competencies playing a crucial role in the success of an individual in charge of a group. If you neglect the development of your emotional intelligence, and the underlying function of your social brain, you undermine the potential growth of your leadership capabilities. In the subsequent sessions, we will explore the details of each emotional intelligence domain, analyzing its typical characteristics and hallmarks.

Self-Awareness

Self-awareness is the capacity of having a deep understanding of one's emotional and behavioral substrate.[xlv] Individuals with significant levels of self-awareness tend to recognize how experiential, behavioral, and physiological elements influence their cognition and use that comprehension to connect meaningfully to their purpose, priorities, and objectives.[xlvi]

Self-aware persons tend to be realistic, achieving a healthy balance between being self-critical and self-complacent, preventing moments of reflection that can potentially damage their self-esteem.[xlvii] Their healthy level of self-esteem allows them to be honest concerning their nature and candid regarding their intentions projecting their perspicuous understanding of their purpose, priorities, and objectives.[xlviii] This condition also enables them to construct a reliable representation of their strengths and weaknesses that they can leverage to support an effective developmental program.[xlix]

The most tangible manifestation of self-awareness is the propensity to invite reflection and reserve time for contemplation to cope with adverse events and ponder significant decisions avoiding impulsive reactions. The capabilities associated with self-awareness are the following:[l]

- **Emotional awareness:** The ability to interpret your own emotions and recognize their impact

- **Capability assessment:** The ability to perceive and articulate one's strengths, limits, and worth

Hallmarks of self-awareness are

- Display self-confidence.

- Define realistic self-assessments.

- Use self-deprecating sense of humor.

Self-Management

Self-management is the capacity of having active control of one's emotional and behavioral substrate.[li] Individuals with solid self-management abilities tend to govern experiential, behavioral, and physiological elements influencing their cognition to propel the behavior necessary to achieve their objectives and remain aligned with their purposes.[lii]

Persons that demonstrate self-management abilities tend to remain adaptable and positive during challenging moments, avoiding jeopardizing the achievement of their objectives by surrendering to irrational and disruptive impulses.[liii] Their healthy way of managing the internal dialogue allows them to naturally project their purpose and priorities, which, in turn, enables them to demonstrate integrity, increasing their credibility and trustworthiness as a leader.[liv] This condition also enables them to engender trust and respect, creating the preconditions for a psychologically safe climate and creative collaboration.[lv]

The most tangible manifestation of self-management is the propensity to preserve intrinsic motivation and maintain high levels of focus. The capabilities associated with self-management are the following:

- **Emotional consciousness:** The ability to keep disruptive emotions and impulses under control and channel them in a constructive direction

- **Achievement orientation:** The ability to be innerly motivated to improve performance and meet internal standards of excellence

- **Adaptability:** The ability to adapt to changing situations and overcome obstacles

- **Optimism:** The ability to expect favorable and desirable outcomes

Hallmarks of self-management are

- Demonstrate integrity and foster credibility.
- Navigate ambiguity.
- Display perseverance.
- Embrace change.

Social Awareness

Social awareness is the capacity of having a deep understanding of others' emotional and behavioral substrate.[lvi] Individuals with significant levels of solid social awareness tend to recognize how experiential, behavioral, and physiological elements influence others' cognition and use that comprehension to connect with them empathetically.[lvii]

Social-aware persons tend to be active listeners considering external perspectives and factoring them into a transparent decision-preparation and decision-making process.[lviii] This condition also enables them to communicate beyond the limited realm of words creating a psychological safety climate and a healthy, sustainable, and productive environment for diversity.[lix]

The most tangible manifestation of social awareness is the propensity to calibrate their communication style to optimize its efficacy, inspiring and motivating the team. The capabilities associated with social awareness are the following:[lx]

- **Empathy:** The ability to sense others' emotions, understanding their perspective, and take an active interest in their concerns
- **Organizational consciousness:** The ability to read organizational currents, decision networks, and politics

Hallmarks of social awareness are

- Display empathy.
- Leverage inclusion.
- Use proficient communication.

Social Management

Social management is the capacity of having a passive control of others' emotional and behavioral substrate.[lxi] Individuals with solid social management abilities tend to govern experiential, behavioral, and physiological elements influencing others' cognition to promote the behavior necessary to achieve the team's objectives and remain aligned with the group's purposes.[lxii]

Persons that demonstrate social management abilities tend to initiate, nurture, and efficiently leverage relationships to collaboratively achieve objectives and, when it is necessary, resolve conflicts.[lxiii] Their healthy way of managing the external dialogue allows them to adapt to different environments and excel in various design contexts, from centralized to distributed teams, from intradepartmental to interdepartmental initiatives.[lxiv] This condition also enables them to build a prolific social system that augments their ability to fulfill ambitious purposes and achieve challenging objectives.[lxv]

The most tangible manifestation of social management is the propensity to develop and retain talents mitigating the primary factor of high turnover within an organization, the absence of personal development. The capabilities associated with social management are the following:[lxvi]

- **Inspiration:** The ability to guide and motivate other individuals

- **Collaboration:** The ability to cooperate with other individuals and groups

- **Influence:** The ability to shape individual and group behavior desirably and favorably

- **Conflict resolution:** The ability to decipher and dissipate misalignments between individuals and groups

- **Talent development:** The ability to bolster the capabilities of individuals and groups through feedback and guidance

Hallmarks of social management are

- Mobilize individuals and groups.

- Use persuasive communication.

- Build successful teams.

- Build and leverage networks.

Conclusively, being aware and in control of the dynamics that shape the way we interact with ourselves and other individuals is an essential milestone on your developmental journey.

Assess Emotional Intelligence

Since psychologists Peter Salovey and John Mayer published their article in 1990, numerous widely used measures of EI have been developed. Within the context of design management and leadership, where the intent is to capture an overall estimation of emotional functioning that can predict professional effectiveness, a *trait-based* measure tends to demonstrate excellent psychometric properties that correlate moderately and meaningfully with a broad set of outcome variables.[lxvii]

The Trait Emotional Intelligence Questionnaire[lxviii] (TEIQue) presented in Table 2-1, created by Konstantinos Petrides, Professor of Psychology and Psychometrics at University College London and the founding director of the London Psychometric Laboratory, can help you to self-assess your level of emotional intelligence.

Table 2-1. Trait-based emotional intelligence questionnaire

Emotional Intelligence

1. Expressing my emotions with words is not a problem for me.
2. I often find it difficult to see things from another person's viewpoint.
3. Overall, I'm a highly motivated person.
4. I usually find it difficult to regulate my emotions.
5. I usually don't find life enjoyable.
6. I can deal effectively with people.
7. I tend to change my mind frequently.
8. I often can't figure out what emotion I'm feeling.
9. I feel that I have several good qualities.
10. I often find it difficult to stand up for my rights.
11. I'm generally able to influence the way other people feel.
12. I often have a gloomy perspective on most things.
13. Those close to me often complain that I don't treat them right.
14. I usually find it difficult to adjust my life according to the circumstances.
15. Overall, I'm able to deal with stress.
16. I usually find it difficult to show my affection to those close to me.
17. I'm normally able to "get into someone's shoes" and experience their emotions.
18. I normally find it difficult to keep myself motivated.
19. I'm usually able to find ways to control my emotions when I want to.
20. Overall, I'm pleased with my life.
21. I would describe myself as a good negotiator.
22. I tend to get involved in things I later wish I could elude.
23. I often pause and think about my feelings.
24. I believe I'm full of personal strengths.
25. I tend to "back down" even if I know I'm right.
26. I don't seem to have any power at all over other people's feelings.
27. I generally believe that things will work out fine in my life.
28. I find it difficult to bond well, even with those close to me.
29. Overall, I'm able to adapt to new environments.
30. Others admire me for being relaxed.

7-Point scale: Strongly disagree, Disagree, Slightly disagree, Neutral, Slightly agree, Agree, Strongly agree.

The Trait Emotional Intelligence Questionnaire also exists in a long form constituted by 153 questions, typically utilized to conduct more formal types of research investigation. Beyond the psychometric limitations of a self-assessment, the short form presented in Table 2-1 can be regarded as sufficient to develop an initial understanding of your level of emotional intelligence maturity and inform the initial phase of your developmental strategy. In Chapter 3, you will familiarize yourself with the concept of 360-degree feedback and a psychometrically valid approach to assess your capabilities using an *ability-based* version of this questionnaire that integrates with a competency framework.

Endnotes

i.   Yukl, Gary A. Leadership in Organizations. Prentice-Hall International, 1998.

ii.   Zenger, John H., and Joseph R. Folkman. The Extraordinary Leader. McGraw-Hill, 2009.

iii.   Zenger, John H., and Joseph R. Folkman. The Extraordinary Leader. McGraw-Hill, 2009.

iv.   Zenger, John H., and Joseph R. Folkman. The Extraordinary Leader. McGraw-Hill, 2009.

v.   Zenger, John H., and Joseph R. Folkman. The Extraordinary Leader. McGraw-Hill, 2009.

vi.   Zenger, Jack, and Joe Folkman. "Key Insights From the Extraordinary Leader: 20 New Ideas about Leadership Development." Zenger | Folkman, 2017, zengerfolkman. com/wp-content/uploads/2019/05/White-Paper-Extraordinary-Leader-Insights-Excerpts-from-The-Extraordinary-Leader.pdf.

vii.   Zenger, John H., and Joseph R. Folkman. The Extraordinary Leader. McGraw-Hill, 2009.

viii.   Zenger, John H., and Joseph R. Folkman. The Extraordinary Leader. McGraw-Hill, 2009.

ix.   Zenger, John H., and Joseph R. Folkman. The Extraordinary Leader. McGraw-Hill, 2009.

x.   Zenger, Jack, and Joe Folkman. "Extraordinary Leader 360-Degree Survey." Zenger | Folkman, 7 Apr. 2019, zengerfolkman.com/white-papers/el360/.

xi. Zenger, Jack, and Joe Folkman. "Key Insights From the Extraordinary Leader: 20 New Ideas about Leadership Development." Zenger | Folkman, 2017, zengerfolkman.com/wp-content/uploads/2019/05/White-Paper-Extraordinary-Leader-Insights-Excerpts-from-The-Extraordinary-Leader.pdf.

xii. Zenger, John H., and Joseph R. Folkman. The Extraordinary Leader. McGraw-Hill, 2009.

xiii. Zenger, John H., and Joseph R. Folkman. The Extraordinary Leader. McGraw-Hill, 2009.

xiv. Zenger, John H., and Joseph R. Folkman. The Extraordinary Leader. McGraw-Hill, 2009.

xv. Zenger, John H., and Joseph R. Folkman. The Extraordinary Leader. McGraw-Hill, 2009.

xvi. Zenger, John H., and Joseph R. Folkman. The Extraordinary Leader. McGraw-Hill, 2009.

xvii. Zenger, Jack, et al. "Leadership Under the Microscope: The Science behind Developing Extraordinary Leaders." Zenger | Folkman, 2017, zengerfolkman.com/wp-content/uploads/2019/06/White-Paper-Leadership-Under-The-Microscope.pdf.

xviii. Zenger, John H., and Joseph R. Folkman. The Extraordinary Leader. McGraw-Hill, 2009.

xix. Zenger, Jack, et al. "Leadership Under the Microscope: The Science behind Developing Extraordinary Leaders." Zenger | Folkman, 2017, zengerfolkman.com/wp-content/uploads/2019/06/White-Paper-Leadership-Under-The-Microscope.pdf.

xx. Zenger, John H., and Joseph R. Folkman. The Extraordinary Leader. McGraw-Hill, 2009.

xxi. Zenger, Jack, and Joe Folkman. "Key Insights From the Extraordinary Leader: 20 New Ideas about Leadership Development." Zenger | Folkman, 2017, zengerfolkman.com/wp-content/uploads/2019/05/White-Paper-Extraordinary-Leader-Insights-Excerpts-from-The-Extraordinary-Leader.pdf.

xxii.   Zenger, Jack, and Joe Folkman. "Key Insights From the Extraordinary Leader: 20 New Ideas about Leadership Development." Zenger | Folkman, 2017, zengerfolkman. com/wp-content/uploads/2019/05/White-Paper-Extraordinary-Leader-Insights-Excerpts-from-The-Extraordinary-Leader.pdf.

xxiii.   Zenger, Jack, and Joe Folkman. "Key Insights From the Extraordinary Leader: 20 New Ideas about Leadership Development." Zenger | Folkman, 2017, zengerfolkman. com/wp-content/uploads/2019/05/White-Paper-Extraordinary-Leader-Insights-Excerpts-from-The-Extraordinary-Leader.pdf.

xxiv.   McClelland, David C. Achievement Motivation Can Be Developed. American Institute of Motivation, 1965.

xxv.   Zenger, Jack, and Joe Folkman. "Key Insights From the Extraordinary Leader: 20 New Ideas about Leadership Development." Zenger | Folkman, 2017, zengerfolkman. com/wp-content/uploads/2019/05/White-Paper-Extraordinary-Leader-Insights-Excerpts-from-The-Extraordinary-Leader.pdf.

xxvi.   Zenger, Jack, and Joe Folkman. "Key Insights From the Extraordinary Leader: 20 New Ideas about Leadership Development." Zenger | Folkman, 2017, zengerfolkman. com/wp-content/uploads/2019/05/White-Paper-Extraordinary-Leader-Insights-Excerpts-from-The-Extraordinary-Leader.pdf.

xxvii.   Zenger, Jack, and Joe Folkman. "Key Insights From the Extraordinary Leader: 20 New Ideas about Leadership Development." Zenger | Folkman, 2017, zengerfolkman. com/wp-content/uploads/2019/05/White-Paper-Extraordinary-Leader-Insights-Excerpts-from-The-Extraordinary-Leader.pdf.

xxviii. Zenger, Jack, and Joe Folkman. "Key Insights From the Extraordinary Leader: 20 New Ideas about Leadership Development." Zenger | Folkman, 2017, zengerfolkman. com/wp-content/uploads/2019/05/White-Paper-Extraordinary-Leader-Insights-Excerpts-from-The-Extraordinary-Leader.pdf.

xxix. Zenger, Jack, and Joe Folkman. "Key Insights From the Extraordinary Leader: 20 New Ideas about Leadership Development." Zenger | Folkman, 2017, zengerfolkman.com/wp-content/uploads/2019/05/White-Paper-Extraordinary-Leader-Insights-Excerpts-from-The-Extraordinary-Leader.pdf.

xxx. Zenger, Jack, and Joe Folkman. "Extraordinary Leader 360-Degree Survey." Zenger | Folkman, 7 Apr. 2019, zengerfolkman.com/white-papers/el360/.

xxxi. Zenger, Jack, and Joe Folkman. "Zenger Folkman Webinar: 6 Keys to Having It All - Delivering Outstanding Results and Creating an Engaged Team." Zenger | Folkman, 7 Aug. 2018, www.youtube.com/watch?v=rDTP0skxP5c.

xxxii. Zenger, Jack, and Joseph Folkman. "How Managers Drive Results and Employee Engagement at the Same Time." Harvard Business Review, 20 Sept. 2017, hbr.org/2017/06/how-managers-drive-results-and-employee-engagement-at-the-same-time.

xxxiii. Zenger, Jack, and Joseph Folkman. "How Managers Drive Results and Employee Engagement at the Same Time." Harvard Business Review, 20 Sept. 2017, hbr.org/2017/06/how-managers-drive-results-and-employee-engagement-at-the-same-time.

xxxiv. Zenger, John H., and Joseph R. Folkman. The Extraordinary Leader. McGraw-Hill, 2009.

xxxv. Zenger, Jack, and Joe Folkman. "Key Insights From the Extraordinary Leader: 20 New Ideas about Leadership Development." Zenger | Folkman, 2017, zengerfolkman.com/wp-content/uploads/2019/05/White-Paper-Extraordinary-Leader-Insights-Excerpts-from-The-Extraordinary-Leader.pdf.

xxxvi. Day, Henry Noble. Elements of Psychology. Nabu Press, 2010.

xxxvii. Day, Henry Noble. Elements of Psychology. Nabu Press, 2010.

xxxviii. Gardner, Howard. Multiple Intelligence: The Theory in Practice. Westview Press, 1993.

xxxix. Salovey, Peter, and John D. Mayer. "Emotional Intelligence." Imagination, Cognition and Personality, vol. 9, no. 3, 1990, pp. 185–211., doi:10.2190/dugg-p24e-52wk-6cdg.

xl. Goleman, Daniel, et al. Primal Leadership: Learning to Lead with Emotional Intelligence. Harvard Business School Press, 2013.

xli. Goleman, Daniel. Emotional Intelligence: Why It Can Matter More than IQ. Bloomsbury, 1996.

xlii. Lieberman, Matthew D. "Social Cognitive Neuroscience: A Review of Core Processes." Annual Review of Psychology, vol. 58, no. 1, 2007, pp. 259–289., doi:10.1146/annurev.psych.58.110405.085654.

xliii. Lieberman, Matthew D. Social: Why Our Brains Are Wired to Connect. Oxford University Press, 2015.

xliv. Lieberman, Matthew D. "Social Cognitive Neuroscience: A Review of Core Processes." Annual Review of Psychology, vol. 58, no. 1, 2007, pp. 259–289., doi:10.1146/annurev.psych.58.110405.085654.

xlv. Goleman, Daniel. Emotional Intelligence: Why It Can Matter More than IQ. Bloomsbury, 1996.

xlvi. Goleman, Daniel. Emotional Intelligence: Why It Can Matter More than IQ. Bloomsbury, 1996.

xlvii. Goleman, Daniel, et al. Primal Leadership: Learning to Lead with Emotional Intelligence. Harvard Business School Press, 2013.

xlviii. Goleman, Daniel, et al. Primal Leadership: Learning to Lead with Emotional Intelligence. Harvard Business School Press, 2013.

xlix. Goleman, Daniel, et al. Primal Leadership: Learning to Lead with Emotional Intelligence. Harvard Business School Press, 2013.

l. Goleman, Daniel. Emotional Intelligence: Why It Can Matter More than IQ. Bloomsbury, 1996.

li. Goleman, Daniel. Emotional Intelligence: Why It Can Matter More than IQ. Bloomsbury, 1996.

lii. Goleman, Daniel. Emotional Intelligence: Why It Can Matter More than IQ. Bloomsbury, 1996.

liii. Goleman, Daniel, et al. Primal Leadership: Learning to Lead with Emotional Intelligence. Harvard Business School Press, 2013.

liv. Goleman, Daniel, et al. Primal Leadership: Learning to Lead with Emotional Intelligence. Harvard Business School Press, 2013.

lv. Goleman, Daniel, et al. Primal Leadership: Learning to Lead with Emotional Intelligence. Harvard Business School Press, 2013.

lvi. Goleman, Daniel. Emotional Intelligence: Why It Can Matter More than IQ. Bloomsbury, 1996.

lvii. Goleman, Daniel. Emotional Intelligence: Why It Can Matter More than IQ. Bloomsbury, 1996.

lviii. Goleman, Daniel, et al. Primal Leadership: Learning to Lead with Emotional Intelligence. Harvard Business School Press, 2013.

lix. Goleman, Daniel, et al. Primal Leadership: Learning to Lead with Emotional Intelligence. Harvard Business School Press, 2013.

lx. Goleman, Daniel. Emotional Intelligence: Why It Can Matter More than IQ. Bloomsbury, 1996.

lxi. Goleman, Daniel. Emotional Intelligence: Why It Can Matter More than IQ. Bloomsbury, 1996.

lxii. Goleman, Daniel. Emotional Intelligence: Why It Can Matter More than IQ. Bloomsbury, 1996.

lxiii. Goleman, Daniel, et al. Primal Leadership: Learning to Lead with Emotional Intelligence. Harvard Business School Press, 2013.

lxiv. Goleman, Daniel, et al. Primal Leadership: Learning to Lead with Emotional Intelligence. Harvard Business School Press, 2013.

lxv. Goleman, Daniel, et al. Primal Leadership: Learning to Lead with Emotional Intelligence. Harvard Business School Press, 2013.

lxvi. Goleman, Daniel. Emotional Intelligence: Why It Can Matter More than IQ. Bloomsbury, 1996.

lxvii. O'Connor, Peter J., et al. "The Measurement of Emotional Intelligence: A Critical Review of the Literature and Recommendations for Researchers and Practitioners." Frontiers in Psychology, vol. 10, 2019, doi:10.3389/fpsyg.2019.011sixteen.

lxviii. Petrides, Konstantinos V, and Adrian Furnham. "Trait Emotional Intelligence: Psychometric Investigation with Reference to Established Trait Taxonomies." European Journal of Personality, vol. 15, no. 6, 2001, pp. 425–448., doi:10.1002/per.4sixteen.

Managing Yourself

Create Your Developmental Program

The 20th century saw the transition from the management age of expertise to the era of empathy; this foundational shift changed the management dynamics and redefined the leadership requirements. Today, being a design manager with leadership responsibilities requires a substantial set of abilities that were inconceivable a few decades ago. These requirements materialize a lifelong journey of constant development that needs to be initiated exceptionally early in your career. Under these conditions, establishing a set of continuous learning habits early in your professional life can set you on a trajectory that can make the difference between expressing your full potential reaching the pinnacle in your field and inhabiting the average dimension of mediocrity. Being the best you can be professionally and ultimately humanly represents a moral imperative that necessitates deliberate and proactive initiatives. You must not leave your chances of success in the hands of fate.

In this chapter, you will learn how to create your developmental program. You will explore the latest insight on leadership development, how your contribution as a design manager and leader evolves during your career,

© Andrea Picchi 2022
A. Picchi, *Design Management*, https://doi.org/10.1007/978-1-4842-6954-1_3

and how that progression affects your developmental needs. You will learn how to assess your competencies and define a leadership archetype based on your organizational needs. You will also learn how to create your personalized developmental program implemented using macro, meso, and micro iterations.

The Four Stages of Contribution

The success of your career is correlated directly to the quantity and quality of value that you add every day in your role; the more value you generate, the higher is the probability that you can be perceived as successful and, consequently, advance in your profession. The discriminant in this scenario is that the way you generate value evolves as you progress through different roles in your career. An effective way to illustrate this phenomenon is to use the Four Stages of Contribution model based on the research of Gene Dalton and Paul Thompson,[i] two former professors at the Harvard Business School. In Figure 3-1, you can see this model adapted to the design context illustrating how the contribution of a designer evolves and traverses four main stages.[ii]

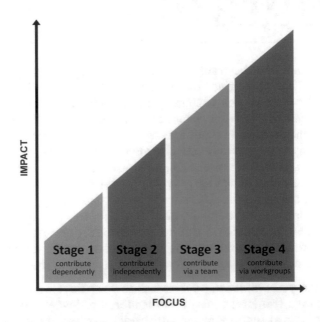

Figure 3-1. The Four Stages of Contribution model adapted to the design context

Stage 1 of contribution is typically inhabited by junior practitioners and characterized by a class of participation that leverages other individuals' knowledge; this is pronominally a dependent type of value generation.[iii]

Stage 2 of contribution is typically inhabited by senior practitioners and characterized by a class of participation that leverages personal knowledge; this is pronominally an independent type of value generation.[iv] Stage 3 of contribution is typically inhabited by Head-level practitioners and characterized by a class of participation that operates via a team; this is pronominally a monodimensional collective type of value generation.[v] Stage 4 of contribution is typically inhabited by Officer-level practitioners and characterized by a class of participation that operates via a workgroup; this is pronominally a multidimensional collective type of value generation.[vi] Figure 3-2 visualizes these different social dynamics.

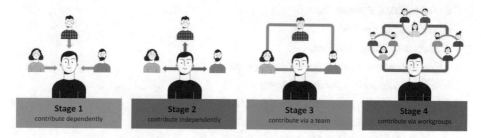

| Stage 1 | Stage 2 | Stage 3 | Stage 4 |
| contribute dependently | contribute independently | contribute via a team | contribute via workgroups |

Figure 3-2. The collaborative dynamics behind the Four Stages of Contribution model

Apart from where explicitly specified, this book will use the terms teams and workgroups interchangeably, referring to the shared peculiarities characterizing these two types of collaborative entities.[1] Irrespective of its domain, a group consists of two or more interdependent individuals driven by needs and goals that cause them to influence each other.[vii] While there is undoubtedly a prevalent job title associated with each of the four stages of contribution, different organizations typically punctuate these four stages using different roles. Furthermore, a job title does not universally capture how an individual contributes to an organization, and it's not uncommon to see an employee generating value across multiple stages.[viii] In Chapter 5, you will learn how building and wielding influence within your organization can help you to go beyond the constraints of your role and the domain of your job description.

The Four Stages of Contribution model provides you with a mental reference that helps you to effectively identify the specific behaviors required to produce the expected value associated with your current or envisioned role. Understanding the primary form of collaboration and value creation of a given position must be considered a prerequisite for self-development and success in that specific role. Being able to manipulate and optimize these variables is the most effective approach to provide continuity to your progress during the

---

[1] A collaborative entity is a group that consists of two or more interdependent individuals driven by needs and goals that cause them to influence each other.

different steps of your career. In Table 3-1, you can see an analysis of the Four Stages of Contribution model with the prevalent behaviors associated with each phase.[ix]

Table 3-1. Behavioral breakdown of the Four Stages of Contribution model

---

Stage 1: Contributing Dependently
Common job title: Junior Designer
Foundational activities:

- Work under supervision, leveraging others' knowledge.
- Accept direction from others.
- Utilize essential resources.
- Fulfill business needs.
- Execute business initiatives.
- Execute on the team on critical strategic issues.
- Achieve success in a portion of a project.
- Follow the design process as a rigid procedure.
- Learn personal technical expertise and ability.
- Initiate the internal social system.
- Establish a professional reputation.

Developmental activities:

- Initiate the development of emotional intelligence.
- Initiate the development of personal character.
- Initiate the development of technical skills.
- Initiate the development of social-oriented skills.

Stage 2: Contributing Independently
Common job title: Senior Designer
Foundational activities:

- Work with no supervision, leveraging personal knowledge.
- Provide direction to the self.
- Utilize essential resources.
- Fulfill business needs.
- Execute business initiatives.
- Execute on the team on critical strategic issues.
- Achieve success in a project.
- Implement the design process with a flexible mindset.
- Master personal technical expertise and ability.
- Consolidate the internal social system.
- Mature a professional reputation.

Table 3-1. *(continued)*

Developmental activities:

- Consolidate the development of emotional intelligence.
- Consolidate the development of personal character.
- Consolidate the development of technical skills.
- Consolidate the development of social-oriented skills.
- Initiate the development of results-oriented skills.
- Initiate the development of organizational change skills.

Stage 3: Contributing via Teams

Common job title: Head of Design

Foundational activities:

- Work collaboratively, leveraging the team's abilities.
- Provide direction to the team.
- Allocate essential resources.
- Support business needs.
- Support business initiatives.
- Represent the team on critical strategic issues.
- Achieve success in several projects.
- Support and protect the design process.
- Develop others' technical expertise and ability.
- Expand the internal and initiate the external social system.
- Prepare reputable and promising individuals for managerial roles.

Developmental activities:

- Expand the development of emotional intelligence.
- Expand the development of personal character.
- Expand the development of technical skills.
- Expand the development of social-oriented skills.
- Consolidate the development of results-oriented skills.
- Consolidate the development of organizational change skills.

(continued)

Table 3-1. (*continued*)

Stage 4: Contributing via Workgroups
Common job title: Chief Design Officer
Foundational activities:
■ Work collaboratively, leveraging departments' abilities.
■ Provide direction to the organization.
■ Obtain essential resources.
■ Define business needs.
■ Drive business initiatives.
■ Represent the organization on critical strategic issues.
■ Achieve success in several programs.
■ Define and establish the design process.
■ Develop others' expertise and ability.
■ Expand the internal and consolidate the external social system.
■ Prepare reputable and promising individuals for leadership roles.
Developmental activities:
■ Expand the development of emotional intelligence.
■ Expand the development of personal character.
■ Expand the development of technical skills.
■ Expand the development of social-oriented skills.
■ Expand the development of results-oriented skills.
■ Expand the development of organizational change skills.

In this analysis, you can notice how some of the 16 leadership abilities are deployed differently across the four stages of contribution. For instance, the "ability to produce outcomes" initially defined at Stage 1 as achieving success in a portion of a project becomes achieving success in a project at Stage 2 and later evolves at Stages 3 and 4 into achieving success in several projects and programs.

The Evolution of Your Contribution

As you traverse the different stages of your career, you must be aware of the exact type of contribution required by your current position and, if necessary, realign your priorities with that expectation. Table 3-2 captures the salient characteristics of this journey.

Table 3-2. The evolution of the area of focus across the Four Stages of Contribution model

Stages 1–2	Stage 3	Stage 4
Leading yourself.	Leading teams.	Leading workgroups.
Drive the achievement of individual and collective objectives.	Drive the achievement of the team's objectives.	Drive the achievement of the workgroups' objectives.
Develop yourself.	Develop the team.	Develop the workgroups.

This incessant shift in expectations is particularly crucial between Stages 2 and 3, where a person initiates the transition from an individual contributor role to a managerial position. This specific change in responsibilities, and consequently expected abilities, is undoubtedly the most challenging moment during the career progression of a designer. The dynamics behind this condition can be analyzed by visualizing the shift in expectations associated with the five clusters of leadership capabilities across the four stages of the model. The four permutations illustrated in Figure 3-3 represent the archetypical configuration of abilities that a designer should strive to reach before contemplating a stage transition. As you will learn later in this next section, the exact blueprint of a successful design manager with leadership capabilities is a function of the organization's culture and needs in which that given person operates.

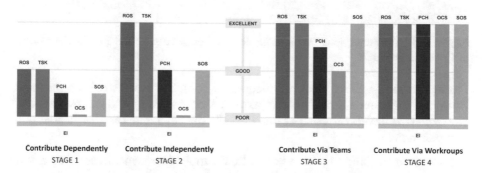

Figure 3-3. The evolution of the five clusters of capabilities across the Four Stages of Contribution model

During Stage 1, a Junior Designer who contributes dependently must be aware that there is probably an expectation regarding the achievement of a good level of competency in results-oriented and technical skills while beginning the development of personal character and social-oriented skills.

During Stage 2, a Senior Designer who contributes independently must be conscious that there is probably an expectation regarding the achievement of an excellent level of competency in results-oriented and technical skills and a good level of competency in personal character and social-oriented skills.

During Stage 3, a Head of Design who contributes via teams must be mindful that there is probably an expectation regarding the achievement of an excellent level of competency in results-oriented, technical, and social-oriented skills and a good level of competency in personal character and organizational change skills. At this point, the challenge arises from the fact that the results-oriented and technical skills of an independent contributor and the ones of a manager are defined by a completely different ability set (Figure 3-4). For this reason, the transition between Stages 2 and 3 presents a significantly higher rate of failure.

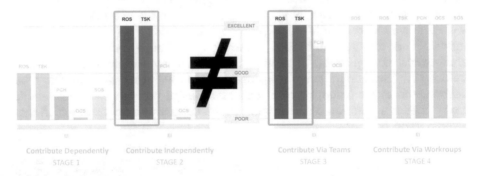

Figure 3-4. The nonobvious diverse ability set required at Stages 2 and 3 of the Four Stages of Contribution model

During Stage 4, a Chief Design Officer who contributes via workgroups must be aware that there is probably an expectation regarding the achievement of an excellent level of competency in all the five clusters of leadership competencies.

Moving from a Junior to a Senior role, an individual contributor has a few years to develop those abilities from a poor to a good and then an excellent level; the same is true from a Head to an Officer role. Instead, a new manager is expected to demonstrate proficiency, if not excellence, from the beginning. Unfortunately, working in a company that is willing to invest in preparing you for a managerial job while still in an individual contributor role is the exception and not the norm. In Chapter 4, you will learn how to overcome this challenge by finding a mentor that can support your transition from Stage 2 to Stage 3 and subsequently Stage 4.

New Research Insights on Leadership Development

As part of their research effort, Jack Zenger and Joe Folkman made a deep dive into the development of leadership capabilities analyzing over 200,000 managers and their relative stakeholders.[x] The data gathered surfaced some essential insights that are important to discuss before concentrating on the creation of your developmental program.[xi]

Leadership abilities can be developed. Managers with leadership capabilities are made, not born. This insight can appear tautological and unnecessary, but the practicality of life demonstrated otherwise, illuminating the fine line between knowing and comprehending. In practice, only a minor fraction of the over 25,000 managers in the study, approximately 10%, had a personal developmental plan, and they regularly dedicated time to it.[xii]

Leadership development and performance improvement connect in a counterintuitive way. The field of human learning can still be considered vastly uncharted. A variety of divergent and sometimes conflicting theories tried to frame this fascinating process. The complexity that permeates the learning process originates from its systemic nature that involves a diversity of dependent and independent variables. Despite that, we can say with certainty that incremental developments in leadership create nonlinear improvements in performance. Your learning process tends to follow a saltatory pattern where the abilities involved must reach critical mass before generating a cumulative leap forward in performance outcome. This process is broadly misunderstood and vastly misinterpreted, with negative repercussions on personal development. Not seeing a linear correlation between the effort allocated to self-development and their performance increase, some individuals tend to jump to premature conclusions. Some may believe that what they are doing is not working or even worse that they are not "born for that thing" and that, therefore, self-development represents a futile endeavor. Others may look at their peers and the time these persons invest in a developmental plan without immediate results, dismissing these initiatives as unworthy. This mindset and behavior typically characterize the average performers inhabiting the 20th–80th percentile bracket of the study.[xiii]

Leadership development requires different strategies. The most efficient approaches to develop an ability from poor to good, and good to excellent, are diametrically different. When a leadership competence is in its infancy or underdeveloped, the best way to improve it is to use a linear and direct developmental approach.[xiv] In this case, the focus is on deepening the knowledge domain of that given ability. For many design managers and leaders, this is the only way they know to develop capabilities. As a result of that limited comprehension, they never achieve an elite level failing to improve their capabilities from good to excellent. When a leadership competence, instead, is already functional, the most efficient strategy to develop it from a good to an excellent level is to use a nonlinear and indirect developmental approach.[xv] In this case, the focus is on deepening the knowledge domain of a set of correlated abilities. You implicitly explored this dynamic in Chapter 1, learning that results-oriented skills can magnify the expression of social-oriented skills and vice versa. These types of competencies are called *companions* and represent a foundational element of the *strength-based developmental approach*[xvi] fostered in this book.

Leadership abilities are intertwined. Leadership competencies rarely operate in isolation, and more often than not, one competence needs to leverage one or more companions to achieve a high degree of proficiency and be perceived by others as a productive skill. This interdependent nature is responsible for known as the *halo effect*. Mastery in one ability often enhances the perception of a correlated one. For instance, if a person displays the ability to communicate powerfully and prolifically, they are more likely to be perceived as proficient at taking initiatives.

Leadership ability synergies can generate superlinear results. The intertwined nature of leadership competencies originates *powerful combinations* that produce performance improvements that surpass the linear tendency.[xvii] Two examples of these productive interactions between *competency companions* are, at a macro-level, the collaboration between the results-oriented and social-oriented clusters of skills and, at a micro-level, the cooperation between the ability to solve problems and analyze issues and the ability to foster collaboration and teamwork.

Leadership abilities have different significance. The five clusters of capabilities are not equally important. Personal character, for instance, has a foundational role. As we introduced in Chapter 2, it's improbable to be perceived as an excellent leader without demonstrating high integrity, honesty, and credibility; this is why personal character occupies the central position in the metaphor depicted by the Leadership Tent model. Within the dimension defined by the Leadership Tent model, organizational factors vastly influence the precise significance associated with each of its five clusters.[xviii] The operational needs determine which actions are expected from a given role, while the social norms underpinning the culture define the exact importance associated with each specific ability establishing which behavior is accepted, rewarded, or disincentivized.

Effective leadership culminates in achieving change. Demonstrating leadership competencies is solely aimed at the achievement of organizational change. Positive change is the bottom line of every manager who manages like a leader. Conversely, if you don't produce any change, you fail. You can achieve change at different levels depending on your current stage of contribution: designing a product or service, improving the creative collaboration, and redefining the social norms underpinning your organization's culture are just a few examples. As you learned in Chapter 2, design leadership is, ultimately, about coping with change.

Effective leadership is not universal. Some competencies overlie and bestride others, but this does not impede design managers from developing their leadership style. While a design manager accredited with "excellent" leadership competencies consistently embodies an efficient distribution of competencies across all the five clusters, many valid permutations can be leveraged to achieve similar results and uniquely express a person's style.

Effective leadership has profound strengths. A competence at that level can catapult you into the elite group at the top of your organization. A *profound strength* is a mission-critical leadership ability to a given role, observable by others, assessed as "excellent," representing a characteristic that distinguishes an individual more than any other competency.[xix] Statistically, a profound strength is an ability at or below the 90th percentile: it falls in the top 90% compared with the same ability assessed in other design leaders.[xx] As discussed in Chapter 2, a leader with no strengths falls in the 30th percentile, with one strength in the 60th percentile, three in the 70th percentile, and five in the 90th percentile.[xxi]

Effective leadership does not imply the absence of weakness. A vast portion of leaders, approximately 84%, while they do not possess any severe weaknesses, they are still not perceived as "excellent" by their manager, peers, and direct reports. These persons navigate the unaspiring and possibly frustrating average bracket of the study.[xxii] This part of the statistical population tends to focus on limiting their vulnerabilities, completely neglecting the development of their strong characteristics.[xxiii] Unfortunately, the absence of weaknesses combined with the absence of any pronounced strength only generates mediocrity. A leader has the moral obligation of improving their weak points, but it is not uncommon that a leader is perceived as excellent while possessing some nonsevere weaknesses.

Effective leadership does imply the absence of severe weakness. While the absence of weakness does not define overall leadership effectiveness, however, it demands the absence of severe weaknesses, also known as *fatal flaws*. A fatal flaw is a mission-critical leadership ability to a given role, observable by others, assessed as "poor," representing a severe weakness requiring significant training.[xxiv] Statistically, a fatal flaw is a weakness at or below the 10th percentile: it falls in the bottom 10% compared with the same ability assessed in other design leaders.[xxv] One or more fatal flaws have the potential to jeopardize a manager's career unless those mission-critical flaws are fixed. Those types of weak points are so dramatic that they can severely hamper an individual's career.

Effective leadership development is influenced by the line manager. Your manager in line, *volens nolens*, is the person that has the highest degree of influence on your developmental journey. That person can sustain or frustrate your developmental effort in various direct and indirect ways: from deciding not to offer coaching sessions to not creating or delegating tasks that expand your area of competence to undermining your attempt to make time in your calendar to develop new abilities.

The Context Dependency of Leadership

We do not see reality; we do not perceive the world as it is.[xxvi] For our brain, information from the physical world is nothing more than a collection of particles with a correspondent energy level, from the photons absorbed by the eye, the phonons detected by the ear, to the ones registered by our nose, our tongue, and skin.[xxvii] Our brain did not evolve to see reality but to survive, precisely, to not die. As you will learn in Chapter 8, survival requires creative problem-solving to acquire, elaborate, and implement information.

Instead of representing the information from the physical world directly, the brain evolved to process and extract meanings from it to elicit a helpful response aimed to resolve uncertainty and increase the likelihood of survival (Figure 3-5). In a state of ambiguity, different interconnected regions of the brain process the information from the five senses, access your memory in search of useful experiential references to interpret the benevolent or malevolent nature of the stimuli, and prepare your body to react in accordance with the resultant level of situational uncertainty.

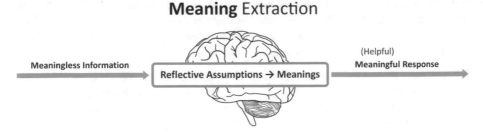

Meaning Extraction

Meaningless Information → Reflective Assumptions → Meanings → (Helpful) Meaningful Response

Figure 3-5. The extraction of meanings from the physical world

In these situations, the brain constructs meaning from the meaningless information from the physical world by engaging empirically in perceptions of the world.[xxviii] During those experiences, the brain creates associations that connect actions and responses to a given context and subsequently assign behavioral meanings to these associations.[xxix] This perceptual history of reality defined by these relationships generates *reflective assumptions* that the brain uses to interpret the world. These assumptions underpin everything we think, believe, and ultimately consider to "know."[xxx]

Given a situation, our brain processes this perceptual history that defines the *empirical significance of information* to identify what is helpful in the hope of increasing the probability of surviving in the future, whichever meaning the word "survive" represents in that specific situation.[xxxi] Ultimately, we do not see reality. We only see what was beneficial to see in the past.[xxxii] As human beings, we understand the world in a predictive way but not necessarily accurate in some absolute, objective sense. Only 10% circa of what we see comes from our eyes, all perceptions and behaviors are a direct manifestation of what was helpful for us to see in the past, and the responses we choose are based on assumptions grounded in our perceptual history.[xxxiii]

The Perceptual Implications of Leadership

While research surfaced the five clusters of capabilities of successful managers with leadership qualities, the empirical significance associated with each of the 16 differentiating design leadership abilities within a company varies based on the perceptual history and reflective assumptions of the population of that given organization.[xxxiv] For the brain, the skills of a person are like any other physical element in the physical world: a source of inherently meaningless information that needs to be interpreted.[xxxv] The meanings associated with those abilities define the domain within which the concept of effective leadership is determined and validated.

This dynamic has two interconnected and critical implications on your developmental journey. Your leadership effectiveness is validated through the perception of the individuals inhabiting your organizational context,[2] and, consequently, what is perceived as effective in a given context can be dismissed as ineffective in a different one.[xxxvi] With the only exception of the "ability to lead by example demonstrating high integrity, honesty, and credibility" that represents a necessary competence for a design manager who wants to manage like a leader, irrespective of the organizational context, the empirical significance associated with other abilities tends to present variations based on the perceived business needs and expected social behaviors of a given company.[xxxvii] Identifying these reflective assumptions and their implications within the context of your organization is necessary to create a developmental program that can maximize your probability of being successful.

---

[2] From a cognitive perspective, the organizational context of a given company is defined by the reflective assumptions underpinning the meanings associated with the perceived business needs and expected social behaviors considered necessary to be successful.

Create Your Developmental Program

In this section, you will lay the foundation for your future development. At the end of this chapter, you will have a complete plan punctuated by multiple events that will delineate the improvement of your leadership abilities. The process is composed of three main activities:

1. Define the competency model.

2. Assess your capabilities.

3. Create your developmental strategy.

The first phase defines a competency model by extrapolating the reflective assumptions permeating your working environment, defining the elements of what is considered "effective leadership" in the context of your organization, and identifying the critical abilities that you need to possess to be successful in your current role. The second phase assesses your capabilities against the competency model by engaging different types of colleagues to capture a complete snapshot of your existing ability profile. The third phase creates your developmental strategy by analyzing your current ability profile identifying your area of improvement and creating a developmental strategy at the intersection of personal and organizational needs to inform an iterative plan of action.

Define the Competency Model

A competency model is identified by a set of abilities and implicitly by their underpinning reflective assumptions that define the successful behaviors to exhibit in order to be successful in a specific organizational context. It is a framework that captures the definition of "effective leadership" approved by a given team, workgroup, or organization in the form of a list of prioritized and quantified abilities deemed essential to a precise role. No model is perfect by definition, but leveraging a competency-based framework constitutes a meritocratic way to manage recruitment, promotion, and compensation, among other things. A competency model also provides a shared language that facilitates and clarifies the conversation around behavioral standards and how those actions propel what is considered high performance in a specific company. The use of a competency model helps to differentiate between what is regarded as a "poor," "good," and "excellent" performance unequivocally.

Define the Statistical Population

In statistics, a population can be seen as the collection of all the possible observations of the feature under study: a set of elements of interest for some questions.[xxxviii] Begin creating a list of individuals from your current seniority level onward that are considered successful managers with leadership qualities in your organization. If your role is Head of Design, recruit all the Head of Department horizontally across the company and all the designer managers senior to you vertically, such as the Chief Design Officer or any other position present in your organization. Ideally, you should be able to recruit the entire *effective leadership population* in your organization. If you work for a large company, you can alternatively decide to proceed with a statistical sample or *sufficient estimator*[3] based on a smaller dataset and subsequently calculate the sample standard deviation to scale the results. The sample standard deviation measures the spread of scores within your set of data. It allows you to use a sample from which you can estimate and generalize to the entire population of successful managers with leadership qualities in your company.

Select the Assessment Method

Depending on the level of rigor that your actual social and technical position within the company allows you to achieve, there are a few methods at your disposal that you can use to create a competency model. The following three sections will introduce three methods presented in order of statistical effectiveness that can cover most scenarios.

Survey Questionnaire: 360-Degree Feedback

If you work for an organization that falls in the circa 85% of the Fortune 500 companies that already use 360-degree feedback as an integral part of their evaluation process,[xxxix] this method should be the first to consider to assess the *effective leadership population* of your organization. This method administrates a questionnaire anonymously to the statistical population to surface their reflective assumptions and indirectly identify their *behavioral configuration*.

In this scenario, you use the already present 360-degree feedback infrastructure to conduct a survey questionnaire to evaluate your statistical population against the 16 differentiating design leadership abilities and understand in which order and to which extent these capabilities contributed to their success. This questionnaire[4] uses a 5+1 Likert scale: Outstanding strength,

---

[3] A sufficient estimator utilizes all the relevant information from a sample in order to approximate a population parameter.

[4] The survey questionnaire used with the 360-degree method is presented in Table 3-3 in the section dedicated to assessing your competencies to avoid repetitions.

Strength, Competent, Needs some improvement, Needs significant improvement, Do not know/Does not apply. Using a 360-degree feedback approach also allows you to expand the dataset, including the direct reports of these managers that demonstrated leadership qualities, with the result of increasing the statistical effectiveness of your investigation.

If you are working in a large company that doesn't already utilize this method to generate performance reviews, it can be challenging to set up in isolation without the active involvement of the human resources department. If you work in a small company, instead, and you built some degree of influence, you may have an opportunity to introduce this method in your company practices.

Survey Questionnaire: Professional Opinion

If a preexisting 360-degree infrastructure is not already present in your organization, sending the same questionnaire directly to each leader composing the statistical population represents a valid alternative. When you forward the survey questionnaire to the respondents, instead of requesting them to do a self-evaluation, you can increase the likelihood of participation, asking them to express their professional opinion on the importance of the 16 leadership abilities.

This class of assessment represents a lower form of evaluation because it introduces an element of subjectiveness that may represent a potential source of inaccuracy. The underlying assumption leveraged by a professional opinion-based assessment is that, given the individual's success, their conception of which ability is more or less decisive is accurate and reliable. Furthermore, this assumption is also correlated with the level of self-awareness of the responder, which is typically sufficiently high in successful leaders, but this is not invariably the case.

Card Sorting

If you don't have a relationship with the leaders in your organization, asking them to complete a survey can be challenging; in this case, card sorting represents your next best alternative. The card sorting method provides a practical and facile way to generate a prioritized list of the 16 differentiating design leadership abilities based on their contextual significance. The categories used to sort the abilities are numbers from 1 to 16, where 1 represents the most important, and 16 denotes the least essential competency.

Card sorting represents another opinion-based assessment that leverages the same knowledge and self-awareness-driven assumption discussed with the "professional opinion" method. Despite that, card sorting provides the best compromise between effort demanded and result achievable; for this reason, it represents the most suitable opportunity for the vast majority of individuals.

While card sorting cannot produce a quantified distribution of capabilities and inform a competency model comprehensively, its prioritized list of abilities is enough to set the overall direction of your developmental strategy.

Define the Leadership Archetype

While research isolated the building blocks of a design manager with leadership capabilities, the contextual dependency of these abilities makes the concept of the "perfect leader" an illusion.[xl] Your profile defined by your level of proficiency across the 16 differentiating leadership abilities can be perceived as paramount in a company and surprisingly unimportant in another. This cognitive misalignment also represents the deciding factor in situations where a successful designer leaves a company with the reputation of being an achiever only to join a new team but, instead, fails to deliver any substantial result.

Those differences tend to disappear inside the same organizational context characterized by an aligned set of reflective assumptions.[xli] This dynamic allows us to identify an archetype that delineates an ideal profile that you can use as a reference point to evolve your unique leadership style and strategize your developmental program accordingly. Depending on the method adopted to define the competency model, this *effective leadership archetype* is delineated using different deliverables.

If you decide to use the survey questionnaire to assess the leadership population with 360-degree feedback or a professional opinion-based assessment, the archetype is delineated by a quantified and prioritized distribution of the 16 leadership abilities. In this case, you can visualize for each ability the dataset fluctuation using the standard deviation (Figure 3-6).

Quantified Distribution

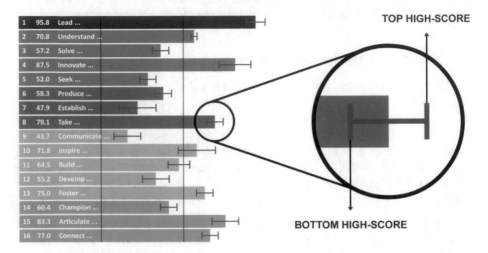

1	95.8	Lead ...
2	70.8	Understand ...
3	57.2	Solve ...
4	87.5	Innovate ...
5	52.0	Seek ...
6	58.3	Produce ...
7	47.9	Establish ...
8	79.1	Take ...
9	43.7	Communicate ...
10	71.8	Inspire ...
11	64.5	Build ...
12	55.2	Develop ...
13	75.0	Foster ...
14	60.4	Champion ...
15	83.3	Articulate ...
16	77.0	Connect ...

TOP HIGH-SCORE

BOTTOM HIGH-SCORE

Figure 3-6. The effective leadership archetype represented using a quantified and prioritized distribution of the 16 leadership abilities

Instead, if you decide to use card sorting to assess the leadership population, the archetype is delineated by a prioritized list of the 16 differentiating design leadership abilities (Figure 3-7).

Prioritized List

1	Lead by example displaying high integrity and credibility
4	Innovate
15	Articulate visions and strategies
8	Take Initiative
16	Connect the group to the outside world
13	Collaborate
10	Inspire and motivate others to high-performance
2	Understand people, business, and technological needs and implications
11	Build relationships
14	Champion change
6	Produce outcomes
3	Solve problems and analyze issues
12	Develop individual and collective skills
5	Seek, process, and implement feedback
7	Establish stretch goals
9	Communicate powerfully and prolifically

Figure 3-7. The effective leadership archetype represented using a prioritized list

A prioritized list cannot quantify the perceptual differences between the 16 differentiating design leadership abilities across the leadership population but can still productively inform your developmental strategy, ranking these competencies in order of significance. Once you defined the competency model, you are ready to assess your leadership and emotional intelligence competencies.

Assess Your Competencies

In the following sections, you will learn how to assess your current leadership capabilities and their substrate, emotional intelligence, and how to analyze and interpret the results to baseline the beginning of your journey and inform your developmental strategy.

Leadership Competencies

Assessing leadership means evaluating its five clusters of capabilities and appraise the maturity of its constituent abilities. As mentioned before, the single most effective way to investigate abilities and their correlated behaviors is to use a questionnaire to gather 360-degree feedback.

Define the Statistical Population

Compared to the work introduced previously, in this context, the statistical population spans across a broader range of positions, internal and external to your team or workgroup, and includes the following categories:

- Line manager (1)
- Peers (≥ 3)
- Direct reports (≥ 3)
- Others (≥ 3)
- Self

"Line manager" represents the person to whom you directly report, your supervisor. The "peers" describes individuals with similar role scope that collaborate with you on some projects; if your role is Head of Design, for instance, and your company has a balanced reporting line structure, you typically include, among others, positions like Head of Product, Head of Development, and Head of Marketing. "Direct reports" captures the individuals in your team that you manage directly. "Others," instead, is a facultative category that can be adapted to encompass specific perspectives depending on whether you are working for an internal team or an external agency, typically indirect reports, clients, partners, and other sorts of collaborators.

While there is no conclusive evidence of an ideal number of respondents when the intention is to assess behaviors,[xlii] in this specific situation, a value between 10 and 16 can be considered a reliable guideline capable of mitigating significant biases and subjectivity. This number does not include your self-assessment, which aims to instigate your self-awareness and not inform your evaluation.

An obvious but essential remark is that, while reaching a significant number of respondents is necessary, who responds to the survey is equally, if not more, critical in terms of statistical significance.[xliii] You have to engage colleagues that had the time and opportunity to interact with you and your abilities and refrain from the temptation to add arbitrary persons to the statistical population. Table 3-3 introduces an ability-based questionnaire that you can use to assess your abilities against the five clusters of leadership capabilities.[xliv]

Table 3-3. Ability-based leadership questionnaire

Personal character[α]

Ability to lead by example demonstrating high integrity and credibility.

- Being approachable by anyone.
- Being adherent to ethical values.
- Being consistent with moral principles.
- Being trustworthy and keeping your commitments.
- Being kind and treating everyone with dignity regardless of their status.
- Being inclusive and transparent when making decisions.
- Being emotionally resilient and tenacious during difficult times.
- Being receptive to constructive criticism.
- Being humble and practicing self-development openly.

Technical skills[α]

Ability to understand persons, business, and technological needs and implications.

- Understand human-centered implications of business and technology.
- Create an overarching vision for the organization.
- Define a strategic direction for the organization.
- Translate the strategic direction into personal objectives for the team.
- Concurrently use a holistic and operational mindset to navigate challenges.
- Allocate resources to multiple teams and/or workgroups adequately.

(continued)

Table 3-3. (*continued*)

Ability to solve problems and analyze issues.

- Understand and contextualize problems and issues holistically.
- Promote group cohesiveness.
- Promote data-informed, insight-driven explorations.
- Protect and nurture diverse points of view.
- Protect and encourage divergent ideas.
- Make decisions leveraging analysis, experience, and judgment.

Ability to innovate.

- Scan the horizon for change, monitoring persons, business, and technological trends.
- Protect the creative nature and approach of problem-solving.
- Continuously refine and improve the way of working and its environment.
- Empower individuals to find efficient new ways to overcome challenges.
- Establish a culture based on continuous learning.
- Create a healthy, sustainable, and productive environment for diversity.

Ability to seek, process, and implement feedback.

- Reserve time to meet persons and request feedback.
- Reserve time to analyze and process feedback.
- Reserve time to work on your development strategy and plan.
- Reserve time to work on your skills and abilities.
- Reserve time to process adverse events.

Results-oriented skills[α]

Ability to produce outcomes.

- Represent customer's needs within the organization.
- Define and pursue high standards of excellence.
- Connect the design effort to the business impact.
- Connect organizational targets to departmental objectives and goals.
- Set objectives and goals for the team.
- Set individual targets for the members of the group.
- Define objectives and goals balancing long-term and short-term targets.
- Pursue projects and tasks until completion.
- Reserve time to develop relationship power.
- Reserve time for reflection.

Ability to establish challenging objectives.

- Promote a mindset of continuous improvement.
- Define measurable standards and metrics for themselves.
- Define measurable standards and metrics for the team.
- Understand the individual and collective potential of the team.
- Articulate the meaning and impact of objectives and goals.
- Encourage individuals and groups to go beyond their comfort zone.

(continued)

Table 3-3. (*continued*)

Ability to take initiative.

- Align your behavior with your priorities.
- Develop comfort with ambiguity.
- Strategize your time and find your focus.
- Initiate programs, projects, and processes.
- Personally sponsor initiatives and campaigns.
- Take accountability for personal outcomes.
- Take accountability for the outcomes of the team.
- Push yourself beyond your comfort zone to exceed expectations.

Social-oriented skills[a]

Ability to communicate powerfully and prolifically.

- Listen actively.
- Articulate new insights.
- Articulate strategic directions.
- Illuminate connections between individual work and collective objectives.
- Exercise influence at every level of the organization.

Ability to inspire and motivate others to high performance.

- Find and activate the team purpose.
- Champion others to grow personally and professionally.
- Reserve time to coach for performance.
- Understand individual abilities, drivers, and ambitions.
- Understand collective abilities, drivers, and ambitions.
- Drive individuals to stretch and exceed their expectations healthily.
- Infuse energy and curiosity into daily activity.

Ability to build relationships.

■ Reserve time to know individuals personally and professionally.

■ Handle stressful situations constructively and tactfully.

■ Balance concern for productivity with sensitivity for personal needs.

■ Recognize and reward individual and collective contributions.

■ Remain approachable, friendly, and constructive during complaints.

Ability to develop individual and collective skills.

■ Make time in the diary to coach for development.

■ Show genuine concern in the work and success of individuals and the group.

■ Allocate responsibility considering individual abilities, needs, and passions.

■ Promote candid conversations.

■ Provide continuous feedback to individuals and groups.

■ Provide an appropriate balance of positive and negative feedback.

■ Identify and activate the team members' superpowers.

■ Integrate the team members' shadow side.

(continued)

Table 3-3. (*continued*)

Ability to foster collaboration and teamwork.

- Build trust and respect.
- Create a psychologically safe climate.
- Institute appropriate social norms.
- Promote a mindset of inclusiveness and cooperation.
- Nurture collaboration with individual accountability.
- Adopt a transparent decision-preparation and decision-making process.
- Adapt the decisional approach to tactical and strategic challenges.
- Develop cooperative working relationships with other groups.
- Ensure healthy collaboration dynamics between groups.
- Create rituals and establish behavioral patterns.
- Frame challenges to face uncertainty with curiosity.
- Guide experiments and support experiential learning.
- Choreograph meetings and conversations.
- Manage creative tensions and frictions.
- Manage and resolve conflicts and crises.

Organizational change skills[α]

Ability to champion change.

- Build your social system.
- Develop and project influence.
- Promote a change mindset within the organization.
- Lead and present initiatives to engage collaborative support in the organization.
- Promote the work and achievements of the group in the organization.
- Represent the work and achievements of the group outside the organization.
- Hire and develop the appropriate talents.

Ability to articulate visions and strategies.

- Develop a strategic perspective on the future of the organization.
- Develop a strategy balancing short- and long-term organizational needs.
- Articulate the contribution of your group to the organization's strategy.
- Connect organizational needs to departmental priorities.

Ability to connect the group to the outside world.

- Reserve time to discuss persons, business, and technological trends.
- Reserve time to discuss external event engagements.
- Organize internal events inviting external contributors.
- Organize internal events extending the invite to the outside community.

Additional comments[β]

- What would you recommend to keep doing?
- What would you recommend to stop doing?

α 5+1-Point scale: Outstanding strength, Strength, Competent, Needs some improvement, Needs significant improvement, Do not know/Does not apply.

β Open answer with text field.

In the next section, you will learn how to assess your emotional intelligence and how this exercise has a more personal and intimate nature that requires a distinctive approach to selecting the respondents.

Emotional Intelligence Competencies

Assessing your emotional intelligence means evaluating its four domains and appraising the maturity of its abilities. Similarly to other forms of intelligence, emotional intelligence represents an enabler. In this specific context, it denotes the substrate of leadership, which means that it exists in the actualization of leadership qualities, and as such, it can only be measured indirectly via the assessment of these abilities.[xlv] This correlation implies that if you assess your leadership capabilities, you also implicitly and indirectly evaluate the maturity of your emotional intelligence capacities.[xlvi]

Define the Statistical Population

Conversing on emotional intelligence is more personal and intimate than discussing leadership qualities, and it requires a more socially prudent approach. Instead of using a 360-degree feedback approach, it's more indicated considering using a survey questionnaire to engage one or possibly a few close colleagues, with whom you can exchange a candid and caring interaction, following the structure of this statistical population:

- Close colleagues (≥ 1)
- Self

A "close colleague" can be any person within the company that developed a close and trustworthy relationship with you and had the time and opportunity to interact with you and your abilities. It is essential to choose someone whom you can consider "safe," a friend with your best interests in mind, who would never harm you using the personal information exposed during this exercise against you. Table 3-4 introduces an ability-based questionnaire that integrates the trait-based one presented in Chapter 2 and that you can use to assess your abilities against the four domains of emotional intelligence capabilities.[xlvii]

Table 3-4. Ability-based emotional intelligence questionnaire

Self-awareness [α]
Emotional awareness: Ability to interpret your own emotions and recognize their impact.

- Interpret emotions in the moment that are experienced.
- Interpret feelings in the moment that are experienced.
- Understand how emotions affect perception.
- Understand how emotions affect feelings.
- Understand how feelings affect behavior.

Capability assessment: Ability to perceive and articulate one's strengths, limits, and worth.

- Understand personal strengths and limits.
- Articulate the personal worth of the organization.

Self-management [α]
Emotional control: Ability to keep disruptive emotions and impulses under control and channel them in a constructive direction.

- Manage stress productively.
- Maintain calmness under pressure and emotional turmoil.
- Redirect strong emotions and impulses in productive ways.
- Control impulses.
- Demonstrate patience.

Achievement orientation: Ability to be innerly motivated to improve performance and meet internal standards of excellence.

- Maintain focus during challenging circumstances.
- Demonstrate tenacity under pressure and emotional turmoil.
- Set and pursue standards of excellence.
- Practice self-development.

Adaptability: Ability to adapt to changing situations and overcome obstacles.

- Demonstrate flexibility when situations change unexpectedly.
- Manage multiple conflicting demands efficiently.
- Adapt goals effortlessly when circumstances change.
- Shift personal priorities rapidly.
- Navigate ambiguity comfortably.

Optimism: Ability to expect favorable and desirable outcomes.

- Expect favorable outcomes during challenging circumstances.
- Focus on opportunities rather than obstacles.
- See persons as good and well-intentioned.
- Look forward to the future.
- Demonstrate hopefulness.

(continued)

Table 3-4. (*continued*)

Social awareness $^{\alpha}$
Organizational awareness: Ability to read organizational currents, decision networks, and politics.

- Interpret collective emotional climate.
- Understand guiding values and behaviors.
- Understand decisional networks and processes.
- Recognize individual and collective norms.
- Understand business model dynamics and ramifications.

Empathy: Ability to sense others' emotions, understand their perspective, and take an active interest in their concerns.

- Interpret other individuals' emotions.
- Interpret other individuals' feelings.
- Understand how other individuals' emotions affect their perception.
- Understand how other individuals' emotions affect their feelings.
- Understand how other individuals' feelings affect their behavior.
- Demonstrate active interest in other individuals' viewpoints.
- Comprehend other individuals' concordant and discordant viewpoints.

Social management $^{\alpha}$
Inspiration: Ability to guide and motivate other individuals.

- Guide and motivate individuals.
- Guide and motivate groups.

Collaboration: Ability to cooperate with other individuals and groups.

- Cooperate with individuals.
- Cooperate with groups.

Influence: Ability to shape individual and group behavior desirably and favorably.

- Shape individual behavior favorably.
- Shape group behavior favorably.

Conflict resolution: Ability to decipher and dissipate misalignments between individuals and groups.

- Decipher and dissipate misalignments between individuals.
- Decipher and dissipate misalignments between groups.

Talent development: Ability to bolster the capabilities of individuals and groups through feedback and guidance.

- Bolster the capabilities of individuals.
- Bolster the capabilities of groups.

Table 3-4. (*continued*)

Additional comments [β]
 • What would you recommend to keep doing?
 • What would you recommend to stop doing?
R Reverse score.
α 5+1-Point scale: Outstanding strength, Strength, Competent, Needs some improvement, Needs significant improvement, Do not know/Does not apply.

β Open answer with text field.

In the next section, you will learn how to analyze and interpret your assessment results, exploring a few concepts that will constitute the backbone of the developmental plan that you will create in the last part of this chapter.

Analyze and Interpret the Assessment Results

As you learned at the beginning of this chapter, information without context is meaningless.[xlviii] Analyzing and interpreting your assessment results allows you to develop a compelling story that can provide context to your current situation and inform your developmental strategy from three distinct and interconnected perspectives.

The first viewpoint is informed by the average score associated with each cluster of capabilities; in the example in Figure 3-8, you can see the initial section of a hypothetical analysis that presents the result relative to the cluster "personal character." This perspective is necessary to develop a holistic understanding of your general leadership maturity and, consequently, inform the overarching trajectory of your developmental program.

The second viewpoint is informed by the average score associated with every ability that contributes to constituting a given cluster of capability; in the example in Figure 3-8, you can see the result relative to the ability "lead by example demonstrating high integrity, honesty, and credibility." This perspective is necessary to develop an atomistic understanding of specific aspects of your leadership maturity and, consequently, inform the identification of the appropriate activities of your developmental plan.

The third viewpoint is informed by the specific score associated with every type of colleague surveyed; in the example in Figure 3-8, they are represented by "line manager," "peers," and "direct reports." This perspective is necessary to develop a social understanding of your general leadership maturity and, consequently, inform specific behavioral changes relative to particular contexts and groups.

Personal Character

RATER	SCORE	POOR		AVERAGE		EXCELLENT			FREQUENCY				
		①	②	③		④	⑤		1	2	3	4	5
Average	3.45					▼ ▼				6%	18%	56%	20%

Lead by example displaying high integrity, honesty, and credibility

RATER	SCORE	POOR		AVERAGE		EXCELLENT			FREQUENCY				
		①	②	③		④	⑤		1	2	3	4	5
Average	3.45					▼ ▼				6%	18%	56%	20%
Line manager	4.00											100%	
Peers	3.20									6%	58%	32%	4%
Direct reports	3.60									8%	34%	46%	2%
Self	3.00											100%	

Being approachable by anyone

RATER	SCORE								1	2	3	4	5
Average	4.05					▼ ▼				2%	26%	62%	10%
Line manager	4.00											100%	
Peers	4.40									2%	6%	78%	14%
Direct reports	3.60									4%	68%	22%	6%
Self	4.20												100%

▼ Bottom and Top High Performer

Figure 3-8. The leadership assessment structure: breakdown of scores per ability and respondent

Investigating the data using these three distinct and interconnected perspectives allows you to move from data to insight, informing your developmental program comprehensively. This investigation does not process the data from your self-assessment, which, instead, will be central during the perceptual gap analysis that you will explore in the next section.

Each of the 16 design leadership abilities assessed is visualized in percentile spanning across three intervals labeled as "poor," "good," and "excellent." In Figure 3-9, you can see three common scenarios.

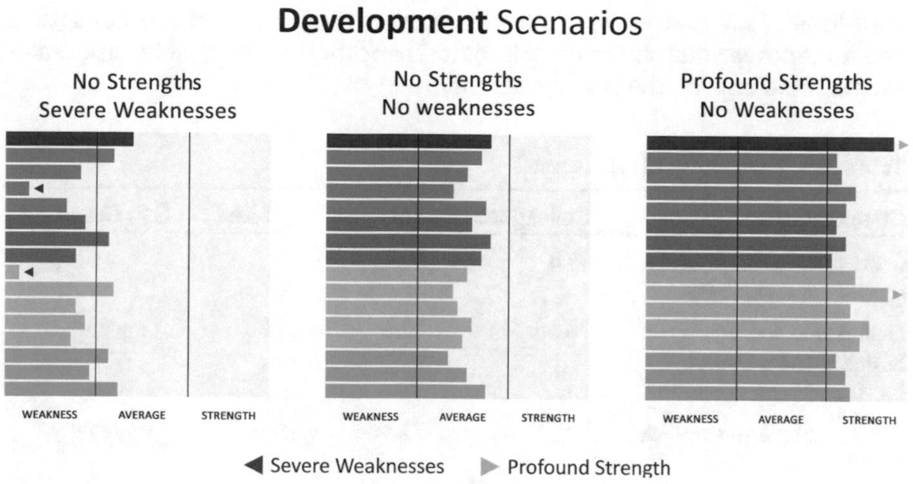

Development Scenarios

No Strengths
Severe Weaknesses

No Strengths
No weaknesses

Profound Strengths
No Weaknesses

WEAKNESS   AVERAGE   STRENGTH   WEAKNESS   AVERAGE   STRENGTH   WEAKNESS   AVERAGE   STRENGTH

◀ Severe Weaknesses   ▶ Profound Strength

Figure 3-9. Three common development scenarios

Abilities evaluated between the 0th and 30th percentile are labeled as "poor" and considered a weakness. When a weakness receives a score at or below the 10th percentile, it is regarded as a "severe weakness" or a "fatal flaw."[xlix] These abilities can be advanced into a "good" level using a linear approach to development.

Abilities evaluated between the 31st and 79th percentile are labeled as "good" and considered average. These abilities can be advanced into an "excellent" level using a combination of linear and nonlinear approaches to development depending on whether they are in the lower or upper part of the interval.

Abilities evaluated between the 80th and 100th are labeled as "excellent" and considered a strength. When a strength receives a score at or above the 90th percentile, it is regarded as a "profound strength." These abilities can be advanced further using a nonlinear approach to development.

Our perceptual process is innately subjective, and our ability to evaluate our capabilities is biased and unreliable.[l] In the next section, you will learn how to use your self-assessment to estimate your perceptual gap and increase your level of self-awareness.

Estimate the Perceptual Gap

Before creating your developmental strategy, it's beneficial for your development to use the data you gather from your colleagues and compare it with your self-assessment to evaluate the difference between your and your colleagues' perception of your leadership capabilities. This exercise can be performed with any assessment and plays an instrumental role in increasing your level of self-awareness. Table 3-5 presents a 16 by 5 matrix that captures and juxtaposes your and your colleagues' hypothetical leadership assessment results, emphasizing the perceptual gap.

Table 3-5. The perceptual gap matrix

Competency	Colleagues	Self	Gap Size	Gap Graph
Demonstrate high integrity, honesty, and credibility.	95.8	90.4	- 5.4	
Understand persons, business, and technical needs and implications.	70.8	82.6	+ 11.8	
Solve problems and analyze issues.	57.2	72.5	+ 15.3	
Innovate.	87.5	85.2	- 2.3	
Seek, process, and implement feedback.	52.0	56.8	- 4.8	
Produce outcomes.	58.3	66.5	- 8.2	
Establish challenging objectives.	47.9	53.7	- 5.8	
Take initiative.	79.1	75.9	+ 3.2	
Communicate powerfully and prolifically.	43.7	51.3	- 7.6	
Inspire and motivate others to high performance.	71.8	67.2	+ 4.6	
Build relationships and networks.	64.5	62.8	+ 1.7	
Develop individual and collective skills.	55.2	59.3	- 4.1	
Foster collaboration and teamwork.	75.0	79.4	- 4.4	
Champion change.	60.4	63.2	- 2.8	
Articulate visions and strategies.	83.4	86.8	- 3.4	
Connect the team to the outside world.	77.0	75.9	+ 1.1	

If you are in a situation where a more granular observation is necessary, you can decompose the average score into four categories visualizing the evaluation of the entire statistical population surveyed: "line manager," "peers," "direct reports," and, if present, "others."

Define Your Developmental Strategy

A research study that analyzed over 85,000 managers globally found that most managers tend to influence the developmental strategy of their direct reports convincing them to improve the capabilities that they believe are essential.[li] When the researcher asked, "What competencies are most important for your direct report to develop?", the vast majority of these managers answered, "deliver results," demonstrating more a self-centered concern for achievements than genuine and altruistic interest to grow the team.

This self-centered attitude is also aggravated by the apprehension of asking more profound questions like "which competency would you love to develop" and the possibility of surfacing a desire to develop a skill potentially misaligned with the individuals' functional requirements.[lii] From the company perspective, it is something that generates value for the individual but not an immediate return on investment for the organization. In the next section, you will explore a model that you can use to reframe this uncomfortable situation productively.

NPR Model

The NPR model aligns your developmental needs, and as you will learn in Chapter 7, the ones of your team, to the functional requirements of your organization. Using this tool, you can prioritize your developmental plan at the intersection of your developmental needs (N) of your competencies, your passions (P), and the functional requirements (R) of your organization (Figure 3-10). The NPR model allows you to maximize your probabilities of success and fulfillment by defining your developmental strategy considering what you need to improve, what you want to grow, and what you have to develop.

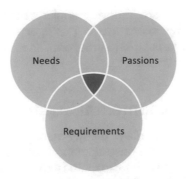

Figure 3-10. The needs (N), passions (P), and requirements (R) model

With the NPR model, you can establish healthy and transparent conversations about developing capabilities that can drive results within the organization without sacrificing the fulfillment of personal passions. Developing an ability, in behavioral terms, demands the creation or modification of an already established conduct. If the new desired ability also fulfills a personal passion, the likelihood of achieving a permanent behavioral change increases significantly.[iiii] This concept applies to self-development and team development equally. As you will learn in Chapter 7, the NPR model represents a central element of coaching that provides the opportunity to identify and activate *superpowers*.[5]

NPR Matrix

The NPR model is implemented using a 16 by 5 matrix. In Table 3-6, you can see an example of this structure visualizing all the 16 differentiating design leadership abilities, your developmental "needs," your "passions," the functional "requirements" of your organization, and the "total" count of instances relative to each leadership ability.

---

[5] A superpower represents the unique contribution that an individual offers to the team.

Table 3-6. The needs (N), passions (P), and requirements (R) matrix

Competency	Needs	Passions	Requirements	Total
Demonstrate high integrity, honesty, and credibility.	✓	✓	✓	3
Understand persons, business, and technical needs and implications.	✓	✓		2
Solve problems and analyze issues.		✓		1
Innovate.	✓			1
Seek, process, and implement feedback.				0
Produce outcomes.			✓	1
Establish challenging objectives.				0
Take initiative.	✓		✓	2
Communicate powerfully and prolifically.		✓		1
Inspire and motivate others to high performance.	✓	✓		2
Build relationships and networks.	✓			1
Develop individual and collective skills.		✓		0
Collaborate.	✓			1
Champion change.				0
Articulate visions and strategies.	✓		✓	2
Connect the team to the outside world.			✓	1

The analysis of your 360-degree feedback identified your areas of strength and weakness. Consider the three abilities that have been assessed with the lower score and the three with the higher score and add these six competencies to the column labeled "needs."

Observe the list of the 16 abilities, understand which one arouses your enthusiasm, and add 6 of these competencies to the column labeled "passions."

The creation of the competency model produced a prioritized list of the critical abilities that you need to possess to succeed within the context of your organization. Add the first six abilities on that list to the column labeled "requirements."

At this point, count the instances of each ability, and add that value to the column labeled "total." The competency associated with the higher value is the ability that will require your attention during the first iteration of your developmental plan. If present, fixing severe weaknesses or fatal flaws has precedence over developing strength. The following pseudocode captures this logic:

```
count the instances
rank the abilities

if there is a severe weakness
select the severe weakness with the higher score
fix the severe weakness

else select the ability with the higher score
develop the ability
end if
```

If the competency selected is a fatal flaw to fix, specifically, an ability that needs to improve from poor to good, you will need to use a linear approach to development. If the competency selected, instead, is a strength to develop further, specifically, an ability that needs to improve from good to excellent, you will need to utilize a nonlinear approach to development. In the next section, you will explore these two methods in practical terms.

From Poor to Good

Developing an ability that has been evaluated as "poor," representing a weakness, requires a linear or direct developmental approach. Using a linear method translates into training the ability that you want to improve directly. The linear development process presents two steps:

1. Select the ability (weakness) from the NPR matrix.

2. Develop the competency linearly.

For instance, suppose the "ability to lead by example demonstrating high integrity, honesty, and credibility" represents a weakness that impedes the display of your "personal character." In that case, you can improve it by developing directly the behaviors associated with that competency. Table 3-7 reports all the 16 differentiating design leadership abilities adapted to a design context, as introduced in Chapter 2.[liv]

Table 3-7. The abilities delineating the 16 differentiating design leadership abilities

Personal character
Ability to lead by example demonstrating high integrity and credibility.

- Be approachable by anyone.
- Be adherent to ethical values.
- Be consistent with moral principles.
- Be trustworthy and keep your commitments.
- Be kind and treat everyone with dignity regardless of their status.
- Be inclusive and transparent when making decisions.
- Be emotionally resilient and tenacious during difficult times.
- Be receptive to constructive criticism.
- Be humble and practice self-development openly.

Technical skills
Ability to understand persons, business, and technological needs and implications.

- Understand human-centered implications of business and technology.
- Create an overarching vision for the organization.
- Define a strategic direction for the organization.
- Translate the strategic direction into personal objectives for the group.
- Concurrently use a holistic and operational mindset to navigate challenges.
- Allocate resources to multiple teams adequately.

Ability to solve problems and analyze issues.

- Understand and contextualize problems and issues holistically.
- Promote data-informed, insight-driven explorations.
- Protect and nurture diverse points of view.
- Protect and encourage divergent ideas.
- Make decisions leveraging analysis, experience, and judgment.

Ability to innovate.

- Scan the horizon for change, monitoring persons, business, and technological trends.
- Protect the creative nature and approach of problem-solving.
- Continuously refine and improve the way of working and its environment.
- Empower individuals to find efficient new ways to overcome challenges.
- Establish a culture of individual and collective learning.
- Encourage and protect alternative approaches.

Ability to seek, process, and implement feedback.

- Make time in the diary to meet persons and request feedback.
- Make time in the diary to analyze and process feedback.
- Make time in the diary to work on your developmental strategy and plan.
- Make time in the diary to work on your skills and abilities.

(continued)

Table 3-7. (*continued*)

Results-oriented skills

Ability to produce outcomes.

- Represent customer's needs within the organization.
- Set and pursue high standards of excellence.
- Set holistic targets for the team.
- Set holistic targets for the member of the team.
- Translate and connect organizational objectives to department and team goals.
- Define goals balancing long-term and short-term objectives.
- Pursue projects and tasks until completion.

Ability to establish challenging objectives.

- Promote a mindset of continuous improvement.
- Define measurable standards and metrics for themselves.
- Define measurable standards and metrics for the group.
- Understand the individual and collective potential of the group.
- Articulate the meaning and impact of objectives and goals.

Ability to take initiative.

- Initiate programs, projects, and processes.
- Personally sponsor initiatives and/or actions.
- Take personal responsibility for personal outcomes.
- Take personal responsibility for groups' outcomes.
- Go spontaneously above and beyond to exceed expectations.

Social-oriented skills

Ability to communicate powerfully and prolifically.

- Listen actively.
- Articulate new insights.
- Articulate strategic directions.
- Illuminate connections between individual work and collective objectives.
- Exercise influence at every level of the organization.

Ability to inspire and motivate others to high performance.

- Champion others to grow personally and professionally.
- Make time in the diary to coach for performance.
- Understand individual abilities, drivers, and ambitions.
- Understand collective abilities, drivers, and ambitions.
- Drive individuals to stretch and exceed their expectations healthily.
- Infuse energy and curiosity into daily activity.

Table 3-7. (*continued*)

Ability to build relationships.

- Make time in the diary to know individuals personally and professionally.
- Handle stressful situations constructively and tactfully.
- Balance concern for productivity with sensitivity for personal needs.
- Recognize and reward individual and collective contributions.
- Remain approachable, friendly, and constructive during complaints.

Ability to develop individual and collective skills.

- Make time in the diary to coach for development.
- Show genuine concern for the work and success of individuals and the group.
- Allocate responsibility considering individual abilities and ambitions.
- Provide continuous feedback to individuals and the group.
- Provide an appropriate balance of positive and negative feedback.

Ability to collaborate.

- Promote a mindset of inclusiveness and cooperation.
- Adopt a transparent decision-preparation and decision-making process.
- Adapt the decisional approach to tactical and strategic challenges.
- Develop cooperative working relationships with other groups.
- Ensure healthy collaboration dynamics between groups.
- Manage and resolve conflicts and crises.

Organizational change skills

Ability to champion change.

- Promote a change mindset within the organization.
- Lead and present initiatives to engage collaborative support in the organization.
- Promote the work and achievements of the group in the organization.
- Represent the work and achievements of the group outside the organization.

Ability to articulate visions and strategies.

- Develop a strategic perspective on the future of the organization.
- Develop a strategy balancing short- and long-term organizational needs.
- Connect the contribution of your group to the organization's strategy.
- Translate the organization's vision into departmental objectives and goals for the group.

Ability to connect the group to the outside world.

- Make time in your diary to discuss persons, business, and technological trends.
- Make time in your diary to discuss external event engagements.
- Organize internal events inviting external contributors.
- Organize internal events extending the invite to the outside community.

Linear development is the form of improvement that we intuitively discover and use during our life. It's an efficient approach to leverage when you need to acquire a new ability or when you need to elevate an underdeveloped one from a "poor" to a "good" level.

From Good to Excellent

Developing a competency that has been evaluated as "good," representing a strength, requires a nonlinear or indirect developmental approach. Using a nonlinear method translates into training the ability that you want to improve indirectly by training one of its *competency companions*. The nonlinear process presents three steps:

1. Select the competency (strength) from the NPR matrix.

2. Select a competency companion that arouses your interest.

3. Develop the competency companion linearly.

For instance, suppose the "ability to lead by example demonstrating high integrity, honesty, and credibility" represents a strength that projects a solid image of your "personal character." In that case, you can develop it further by directly improving the behaviors defined by its *competency companions*. Table 3-8 reports all the competency companions adapted to a design context relative to the 16 differentiating design leadership abilities.[lv]

Table 3-8. The competency companions associated with the 16 differentiating design leadership abilities

---

Personal character
Ability to lead by example demonstrating high integrity and credibility.

- Be assertive.
- Be optimistic in the face of failure.
- Be sensitive to cross-cultural differences.
- Be open-minded and assume good intentions in other individuals.
- Be aware and respectful of other persons' emotions and feelings.
- Be comfortable with ambiguity and adaptable to environmental changes.
- Be a team player and work collaboratively to achieve shared goals.
- Show genuine concern for the work and success of individuals and the team.
- Inspire and motivate others to high performance.
- Make decisions leveraging analysis, experience, and judgment.
- Drive for results.

Technical skills
Ability to understand persons, business, and technological needs and implications.

- Take initiative.
- Communicate powerfully and prolifically.
- Solve problems and analyze issues.
- Build relationships and networks.
- Set and pursue high standards of excellence.
- Develop individual and collective skills.
- Protect the team and act in its best interest.
- Lead by example demonstrating high integrity, honesty, and credibility.

Ability to solve problems and analyze issues.

- Understand persons, business, and technical needs and implications.
- Articulate visions and strategies.
- Take initiative.
- Act independently.
- Communicate powerfully and prolifically.
- Organize and plan meticulously.
- Be open to challenges.
- Innovate.

(continued)

Table 3-8. (*continued*)

Ability to innovate.

- Take initiative.
- Articulate visions and strategies.
- Solve problems and analyze issues.
- Take pondered risks and challenge the status quo.
- Support others in taking pondered risks.
- Learn rapidly from success and failure.
- Champion change.

Ability to seek, process, and implement feedback.

- Be humble and practice self-development openly.
- Protect and encourage divergent ideas.
- Be aware and respectful of other persons' emotions and feelings.
- Listen actively.
- Take initiative.
- Lead by example demonstrating high integrity, honesty, and credibility.
- Take pondered risks and challenge the status quo.
- Inspire and motivate others to high performance.
- Provide continuous feedback and development to individuals and the team.

Results-oriented skills

Ability to produce outcomes.

- Anticipate problems.
- Establish challenging objectives.
- Inspire and motivate others to high performance.
- Solve problems and analyze issues.
- Organize and plan meticulously.
- Allocate resources to multiple teams adequately.
- Operate with speed and intensity.
- Articulate the meaning and impact of objectives and goals.
- Provide continuous feedback and development to individuals and the team.
- Take personal responsibility for personal outcomes.
- Take personal responsibility for groups' outcomes.
- Recognize and reward individual and collective contributions.
- Lead by example demonstrating high integrity, honesty, and credibility.
- Innovate.

Table 3-8. (*continued*)

Ability to establish challenging objectives.

- Understand persons, business, and technical needs and implications.
- Articulate visions and strategies.
- Inspire and motivate others to high performance.
- Take pondered risks and challenge the status quo.
- Drive for results.
- Champion change.
- Make decisions leveraging analysis, experience, and judgment.
- Lead and present initiatives to engage collaborative support in the organization.

Ability to take initiative.

- Anticipate problems.
- Establish challenging objectives.
- Operate with speed and intensity.
- Organize and plan meticulously.
- Pursue projects and tasks until completion.
- Inspire and motivate others to high performance.
- Champion others to grow personally and professionally.
- Lead by example demonstrating high integrity, honesty, and credibility.
- Be comfortable with ambiguity and adaptable to environmental changes.

Social-oriented skills

Ability to communicate powerfully and prolifically.

- Connect the team to the outside world.
- Develop cooperative working relationships with other teams.
- Be trustworthy and keep your commitments.
- Inspire and motivate others to high performance.
- Articulate visions and strategies.
- Translate and connect organizational objectives to department and team goals.
- Establish challenging objectives.
- Solve problems and analyze issues.
- Develop individual and collective skills.
- Take initiative.
- Innovate.

(*continued*)

Table 3-8. (*continued*)

Ability to inspire and motivate others to high performance.

- Lead by example demonstrating high integrity, honesty, and credibility.
- Communicate powerfully and prolifically.
- Establish challenging objectives.
- Develop individual and collective skills.
- Foster collaboration and teamwork.
- Be aware and respectful of other persons' emotions and feelings.
- Translate the organization's vision into departmental objectives and goals for the team.
- Take initiative.
- Champion change.
- Innovate.

Ability to build relationships.

- Listen actively.
- Communicate powerfully and prolifically.
- Promote a mindset of inclusiveness and cooperation.
- Foster collaboration and teamwork.
- Recognize and reward individual and collective contributions.
- Develop individual and collective skills.
- Lead by example demonstrating high integrity, honesty, and credibility.
- Be humble and practice self-development openly.
- Be optimistic in the face of failure.

Ability to develop individual and collective skills.

- Champion others to grow personally and professionally.
- Inspire and motivate others to high performance.
- Recognize and reward individual and collective contributions.
- Promote a mindset of inclusiveness and cooperation.
- Articulate visions and strategies.
- Be humble and practice self-development openly.
- Lead by example demonstrating high integrity, honesty, and credibility.
- Innovate.

Table 3-8. (*continued*)

Ability to collaborate.

- Be trustworthy and keep your commitments.
- Be comfortable with ambiguity and adaptable to environmental changes.
- Communicate powerfully and prolifically.
- Articulate visions and strategies.
- Inspire and motivate others to high performance.
- Establish challenging objectives.
- Develop individual and collective skills.
- Lead by example demonstrating high integrity, honesty, and credibility.

Organizational change skills

Ability to champion change.

- Articulate visions and strategies.
- Take pondered risks and challenge the status quo.
- Build relationships and networks.
- Inspire and motivate others to high performance.
- Promote a mindset of inclusiveness and cooperation.
- Recognize and reward individual and collective contributions.
- Develop individual and collective skills.
- Drive for results.
- Innovate.

Ability to articulate visions and strategies.

- Champion customer's needs within the organization.
- Understand persons, business, and technical needs and implications.
- Solve problems and analyze issues.
- Communicate powerfully and prolifically.
- Inspire and motivate others to high performance.
- Establish challenging objectives.
- Champion change.
- Innovate.

Ability to connect the group to the outside world.

- Build relationships and networks.
- Communicate powerfully and prolifically.
- Understand and contextualize problems and issues holistically.
- Develop holistic perspectives.
- Articulate visions and strategies.
- Inspire and motivate others to high performance.
- Take initiative.
- Champion change.

Nonlinear development is a counterintuitive but highly efficient form of improvement. It's an efficient approach to implement when you need to improve an ability further or when you need to elevate a strength from a "good" to an "excellent" level.

Competency Companion

As discussed previously, the perceived proficiency level associated with your design leadership competencies is a function of the persons' reflective assumptions permeating your working environment; your abilities do not and cannot exist objectively.[lvi] This interdependence represents the most foundational learning from cognitive psychology; nothing in life, including your expertise, can exist objectively.

If a competence only manifests itself in the observers' minds and does not exist as an objective entity, but only as a subjective interpretation of your behavior in a given context, in order to maximize the impact of our developmental program, we must ask ourselves:

How might we enhance the perception of a specific ability?

Research from Zenger Folkman illuminated that some of the comportments propelling the leadership abilities rise and fall in association with other, often nonobvious, behaviors called *companions*.[lvii] These behaviors are functionally interconnected, and they project a degree of reciprocal influence on one another, enhancing your ability to perform and other individuals' perception of your proficiency. Their relationship tends to be complementary and typically nonexclusive.

The *competency companion* concept represents a tacit and cardinal element of many environments, like the world of academia and sport. In the university, you have to learn mathematics before taking a physics class, and in sports, athletes utilize weight lifting to improve functional movements. Michael Jordan said that he developed his vertical jump to increase his ability to jump shot, and Ronaldo developed his performances on the 20-meter dash to improve his ability to dribble at speed. In the nuanced dimension of reality, a capability is always built on the foundation of other competencies and perceived as a function of that relationship.

While researchers in the leadership field are still framing the specific peculiarities of this precise correlation between leadership competencies, a few patterns have already emerged:

- **Pattern 1:** A competency can contribute to building the foundation of another one.

  Example: Practicing self-development represents a propaedeutic activity that builds your ability to develop other individuals.

- **Pattern 2:** A competency can contribute to increasing the proficiency of another one.

  Example: Establishing challenging objectives represents a synergistic activity that enhances your ability to foster collaboration and teamwork.

- **Pattern 3:** A competency can contribute to increasing the perception of another one.

  Example: Communicating powerfully and prolifically represents an instrumental activity that enables your ability to solve problems and analyze issues.

Developing competency companions can allow you to build your strengths to an "excellent" level and, when present, overcome a plateau in your career progression.

Create Your Developmental Plan

Implementing your developmental strategy requires creating a plan of action punctuated by quarterly, semiannual, and annual reviews (Figure 3-11). While everyone improves at different rates based on the retention rate of the information absorbed during the activities practiced to grow a specific competency,[lviii] the typical time frame to improve an ability varies between 6 and 12 months. Figure 3-11 exemplifies this scenario.

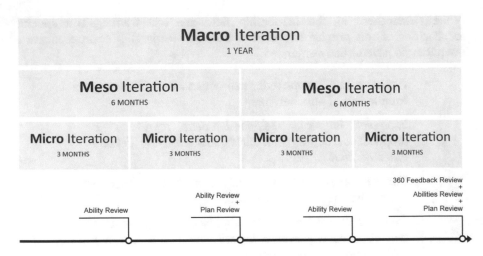

Figure 3-11. The different events that punctuate the developmental plan

Select the ability you prioritized using the NPR matrix and allocate six months to its development on your timeline. You will have the first quarterly review after three months to analyze the initial result of your training, while after six months, you will have a semiannual review to decide if you need three or six more months to work on that given competency, or you are in a position to begin the training of another one. In the average favorable scenario, you will produce noticeable improvements in two abilities per year. After one year, you will gather another 360-degree feedback to capture a complete and updated snapshot of your competency profile and inform the next iteration of the process.

Identify Learning Experiences

After you prioritized the competency to develop and created the plan to implement your developmental strategy, you need to identify which activities you can leverage to train that specific ability. Answering the following questions can assist you in your decision-preparation and decision-making process:

Which activity will allow me to learn and evolve?

Which activity will allow me to develop that specific ability?

How have others at this company developed new abilities?

How have others at this company expanded their responsibilities?

In Chapter 9, you will learn how to calibrate *exogenous stress* to promote learning, a concept that will enhance your capacity to identify the most efficacious learning experiences for your developmental plan. In this section, while you may have access to different resources, depending on your role and the budget allocated to your team, there is a critical concept to absorb before assigning these learning experiences to your plan: the *retention rate* of information.

Retention Rate of Information

Learning is a complex evolutionary event underpinned by several social and cognitive elements.[lix] One of these significant elements to consider when you identify your developmental activities is the retention rate of information that characterizes those learning experiences (Figure 3-12). Different rates of information retention produce a variable degree of influence on your learning curve.[lx]

Xunzi, a Confucian scholar, expressed the same concept more than 2000 years ago, saying, "I hear, and I forget. I see, and I remember. I do, and I understand," or using a more appropriate translation from Chinese, "not hearing is not as good as hearing, hearing is not as good as seeing, seeing is not as good as knowing, knowing is not as good as acting; true learning continues until it is put into action."

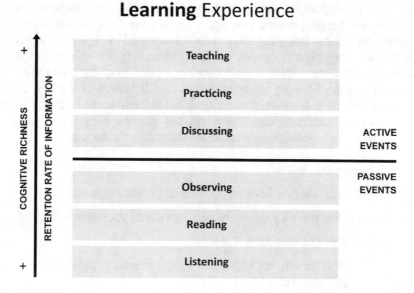

Figure 3-12. The different types of learning experiences and their diverse rate of information retention

Discerning the intricate process of learning in humans goes beyond the scope of this book, but applying some degree of simplification, we can isolate a foundational concept: the *cognitive richness* of an experience is directly correlated with the retention rate of the information associated with that given experience.[lxi] The richer is the cognitive experience, the strongest and more diversified are the neuronal associations formed in our brain.[lxii] The consensus view in neuroscience is that the sorts of memory involved during complex behaviors are distributed among various neural systems, even if certain types of knowledge may be processed and contained in specific regions of the brain.[lxiii] The act of being actively involved in an event tends to recruit more cognitive resources originating an increase in the retention rate of information.[lxiv]

Comprehending this concept allows you to control the tendency of your learning curve proactively.[lxv] An efficient developmental program includes passive and active learning activities to maximize learning during the different scenarios characterizing your week. In Chapter 4, you will investigate how to strategize your week and reserve time for personal development. In the next section, you will learn how to explore and allocate these activities to your developmental plan.

Explore and Allocate Learning Experiences

Exploring and allocating possible activities actualizes your plan of action and implements your developmental strategy. The exploration phase is executed as a divergent session where you brainstorm possible learning experiences and categorize them following the model introduced in Figure 3-12: listening, reading, observing, discussing, practicing, and teaching. The allocation phase is instead executed as a convergent session where you estimate the effort demanded by each learning experience and decide which of them you can or cannot allocate to your developmental plan. Similarly to any other divergent and convergent exercise, these two phases can be structured using one session, following a possible agenda based on a 45-minute time slot:

1. Preparation:

    a. Free up one hour of your time on your agenda.

    b. Prepare a digital or printed version of your yearly plan.

2. Diverge (15 minutes):

    a. Brainstorm activities for the selected capability (10 minutes).

    b. Categorize each activity using the learning categories (5 minutes).

    c. Iterate if necessary.

3. Converge (15 minutes):

   a. Estimate resources demanded by each activity (10 minutes).

   b. Decide which activity to add to your developmental plan (5 minutes).

   c. Iterate if necessary.

4. Summary (10 minutes):

   a. Allocate the activities to your developmental plan (5 minutes).

   b. Allocate time for these activities in your calendar (5 minutes).

During this step, you may realize that you need to discuss a specific activity with someone else; it can be your line manager, the human resources department, or any other entity that owns a resource that your event demands. In this case, make a note annotating the person you need to contact and then move on with the session. If necessary, you can amend the plan after that conversation with that specific person.

Endnotes

i.   Dalton, Gene W., and Paul Thompson. Novations: Strategies for Career Management. G. Dalton and P. Thompson, 1993.

ii. Dalton, Gene W., and Paul Thompson. Novations: Strategies for Career Management. G. Dalton and P. Thompson, 1993.

iii. Dalton, Gene W., and Paul Thompson. Novations: Strategies for Career Management. G. Dalton and P. Thompson, 1993.

iv. Dalton, Gene W., and Paul Thompson. Novations: Strategies for Career Management. G. Dalton and P. Thompson, 1993.

v.   Dalton, Gene W., and Paul Thompson. Novations: Strategies for Career Management. G. Dalton and P. Thompson, 1993.

vi. Dalton, Gene W., and Paul Thompson. Novations: Strategies for Career Management. G. Dalton and P. Thompson, 1993.

vii. Ashforth, Blake E., and Fred Mael. "Social Identity Theory and the Organization." The Academy of Management Review, vol. 14, no. 1, 1989, p. 20., doi:10.2307/258189.

viii. Terry, Paul, et al. "The Four Stages of Contribution." Korn Ferry Institute, 22 Apr. 2014, www.kornferry.com/institute/the-four-stages-of-contribution.

ix. Dalton, Gene W., and Paul Thompson. Novations: Strategies for Career Management. G. Dalton and P. Thompson, 1993.

x. Zenger, John H., and Joseph R. Folkman. The Extraordinary Leader. McGraw-Hill, 2009.

xi. Zenger, Jack, and Joe Folkman. "Key Insights From the Extraordinary Leader: 20 New Ideas about Leadership Development." Zenger | Folkman, 2017, zengerfolkman.com/wp-content/uploads/2019/05/White-Paper-Extraordinary-Leader-Insights-Excerpts-from-The-Extraordinary-Leader.pdf.

xii. Zenger, Jack, and Joe Folkman. "Key Insights From the Extraordinary Leader: 20 New Ideas about Leadership Development." Zenger | Folkman, 2017, zengerfolkman.com/wp-content/uploads/2019/05/White-Paper-Extraordinary-Leader-Insights-Excerpts-from-The-Extraordinary-Leader.pdf.

xiii. Zenger, John H., and Joseph R. Folkman. The Extraordinary Leader. McGraw-Hill, 2009.

xiv. Zenger, John H., and Joseph R. Folkman. "Developing Strengths or Weaknesses."

xv. Zenger, John H., and Joseph R. Folkman. "Developing Strengths or Weaknesses." Zenger | Folkman, 27 Apr. 2019, zengerfolkman.com/wp-content/uploads/2019/08/Developing-Strengths-or-Weaknesses_WP-2019.pdf.

xvi. Zenger, John H., and Joseph R. Folkman. "Developing Strengths or Weaknesses." Zenger | Folkman, 27 Apr. 2019, zengerfolkman.com/wp-content/uploads/2019/08/Developing-Strengths-or-Weaknesses_WP-2019.pdf.

xvii. Zenger, John H., and Joseph R. Folkman. "Developing Strengths or Weaknesses." Zenger | Folkman, 27 Apr. 2019, zengerfolkman.com/wp-content/uploads/2019/08/Developing-Strengths-or-Weaknesses_WP-2019.pdf.

xviii.   Zenger, John H., and Joseph R. Folkman. How to Be Exceptional: Drive Leadership Success by Magnifying Your Strengths. McGraw-Hill, 2013.

xix.   Zenger, John H., and Joseph R. Folkman. How to Be Exceptional: Drive Leadership Success by Magnifying Your Strengths. McGraw-Hill, 2013.

xx.   Zenger | Folkman, 27 Apr. 2019, zengerfolkman.com/wp-content/uploads/2019/08/Developing-Strengths-or-Weaknesses_WP-2019.pdf.

xxi.   Zenger, John H., and Joseph R. Folkman. How to Be Exceptional: Drive Leadership Success by Magnifying Your Strengths. McGraw-Hill, 2013.

xxii.   Zenger, John H., and Joseph R. Folkman. How to Be Exceptional: Drive Leadership Success by Magnifying Your Strengths. McGraw-Hill, 2013.

xxiii.   Zenger, John H., and Joseph R. Folkman. How to Be Exceptional: Drive Leadership Success by Magnifying Your Strengths. McGraw-Hill, 2013.

xxiv.   Zenger, John H., and Joseph R. Folkman. How to Be Exceptional: Drive Leadership Success by Magnifying Your Strengths. McGraw-Hill, 2013.

xxv.   Zenger | Folkman, 27 Apr. 2019, zengerfolkman.com/wp-content/uploads/2019/08/Developing-Strengths-or-Weaknesses_WP-2019.pdf.

xxvi.   Lotto, Beau. Deviate: The Creative Power of Transforming Your Perception. Weidenfeld & Nicolson, 2018.

xxvii.   Gazzaniga, Michael S., et al. Cognitive Neuroscience: The Biology of the Mind. Norton, 2014.

xxviii.   Lotto, Beau. Deviate: The Creative Power of Transforming Your Perception. Weidenfeld & Nicolson, 2018.

xxix.   Lotto, Beau. Deviate: The Creative Power of Transforming Your Perception. Weidenfeld & Nicolson, 2018.

xxx.   Lotto, Beau. Deviate: The Creative Power of Transforming Your Perception. Weidenfeld & Nicolson, 2018.

xxxi.   Lotto, Beau. Deviate: The Creative Power of Transforming Your Perception. Weidenfeld & Nicolson, 2018.

xxxii.   Lotto, Beau. Deviate: The Creative Power of Transforming Your Perception. Weidenfeld & Nicolson, 2018.

xxxiii. Lotto, Beau. Deviate: The Creative Power of Transforming Your Perception. Weidenfeld & Nicolson, 2018.

xxxiv. Zenger, John H., and Joseph R. Folkman. The Extraordinary Leader. McGraw-Hill, 2009.

xxxv. Gazzaniga, Michael S., et al. Cognitive Neuroscience: The Biology of the Mind. Norton, 2014.

xxxvi. Baron, Robert A., and Nyla R. Branscombe. Social Psychology. Pearson, 2017.

xxxvii. Zenger, John H., and Joseph R. Folkman. The Extraordinary Leader. McGraw-Hill, 2009.

xxxviii. Freedman, David A., et al. Statistics. W.W. Norton Company, 2007.

xxxix. Folkman, Joseph R., and John H. Zenger. The Power of Feedback: 35 Principles for Turning Feedback from Others into Personal and Professional Change. Wiley, 2006.

xl. Zenger, Jack, and Joe Folkman. "Key Insights From the Extraordinary Leader: 20 New Ideas about Leadership Development." Zenger | Folkman, 2017, zengerfolkman.com/wp-content/uploads/2019/05/White-Paper-Extraordinary-Leader-Insights-Excerpts-from-The-Extraordinary-Leader.pdf.

xli. Zenger, Jack, and Joe Folkman. "Key Insights From the Extraordinary Leader: 20 New Ideas about Leadership Development." Zenger | Folkman, 2017, zengerfolkman.com/wp-content/uploads/2019/05/White-Paper-Extraordinary-Leader-Insights-Excerpts-from-The-Extraordinary-Leader.pdf.

xlii. Garavan, Thomas N., et al. "360 Degree Feedback: Its Role in Employee Development." Journal of Management Development, vol. 16, no. 2, 1997, pp. 134–147., doi:10.1108/02621719710164300.

xliii. London, Manuel, et al. "A Feedback Approach to Management Development." Journal of Management Development, vol. 9, no. 6, 1990, pp. 17–31., doi:10.1108/02621719010001343.

xliv. Zenger, John H., and Joseph R. Folkman. How to Be Exceptional: Drive Leadership Success by Magnifying Your Strengths. McGraw-Hill, 2013.

xlv. Goleman, Daniel. Emotional Intelligence: Why It Can Matter More than IQ. Bloomsbury, 1996.

xlvi. Goleman, Daniel. Emotional Intelligence: Why It Can Matter More than IQ. Bloomsbury, 1996.

xlvii. Goleman, Daniel. Emotional Intelligence: Why It Can Matter More than IQ. Bloomsbury, 1996.

xlviii. Gazzaniga, Michael S., et al. Cognitive Neuroscience: The Biology of the Mind. Norton, 2014.

xlix. Zenger, John H., and Joseph R. Folkman. How to Be Exceptional: Drive Leadership Success by Magnifying Your Strengths. McGraw-Hill, 2013.

l. Aronson, Elliot, et al. Social Psychology. Pearson, 2016.

li. Zenger, Jack, et al. "Discovering and Developing Hidden Reservoirs of Talent." Zenger | Folkman, 25 Apr. 2019, zengerfolkman.com/wp-content/uploads/2019/08/Discovering-and-Developing-Hidden-Reservoirs_WP-2019.pdf.

lii. Zenger, Jack, et al. "Discovering and Developing Hidden Reservoirs of Talent." Zenger | Folkman, 25 Apr. 2019, zengerfolkman.com/wp-content/uploads/2019/08/Discovering-and-Developing-Hidden-Reservoirs_WP-2019.pdf.

liii. Bandura, Albert, et al. "Factors Influencing Behavior and Behavioral Change." National Institute of Mental Health, 1992.

liv. Zenger, John H., and Joseph R. Folkman. How to Be Exceptional: Drive Leadership Success by Magnifying Your Strengths. McGraw-Hill, 2013.

lv. Zenger, John H., and Joseph R. Folkman. How to Be Exceptional: Drive Leadership Success by Magnifying Your Strengths. McGraw-Hill, 2013.

lvi. Guadagnoli, Mark. Human Learning: Biology, Brain, and Neuroscience. Elsevier Science, 2008.

lvii. Zenger, John, et al. "Making Yourself Indispensable." Harvard Business Review, 9 Sept. 2016, hbr.org/2011/10/making-yourself-indispensable.

lviii. Guadagnoli, Mark. Human Learning: Biology, Brain, and Neuroscience. Elsevier Science, 2008.

lix. Guadagnoli, Mark. Human Learning: Biology, Brain, and Neuroscience. Elsevier Science, 2008.

lx. Guadagnoli, Mark. Human Learning: Biology, Brain, and Neuroscience. Elsevier Science, 2008.

lxi. Guadagnoli, Mark. Human Learning: Biology, Brain, and Neuroscience. Elsevier Science, 2008.

lxii. Guadagnoli, Mark. Human Learning: Biology, Brain, and Neuroscience. Elsevier Science, 2008.

lxiii. Guadagnoli, Mark. Human Learning: Biology, Brain, and Neuroscience. Elsevier Science, 2008.

lxiv. Guadagnoli, Mark. Human Learning: Biology, Brain, and Neuroscience. Elsevier Science, 2008.

lxv. Koh, Aloysius Wei Lun, et al. "The Learning Benefits of Teaching: A Retrieval Practice Hypothesis." Applied Cognitive Psychology, vol. 32, no. 3, 2018, pp. 401–410., doi:10.1002/acp.3410.

Establish Your Core Practices

Some of the most salient aspects of being a designer, such as thinking creatively and critically, rely significantly on a region of the brain called the prefrontal cortex. The prefrontal cortex is formed by the area of the anterior pole of the mammalian brain that receives projections from the mediodorsal nucleus of the thalamus.[i] With its highly evolved neuronal networks, the prefrontal cortex (Figure 4-1) is capable of generating mental representation without sensory stimulation from the environment and responsible for your high-order cognition.[ii]

Prefrontal Cortex

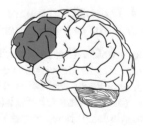

Figure 4-1. The prefrontal cortex

© Andrea Picchi 2022
A. Picchi, *Design Management*, https://doi.org/10.1007/978-1-4842-6954-1_4

This region of the brain was one of the last structures to develop in the course of evolution, it constitutes more than one quarter of the entire cerebral cortex, and its primary function is the representation and execution of new forms of organized goal-directed action.[iii] This capability of the prefrontal cortex serves a superordinate functionality that underpins eight cognitive abilities known as *executive functions*:[iv]

- **Attention:** The ability to allocate and maintain neural resources necessary to process one given item of information, inhibiting the potential influence of other elements[v]

- **Memory:** The ability to maintain networks made of connective associations between neuronal assemblies representing items of information[vi]

- **Working memory:** The ability to retain an item of information for the prospective execution of an action that is dependent on that given neuronal assembly[vii]

- **Planning:** The ability to form and execute goal-directed schemes of action guided by internal cues[viii]

- **Decision-making:** The ability to assess and evaluate schemes of action[ix]

- **Temporal integration:** The ability to complete temporally extended and goal-directed behaviors, speeches, or reasonings[x]

- **Monitoring:** The ability to test internal and external schemes of action to gauge their effect against goal-directed expectations[xi]

- **Inhibitory control:** The ability to suppress inappropriate schemes of action or items of information[xii]

Establishing your core practices fulfills the precondition for your prefrontal cortex to operate in an optimal social and cognitive configuration that maximizes its executive functions, especially the ones oriented to promote creative and critical thinking in the service of decision-making and adaptive behavior.[xiii] In this chapter, you will learn how to establish your core practices. You will learn how to fulfill the precondition for your prefrontal cortex to operate in an optimal social and cognitive configuration that maximizes its executive functions. You will analyze how to control these conditions to promote creative and critical thinking in the service of decision-making and adaptive behavior. You will also learn how to connect with your purpose, identify your priorities, and strategize your week, creating a schedule that allows you to find your focus and achieve a high level of performance.

Connect with Your Purpose

Your ability to connect and remain connected with a clear sense of purpose during your professional journey is necessary to sustain your core practices and achieve high levels of performance.[xiv] When you have an objective that you are pursuing with purpose-congruent actions, your brain maintains a neural representation of that given aspiration that drives your perception accordingly, allowing you to progress on your journey, inhibiting the perception of competing stimuli in the form of diverging priorities, objectives, and goals.[xv] Your holistic purpose transcends three dimensions: personal, professional, and societal (Figure 4-2). This chapter will concentrate on your professional aspiration, which can be defined as *your reason to exist beyond making profits*; it represents your *why* and the impact that you want to have with your work.

Figure 4-2. The three dimensions of purpose

The internalized cognitive structures forming your purpose guide your perception and, consequently, your choices by evoking an underlying sense of "right" and "wrong," making meanings, isolating patterns, and developing internal priorities that motivate and drive you to its activation.[xvi] These priorities motivate your actions and serve as your guiding principles.[xvii] In the next paragraph, you will learn how to find and activate your professional purpose.

Find and Activate Your Professional Purpose

Connecting with what genuinely motivates you can be challenging, but it represents a contributory step to bring your authentic self to work and unlock your talent. The following three steps can guide you through this journey:

1. See it.

2. Name it.

3. Do it.

In the first step, you "see it." You look back in your past at what you have done, trying to identify moments of fulfillment; those are the experiences that remind you of your passion for design. You explore these events by answering the following three questions:

What are the moments that gave me a sense of fulfillment?

What are the things that I would do, even without being paid?

What is the contribution that I can give to the design community?

Think about three moments that you feel had a significant impact on your professional life. Describe those moments writing down, or sketching out, the scenario where you think you expressed yourself at your best. Now select the moment that generated the more profound sense of fulfillment and think about the persons you served, the type of impact you had, and annotate your reflection down.

In the second step, you "name it." You look forward to your future to craft the first iteration of your purpose statement. This declaration needs to be simultaneously broad enough to be aspirational and sufficiently specific to be meaningful. Capture your purpose statement using the following structure:

I exist to [desired impact] in order to serve [intended audience].

In the third step, you "do it." You activate and substantiate your purpose by translating it into priorities and connecting them to objectives, goals, and actions (Figure 4-3).

Figure 4-3. The purpose activation process

The following questions will help you to explore short-term and long-term objectives to define an effective combination of quick wins and bold moves across three different time scales:

Time scale: week

What conversation might I initiate?

What ritual might I test?

Time scale: quarter

What project might I start?

What initiative might I start?

Time scale: year

What objective might I accomplish?

What impact might I have?

In the next paragraph, you will learn how to periodically reflect on your professional purpose as a method to control the trajectory of your journey proactively.

Reflect on Your Professional Purpose

Your life evolves continuously, and there is the eventuality that, at some point, your professional purpose and role responsibilities will misalign. This misalignment can develop from internal factors contributing to an evolution of your priorities or external ones deriving from the development of your position. On some occasions, this situation can also originate from accepting a job offer intrigued by determinants such as money or responsibilities potentially disconnected from your professional purpose. Your priorities will also experience an organic evolution as you traverse the four stages of contribution: shifting your focus naturally from contributing independently to contributing via a team and workgroups.

Based on the structure of the developmental program introduced in Chapter 3, you can review your purpose statement yearly at the end of each macro iteration to verify its alignment with your current role and responsibilities. When you evaluate your purpose statement, invite reflection answering the following question:

What is the legacy that I want to leave behind?

Answering this question will help you investigate your current priorities connected with your purpose and juxtapose them with those defined by your current role and responsibilities. The following five additional questions can help you to assess this alignment further. Use a binary scale to answer them: agree or disagree.

The work I do has a special meaning, and it is more than a job.

The work I do provides me a sense of accomplishment.

I'm proud of the work I have done for my company.

I'm proud to share my work with other persons.

I'm proud to share where I work with other persons.

If you sense that your internal priorities shifted, reflect on them, and update your purpose statement accordingly. Practice reflection is essential to maximize your experiential learning.[i] In the last section of this chapter, you will learn how to invite reflection by exploring the reflective process and its different forms of expression. Assuming that you are in a role aligned with your professional purpose, in the next section, you will learn how to capture your role priorities and align them with your behavior to purposefully and proactively place yourself on a successful trajectory.

Align Your Behavior and Priorities

One of your prefrontal cortex's unique abilities is to structure the present in order to serve the future by apparently inverting the temporal direction of causality.[xviii] This ability allows you to perform objectives and goal-directed reasoning and behaviors.[xix] The neural representation of an objective with its connected goals and schemes of actions antecede and cause those actions to occur via the agency of the prefrontal cortex.[xx] In this section, you will learn how to identify and capture your priorities and discern and assess your behavioral configuration derived from your schemes of actions to direct your attention and drive your behavior purposefully.

Identify and Capture Your Role Priorities

Your role priorities are defined by what your company expects from you, predominantly via your line manager. Identifying and capturing these expectations represents a crucial preparatory phase to control your behavior purposefully, following, in order of reliability, three conventional sources that you can access to investigate them.

Formal performance review. This document officially defines your current and future expected performance, and it represents your most reliable source to understand your priorities. Your performance review is typically managed by the human resources department or, in small companies, directly from your line manager. The performance review document represents the most reliable source because it's an official deliverable, and it's continuously updated.

Job description. This document officially defines your expected performance updated to the day you signed your job contract, and it represents your second most reliable source to understand your priorities. Given its nature,

---

[i] Experiential learning is the process of learning through reflection on doing.

it may be outdated when you consult it and not reflect your current activities comprehensively. The human resources department updates your job description when your role officially evolves in some companies, but this practice is neglected in most organizations.

Line manager. Talking to your manager in line represents the third most reliable source to understand your priorities. The only reason why this source must be considered a less reliable one compared to the previous two is that informally conversing with your manager represents an unverifiable form of interaction. Despite its unofficial nature, your investigation must always begin with an exploratory conversation with your line manager because this approach will help you to fortify your relationship with that person. After that conversation, utilize your performance review document or your job description to continue your exploration.

Irrespective of the source that you decide to access, at the end of this phase, produce a list of role priorities and validate them with your line manager before moving forward with the process.

The Configuration of Your Behavior

Your behavioral configuration is delineated by the activities that you decided to pursue, and it's a function of your mental and physical activities: from their inner intention to their outer actualization. These activities represent the only variable that you can directly control to achieve your objectives and accomplish your goals.[xxi] Beyond the concrete actions that shape your behavior, every other element inside your working environment can only be managed indirectly by projecting influence. If you don't develop a continuous awareness of the time that you proactively or reactively consume with your actions, your probability of remaining aligned with your priorities is almost nonexistent and entirely outside your control.

Assess Your Behavioral Configuration

As a design manager and leader, you need to proactively allocate time to cultivate creative collaboration and optimize design operations concurrently, dedicating time to the creation of design and its management. Assessing your behavioral configuration allows you to identify the actions underpinning your conduct and determine their alignment with your role priorities. In Figure 4-4, you can see an example of possible design and operational priorities spanning over a period of one year.

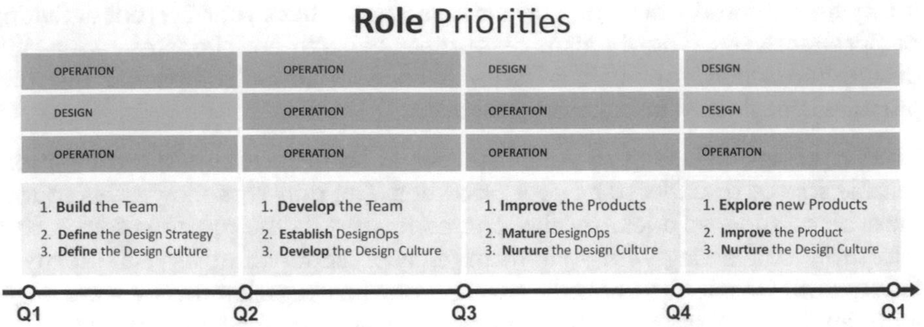

Role Priorities

OPERATION	OPERATION	DESIGN	DESIGN
DESIGN	OPERATION	OPERATION	DESIGN
OPERATION	OPERATION	OPERATION	OPERATION
1. **Build** the Team	1. **Develop** the Team	1. **Improve** the Products	1. **Explore** new Products
2. **Define** the Design Strategy	2. **Establish** DesignOps	2. **Mature** DesignOps	2. **Improve** the Product
3. **Define** the Design Culture	3. **Develop** the Design Culture	3. **Nurture** the Design Culture	3. **Nurture** the Design Culture
Q1	Q2	Q3	Q4	Q1

Figure 4-4. An example of design and operational priorities over a period of one year

The most effective approach to capture the actions delineating your current behavioral configuration is to perform a calendar audit completing the following three steps:

1. Identify the time frame.

2. Analyze the daily workloads.

3. Visualize the activities.

In the first step, you identify the time frame of the audit. Three months represent a reasonable interval for a calendar audit: select two weeks for each of the past three months.

In the second step, you analyze your schedule. Look at your calendar and identify the activities for those given weeks. Decompose all the tasks in working units using blocks of time formed by 30 minutes, approximating and rounding when necessary. If you spent additional time working outside your calendar or working hours, you need to factor that time into the analysis. Add all the tasks to a list and organize them by activity. For instance, if you made a phone interview, label it as "recruitment," or if you run a workshop, mark it as "product design." Once you have all your tasks organized, divide them into two categories: design and management. The following list captures some examples of possible activities:

Product design (e.g., workshops).

Team development (e.g., one-on-one).

Team administration (e.g., budget review).

Recruitment (e.g., interviews).

Personal growth (e.g., seminars).

Free time.

...

After you extrapolated a snapshot of your activities, count the working units of 30 minutes allocated to them. In the following, you can see an example of this calculation:

[category] Management

[Activity] Recruiting: 8 units (4 hours).

[Task] Creating a candidate evaluation framework: 2 units (1 hour).

[Task] Writing job descriptions: 2 units (1 hour).

[Task] Creating behavioral questions: 2 units (1 hour).

[Task] Making phone screenings: 2 units (1 hour).

[category] Design

[Activity] Product development 12 units (6 hours)

[Task] Design ideation: 6 units (3 hours)

[Task] Design review: 2 units (1 hour)

[Task] ...

In the third step, you add the working units to a spreadsheet, and you visualize the average time per activity using a nested donut chart to visualize the design vs. management activity ratio, with their relative breakdown (Figure 4-5).

Figure 4-5. Visualization of the design vs. management activity ratio

Your *current most effective* behavioral configuration, and consequently your optimal form of contribution, is a function of your current challenges and will evolve as you progress into a more senior role. Knowing how to perform a calendar audit will allow you to determine at any time the efficacy of your actions against your current role priorities.

Strategize Your Week

With a clear trajectory in front of you identified by your role priorities, you can begin to strategize your week. While there are different temporal lenses that you can utilize to manage your calendar, the week represents the most effective time frame to balance strategy and execution. As Peter Drucker reminded us in his book *The Effective Executive*, while there is value in processing your priorities in terms of days and months, the week represents the most pragmatic temporal unit of focus.[xxii]

Create Your Backlog

Your backlog captures your role priorities in the form of overarching role objectives. Begin by analyzing your role priorities, reflect on them, and isolate a list of necessary steps delineating a tentative journey aimed to fulfill them. These steps represent your objectives, which, consequently, will compose your backlog (Figure 4-6).

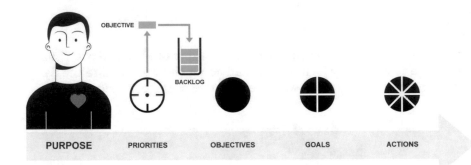

Figure 4-6. The backlog creation process

In the following list, you can see an example of this process elaborating a priority revolving around the creation of a new design team:

Priority: Create the design team.

Objective (step): Define team headcount and budget.

Objective (step): Define team structure and responsibilities.

Objective (step): Create a candidate evaluation framework.

Objective (step): Create a career section on the website.

Objective (step): Create job descriptions.

Objective (step): ...

After you have created the first draft, reflect on it, and then validate it with your line manager if some of the steps involved require a form of approval. Once you have finalized your list, your backlog will present a record of objectives categorized by priority. As you will explore in detail later in this chapter, you must allocate your attention selectively to increase your ability to focus, your capacity to implement a plan of action, and, ultimately, your opportunities to maximize your performance.[xxiii] Allocating your attention selectively requires you to understand your priorities to inform a prioritization of your undivided cognitive recourses toward specific task-relevant stimuli while temporarily ignoring any other agent, event, or situation.[xxiv] Achieving this awareness requires you to rank the priorities and objectives on your list, identifying their level of importance and urgency. In Figure 4-7, you can see a 2x2 matrix that you can use to investigate these two characteristics. Once you ranked all your priorities, isolate your currently three most important ones.

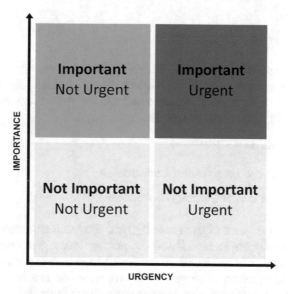

Figure 4-7. The Urgent-Important Matrix, also known as the Eisenhower Matrix

There are many factors to consider when you commit to the delicate decision of identifying your current top three priorities because, in some cases, focusing on the wrong ones can cost you your professional reputation or even your job. Your current top three priorities represent your *current area of focus*; they represent a cognitive lens that drives your perception and directly affects your performance, as you will learn later in this chapter.

Define Your Weekly Objectives

A weekly objective captures the most significant achievement that you can obtain over a week aimed at fulfilling a given priority. Defining your weekly objectives allows you to decompose your top three overarching role objectives into three weekly deliverables. These three weekly objectives will determine your three weekly areas of focus.

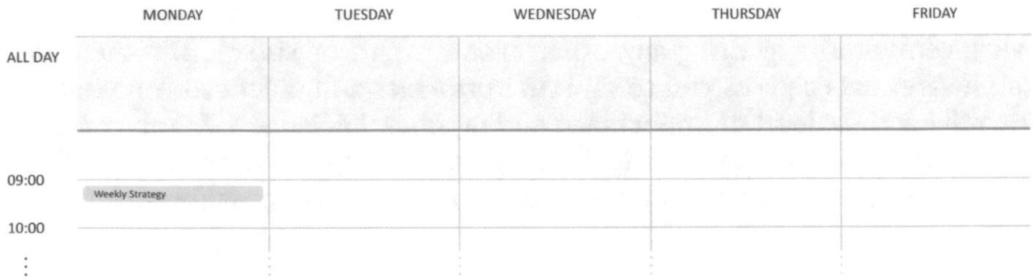

Figure 4-8. The weekly strategy session

Block 30 minutes early on Monday morning to run a weekly strategy session (Figure 4-8) to decide your three weekly focus areas, defining one weekly objective for each of your top three priorities; the following are the two points composing this meeting agenda.

Weekly strategy (30 minutes)

- Review the past week (5 minutes).
- Plan the current week (25 minutes).

During the "review" part of the meeting, you examine the status of your past week's three weekly objectives. If you did not achieve a given objective, reflect on the root causes of the situation, generate new insights, and leverage your augmented awareness to explore new approaches to reach that objective in the current week, making the necessary adjustments to your strategy if necessary. Instead, if you achieved your past week's objective, take a moment to reflect on your action to celebrate your success and understand if, despite the positive result, you still have margins for improvement.

During the "plan" part of the meeting, you look at your backlog, reflect on your top three priorities, and answer the following question:

What can I do this week to get one step closer to fulfill my role priorities?

The response to that question typically underpins results-oriented and social-oriented outcomes. Capture your answer with one weekly objective for each of your top three priorities. Add these three weekly objectives to your calendar in the form of seven-day-long events (Figure 4-9); use three different color labels to improve glanceability.

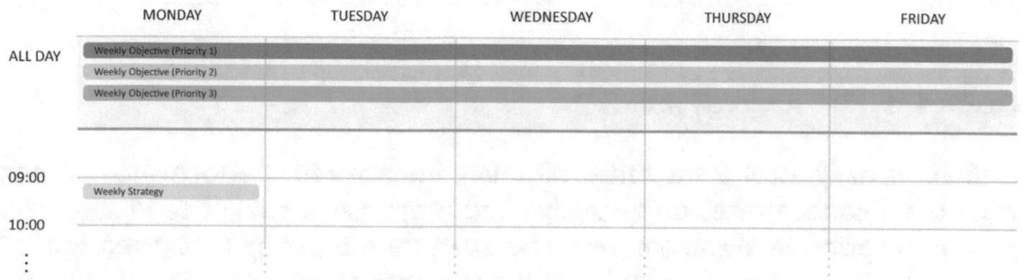

Figure 4-9. The three weekly objectives

It is important to emphasize that all your weekly objectives need to be cumulatively achievable in a weekly unit of time; if one of them is incredibly demanding, the other two must compensate for that fact to respect your weekly capacity.

Define Your Daily Goals for the Week

A daily goal captures the most significant accomplishment that you can obtain over a day aimed at the achievement of a given weekly objective. Defining one daily goal allows you to decompose your three weekly objectives into five daily deliverables. These five daily goals will determine your five daily areas of focus.

During the "plan" part of the meeting, after you defined your weekly objectives, reflect on them, and answer the following question:

What can I do today to get one step closer to achieving one of my weekly objectives?

The response to that question reflects the belief that, while you may have to initiate different daily interactions, only a subset of those actions represent your priority and connect to a specific and measurable daily goal. Capture your answer with one daily goal connected to one of your top three priorities (Figure 4-10). Add these five daily goals to your calendar in the form of all-day events; use the color labels corresponding to the connected weekly objective to improve glanceability. Each goal needs to be accomplishable in a daily unit of work.

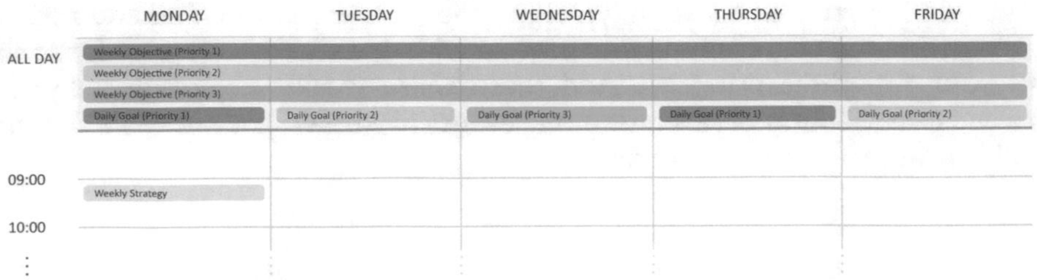

	MONDAY	TUESDAY	WEDNESDAY	THURSDAY	FRIDAY
ALL DAY	Weekly Objective (Priority 1)				
	Weekly Objective (Priority 2)				
	Weekly Objective (Priority 3)				
	Daily Goal (Priority 1)	Daily Goal (Priority 2)	Daily Goal (Priority 3)	Daily Goal (Priority 1)	Daily Goal (Priority 2)
09:00	Weekly Strategy				
10:00					

Figure 4-10. The seven daily goals

Defining a daily focal point can profoundly impact your performance. When your brain concentrates on a specific and measurable goal, it tends to drive your perception in alignment with the attitude framed by that given goal.[xxv] Under this condition, your daily focal point originates a cognitive filter that directs your attention throughout the day, enhancing the perception of what is considered potentially instrumental to the achievement of your goal and deprioritizing what instead is considered potentially distracting.[xxvi] This cognitive dynamic concretely affects what you "see" and what you "don't see" in your working environment.[xxvii]

The deliverable that you obtain at the end of each one of these meetings is a clear weekly strategy driven by three weekly objectives and five specific and measurable daily goals that you can track with your goal-setting framework of choice. Adopting a methodical approach to planning and executing your behavioral strategy will also facilitate the conversation with your line manager during your performance review.

Review Your Progress

The last step of every experience must always be punctuated with a moment of reflection to promote experiential learning. At the end of each day, take a moment to review your progress, investigate the consequences of your actions, and answer the following questions:

Did I achieve my daily goal, and why?

What challenges did I face during the day?

How can I improve my behavioral configuration?

Do I need to reframe or change my weekly objective?

Annotate your answers and decide the most appropriate course of action to factor them into your current schedule. With a precise understanding of your top three priorities, and a backlog able to effectively guide the production of your weekly and daily outcomes, in the next section, you will learn how to leverage your calendar to allocate time to support these areas of focus.

Block Time for the Top Three Priorities

When you align your behavior with your priorities, your mindset must change from *getting more done* to *getting the most important things done*, and your calendar must reflect this cognitive shift. The following is a possible structure that allocates a proportionated amount of weekly time to your top three priorities:

- **Priority 1:** Three slots of 60 minutes a week
- **Priority 2:** Two slots of 60 minutes a week
- **Priority 3:** One slot of 60 minutes a week

As long as you maintain the relationship between the number of slots allocated to each priority, you can also consider using different frequencies and combinations of 90-, 60-, and 30-minute slots if you feel that this approach better suits your context and necessities. These time slots form the backbone of your weekly behavioral strategy, ensuring that you will always be in a position to make weekly progress toward your priorities under any circumstance. These slots do not capture the entire time that you will spend working on your strategic aims, but they will secure you enough time to plan, execute, and review the strategic actions necessary to generate and maintain a positive momentum toward the achievement of your top three priorities. You can use these time slots to have formal meetings or informal conversations or, more commonly, do individual work and think in all its forms: from diverging to converging and reflecting.

Begin from your priority number 1 and assign three weekly 60-minute slots to it using a colored label to distinguish it from the other events in your calendar visually. Regardless if your priority is a design- or management-oriented activity, a good practice is to use Monday, Wednesday, and Friday to uniformly cover your week. Repeat the process for your priority number 2, preferring Tuesday and Thursday as dedicated days, and for your priority number 3, allocating it, preferably on Wednesday, to balance the workload (Figure 4-11).

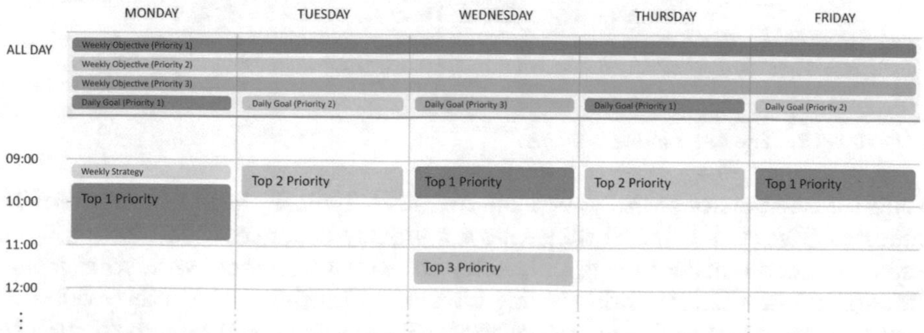

Figure 4-11. The top three priorities

This weekly pattern creates a cadence that allows you to focus on your priorities in order of significance, optimizing your weekly cognitive capacity.[xxviii] When an unforeseen event perturbs your strategy and requires a time slot in your current weekly schedule, consider moving other meetings before pondering rescheduling a session blocked to support one of your top three priorities. If you have to do it in a critical situation, you must reschedule on the same day and never considering removing it from your current weekly schedule.

Block Time for One-on-One Meetings

If you operate at Stage 3 or 4 of contribution and fulfill your role priorities via a team or workgroup, establishing healthy relationships with the members of these groups must be your number zero unwritten priority. Whether you work in a collocated or distributed environment with a physical or remote configuration, one-on-one meetings represent your primary tool to develop a trustworthy and respectful relationship with your collaborators, helping you to build relationship power and influence within the organization.

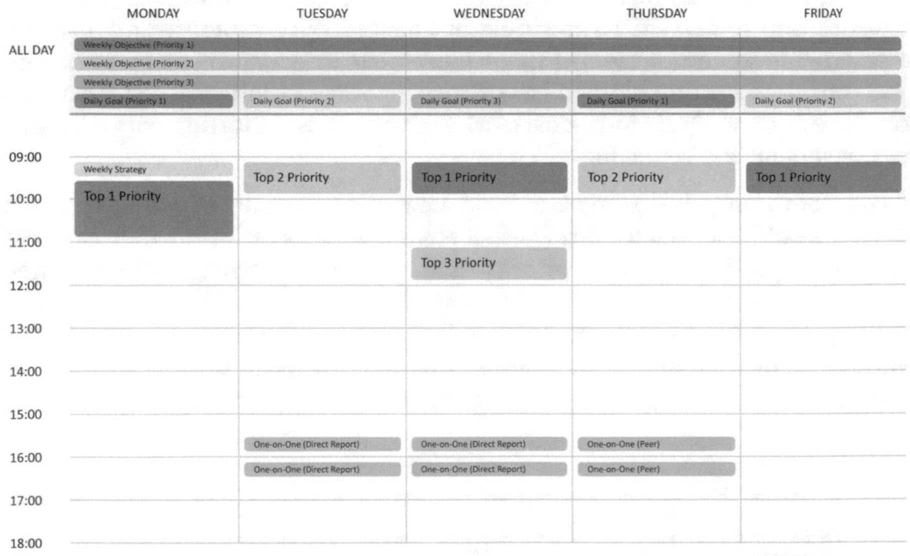

	MONDAY	TUESDAY	WEDNESDAY	THURSDAY	FRIDAY
ALL DAY	Weekly Objective (Priority 1)				
	Weekly Objective (Priority 2)				
	Weekly Objective (Priority 3)				
	Daily Goal (Priority 1)	Daily Goal (Priority 2)	Daily Goal (Priority 3)	Daily Goal (Priority 1)	Daily Goal (Priority 2)
09:00	Weekly Strategy	Top 2 Priority	Top 1 Priority	Top 2 Priority	Top 1 Priority
10:00	Top 1 Priority				
11:00			Top 3 Priority		
12:00					
13:00					
14:00					
15:00					
16:00		One-on-One (Direct Report)	One-on-One (Direct Report)	One-on-One (Peer)	
		One-on-One (Direct Report)	One-on-One (Direct Report)	One-on-One (Peer)	
17:00					
18:00					

Figure 4-12. The one-on-one sessions

One-on-one sessions with your team members typically have a duration of 30 minutes (Figure 4-12), and they follow a weekly cadence; if you have a large team, you can consider using a biweekly cadence to connect with your group. In addition to that, consider having biweekly, monthly, or quarterly one-on-one conversations with the key collaborators in your social system to nurture the relationship with these persons, whether peers or more senior personalities. Furthermore, if your line manager is favorable to one-on-one meetings, agree on a recurring day and time, and add it to the calendar. Analogous to the time slots allocated to your top three priorities, one-on-one meetings can be rescheduled to manage unexpected events but never canceled for the week; when unavoidable, reschedule them preferably on the same day and, in any case, within the same week. In Chapter 7, you will explore how to connect with your team using one-on-one meetings in detail.

Block Time to Manage Emails

Among all the activities not implicitly specified in your job description, reading and writing emails represent an activity that, if not properly structured and organized, can absorb a vast cumulative amount of time during the day, having grave repercussions on your efficiency and, in some cases, mental health.[xxix]

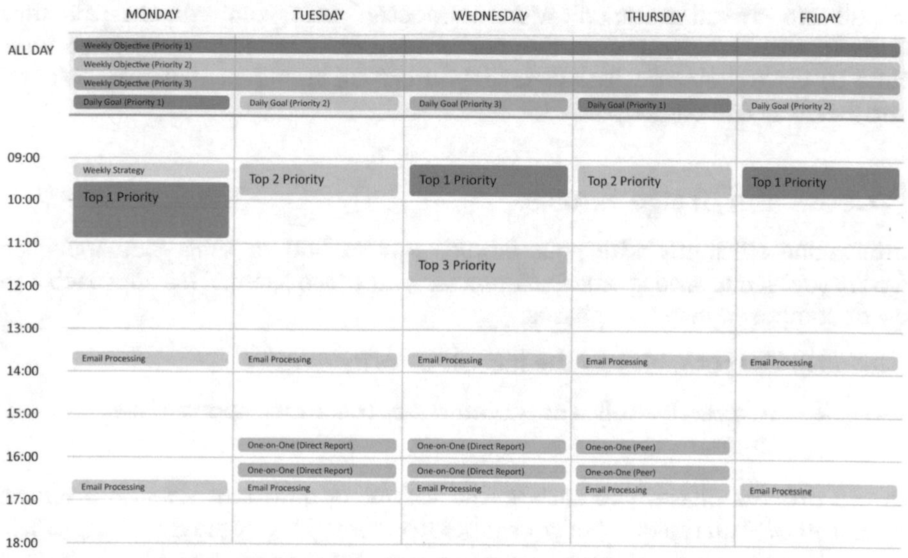

	MONDAY	TUESDAY	WEDNESDAY	THURSDAY	FRIDAY
ALL DAY	Weekly Objective (Priority 1)				
	Weekly Objective (Priority 2)				
	Weekly Objective (Priority 3)				
	Daily Goal (Priority 1)	Daily Goal (Priority 2)	Daily Goal (Priority 3)	Daily Goal (Priority 1)	Daily Goal (Priority 2)
09:00	Weekly Strategy	Top 2 Priority	Top 1 Priority	Top 2 Priority	Top 1 Priority
10:00	Top 1 Priority				
11:00			Top 3 Priority		
12:00					
13:00					
14:00	Email Processing	Email Processing	Email Processing	Email Processing	Email Processing
15:00					
16:00		One-on-One (Direct Report)	One-on-One (Direct Report)	One-on-One (Peer)	
		One-on-One (Direct Report)	One-on-One (Direct Report)	One-on-One (Peer)	
17:00	Email Processing	Email Processing	Email Processing	Email Processing	Email Processing
18:00					

Figure 4-13. The email processing sessions

Block two slots of 15 or 30 minutes each (Figure 4-13), one after lunch and one in the evening at the end of your working day where your cognitive resources and energy are more likely to be depleted. Answering your email during your low energy moments will also have the additional benefits of reducing your tendency to overcommit to demands, be more concise with your answers, and be more inclined to say no to nonprioritized work requests. Moreover, limiting your email management to two times per day outside the most prolific moments has the effect of protecting your efficiency levels and reducing your stress level.[xxx]

From a performance evaluation perspective, communication is subordinate to task completion. If you check your email instinctually during the day concurrently with other activities, you are doing a perilous disservice to yourself, dispersing cognitive resources like attention and resetting others like working memory.[xxxi] If allocating two slots a day to email processing requires you to reframe expectations around your time to respond within your organization, you can say something similar to the following phrases using its autoresponder functionality:

> *I check my email after lunch and late in the evening. If your message is urgent, visit me in the studio; otherwise, I'll respond to you before the end of the day.*

> *I'm slow to respond because I need to prioritize some important projects. If your message is urgent, visit me in the studio; otherwise, I'll respond to you before the end of the day.*

Two timeboxed email sessions during the day are sufficient to remain informed on all the critical communication connected with your current priorities without sacrificing your cognitive performances. If other persons need your time to drive their priorities instead, you can be confident that they will find another way to contact you.

Process and Triage Emails

Interacting efficiently with your inbox requires you to shift your approach from *reading and writing* emails to *process and triage* them. This approach can be decomposed into two phases:

1. **Process:** Identify the highest-priority email.

2. **Triage:** Identify and commit to the most appropriate action.

In the process phase, you answer the following question: *What is the most critical email?* You rapidly scan your inbox to identify the highest-priority email, mentally assigning it a degree of importance and urgency, factoring into your decision the sender, the title, and the content preview showed by your email client. Email related to your top three priorities must be considered having the highest degree of importance and urgency. Once you have isolated the highest-priority email, you begin to read it to understand how to respond.

Email Processing

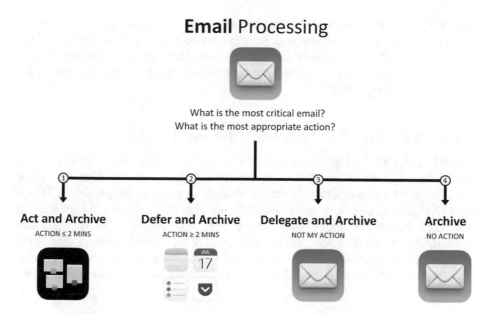

Figure 4-14. The email processing decisional tree

In the triage phase, you answer the question: *What is the most appropriate action?* You identify and commit to the most appropriate action based on four different conditions (Figure 4-14).

Act and archive: When there is an appropriate action from your side that takes less than two minutes. In this scenario, you complete the task and archive the email.

Defer and archive: When there is an appropriate action from your side that takes more than two minutes. In this scenario, always within the two minutes, use one of the following applications to postpone your next step, and then archive the email:

- **To Do:** If the task requires a tangible action
- **Calendar:** If the task requires a meeting
- **Note:** If the task requires answering more questions
- **Read-it-later:** If the action requires reading more information

Delegate and archive: When there is an appropriate action but you are not the most appropriate person to execute it. In this scenario, always within the two minutes, send an email to the delegated person explaining the matter, or requesting more time to explain the situation, then archive the email.

Archive it: When there is no appropriate action from your side that can add value to the conversation. In this scenario, archive the email directly.

You may have noticed that writing long emails is not contemplated, and for a valid reason, emails are inadequate for articulated conversations because their nature is asymmetric, inflexible, and predisposed to misunderstanding. If your answer requires a long email, request a face-to-face discussion, and move on with your triaging process.

Reserve Time for Networking Activities

Breakfast and lunches are great opportunities to spend time with your internal and external connections of any level of proximity; reserve at least two time slots of one hour per week for your networking activity (Figure 4-15). When you plan your networking activities with external connections, always initiate the conversation a few weeks before the intended day to increase the probability of acceptance. When you plan your conversations, you may experience rescheduling, cancellation, and sometimes misses, especially with external connections; don't take them personally and propose alternative moments.

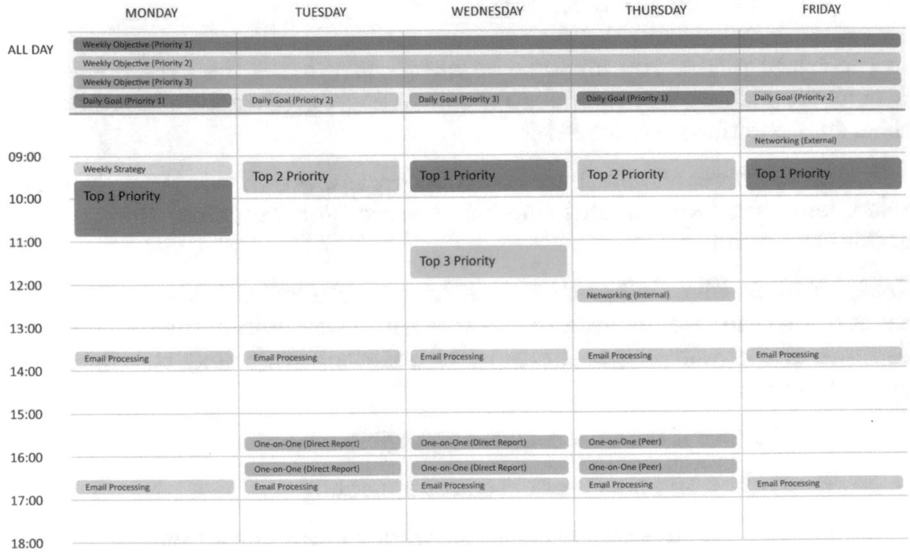

	MONDAY	TUESDAY	WEDNESDAY	THURSDAY	FRIDAY
ALL DAY	Weekly Objective (Priority 1)				
	Weekly Objective (Priority 2)				
	Weekly Objective (Priority 3)				
	Daily Goal (Priority 1)	Daily Goal (Priority 2)	Daily Goal (Priority 3)	Daily Goal (Priority 1)	Daily Goal (Priority 2)
09:00					Networking (External)
	Weekly Strategy	Top 2 Priority	Top 1 Priority	Top 2 Priority	Top 1 Priority
10:00	Top 1 Priority				
11:00					
			Top 3 Priority		
12:00				Networking (Internal)	
13:00					
	Email Processing	Email Processing	Email Processing	Email Processing	Email Processing
14:00					
15:00					
		One-on-One (Direct Report)	One-on-One (Direct Report)	One-on-One (Peer)	
16:00		One-on-One (Direct Report)	One-on-One (Direct Report)	One-on-One (Peer)	
	Email Processing	Email Processing	Email Processing	Email Processing	Email Processing
17:00					
18:00					

Figure 4-15. The networking sessions

Despite being challenging at times, networking provides long-term social benefits. Across the span of 12 months, having just one meeting per week with an internal connection will considerably solidify your internal social salience, and only one gathering per week with an external contact will significantly increase your social reach.

Reserve Time for Personal Development

Depending on the types of experiences, active or passive, that you added to your developmental plan, the time blocked in your calendar can be weekly, biweekly, or monthly. In Figure 4-16, you can see an example of a week that encompasses one learning activity.

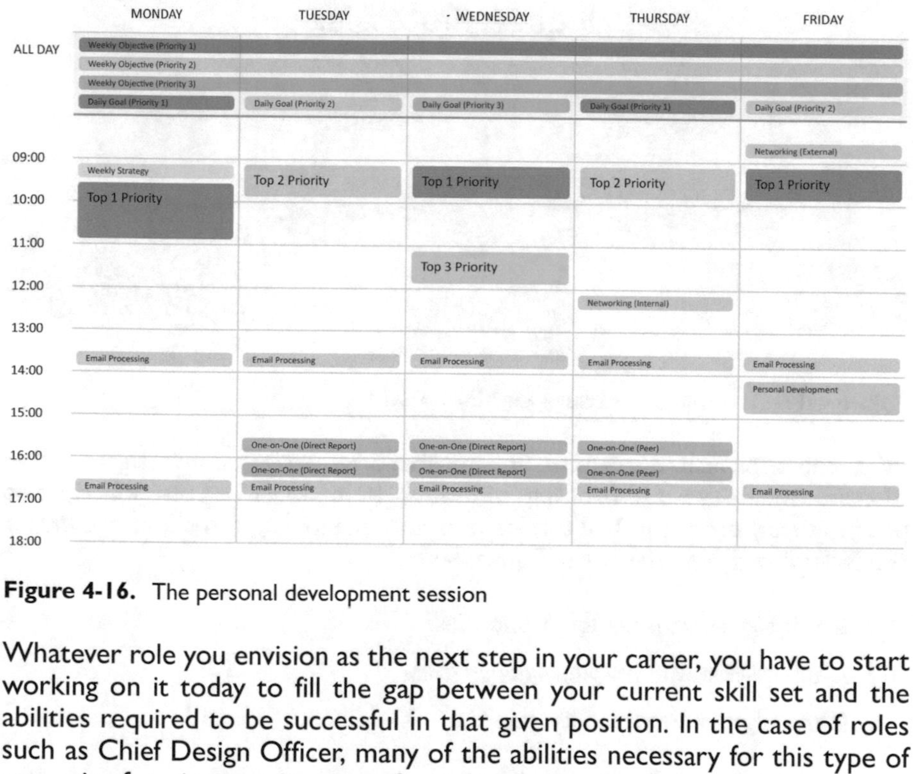

Figure 4-16. The personal development session

Whatever role you envision as the next step in your career, you have to start working on it today to fill the gap between your current skill set and the abilities required to be successful in that given position. In the case of roles such as Chief Design Officer, many of the abilities necessary for this type of executive function require years if not decades to develop, and it's, therefore, crucial for you to ensure that your developmental program always has a reserved place in your calendar to generate a continuous growth.

Govern the Residual Time

After you have allocated time for your prioritized activities, you need to concentrate on governing the residual time of your working week. Managing this portion of your time is an exercise in saying "yes" and "no" in a professional way to the multitude of expected and unforeseen time requests that you receive during the day from various angles of your organization. In the following sections, you will learn how to process and respond to different types of incoming time requests misaligned with your priorities.

Remain Aligned with Your Priorities

Despite all your best effort to strategize the "perfect" week, there is always a certain amount of unforeseen time requests that you cannot prevent and that you have to handle effectively without merely ignoring them. When you process an incoming request, you must switch your default mindset to decline.

Request Processing

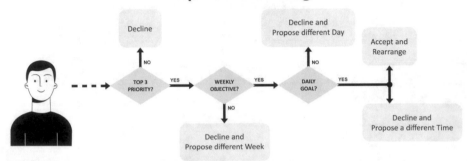

Figure 4-17. The request processing decisional model

With this approach, every time you receive a request, you investigate if that demand meets your *scheduling threshold*; if it doesn't, you politely and professionally decline it. You can inspect your scheduling threshold, answering the following three questions (Figure 4-17):

1. *Is this aligned with my top 3 priorities?*

2. *Is this aligned with my weekly objectives?*

3. *Is this aligned with my daily goal?*

If the request is not aligned with your current top three priorities, you decline it. If the request is aligned with your top three priorities but not with your three weekly objectives, you decline it and propose a different moment in the following weeks. If the request is aligned with your three weekly objectives but not with your daily goal, you decline it and propose a different moment within the current week. If a request is aligned with your daily goal but targets an inconvenient moment, you propose a different time during the same day, or you accept it and rearrange your schedule accordingly.

Declining requests from peers and direct reports. When you receive a request from a peer or a direct report, you need to process it considering its alignment with your role priorities. This scenario accounts for the vast majority of the requests that you may receive during your week. The following examples illustrate some of the approaches that you can use in this case:

Alignment: top 3 priorities (not clear, no agenda, no details)

... can you please share the agenda and help me to understand how I can contribute to the conversation?

Alignment: top 3 priorities (not clear, no details)

... can you please tell me more about what you intend to discuss and how I can contribute to the conversation?

Alignment: top 3 priorities (not aligned)

... unfortunately, I don't see that as something that can benefit from my contribution.

... unfortunately, I don't see that as something aligned with my current priorities.

... unfortunately, I don't see that as something that I can justify to [your line manager].

Alignment: top 3 priorities (yes), weekly objective (no)

... I'm not available [requested day], would [day, future week] work?

Alignment: top 3 priorities (yes), weekly objective (yes), daily goal (no)

... I'm not available [requested day], would [day, same week] work?

When the alignment is not apparent, you must always request more information before processing it. Instead, when a request is aligned with your priorities but not with your weekly objectives or daily goals, the logic is to postpone the conversation to a more appropriate moment for you.

Declining requests from more senior persons. When you receive a request from a more senior person, especially when that arrives from your line manager, you need to process it using a different approach. In this edge case, your mindset needs to switch from default-to-no to default-to-yes because while they may not be directly aligned with your role priorities, they almost certainly align with a broader priority for the company. This scenario accounts for the minority of the requests that you may receive during your week, and while this condition does not imply that you must accept every request passively, it requires you to be prepared to make an additional effort to accommodate them. The following examples illustrate some of the approaches that you can use in this case, considering that some of those requests can disrupt your work and potentially impede you from achieving your weekly or daily goals:

Alignment: top 3 priorities (no)

… [day] I needed to [task], but I can make it happen. I'm happy to make myself available and worry about my schedule later. See you then.

Alignment: top 3 priorities (no)

… [day] I needed to [task], but I can reorganize my calendar and make it happen.

In this situation, it is essential to mention the potential implications of a request on your role priorities to prevent misunderstandings and preserve healthy relationships. Furthermore, when your line manager demands your time without a comprehensive understanding of its implications, that person can inadvertently consider you directly responsible for the consequences of your modified behavioral configuration.

Find Your Focus

When human beings perform a behavior, they not only react to stimuli but also make flexible use of cognitive resources, such as attention, working memory, and internal triggers for action, to comply with task requirements.[xxxii] When you attempt to switch between an interrupted and an interrupting task, your cognitive resources allocated to the interrupted one do not immediately reconfigure on the interrupting one originating *residual switching costs*.[xxxiii] Depending on variables like your levels of intrinsic motivation and significance associated with the interrupted task, the attention residue generated can persist for a considerable amount of time.[xxxiv]

The demanding work you completed previously in this chapter to connect with your purpose and align your behavior with your priorities is part of a preparatory effort that will be futile if you don't learn how to find your focus and concentrate during the sessions that you carefully crafted and allocated in your calendar. Producing high-quality work and achieving peak performance requires you to operate for extended periods with deep concentration on a single task free from distraction.[xxxv] In order to achieve deep states of concentration, you must comprehend the role of your intensity of focus within the dynamics of productivity and go out of your way to minimize the effect of attention residue and maximize your concentration levels. Even Jonathan Ive, former Chief Design Officer at Apple, reminded us of this vital concept in one of his rare interviews, saying that the most important lesson he learned from Steve Jobs is that "you can achieve so much when you truly focus."[xxxvi] The following equation captures the relationship between quality of work, the intensity of focus, and time:[xxxvii]

High-Quality Work = (Intensity of Focus) × (Time Spent)

Your potential intensity of focus is also influenced by your ability to disengage cognitively from other activities and the magnitude of perturbation generated by your environment:

Intensity of Focus = (Ability to Concentrate) - (Ability to Disengage Cognitively) / Environmental Perturbation

In the following sections, you will learn how to maximize the work that you produce per unit of time spent taking decisive actions to address the critical variables of productivity.

Create Your Private Environment

Your private environment needs to promote deep concentration during your focused sessions; it needs to be a tranquil and single-minded space with solely the tools that you need to achieve your goal to prevent distractions.[xxxviii] A quiet and single-minded area is necessary to protect your working memory that, similarly to other forms of sustained attention, is inherently liable to interferences.[xxxix] This type of environment optimizes the inhibitory ability of your prefrontal cortex to protect the working memory from distractions and consequently preserve the temporal structures of behavior that this form of attention mediates.[xl] On a typical day, the following checklist can guide you to prepare a single-minded private environment capable of limiting interruptions, including the temptation of multitasking:

- Activate Do Not Disturb on all your devices.

- Close unnecessary applications on your computer.

- Check the presence of the necessary information.

- Check the presence of the necessary tools.

- Remove unnecessary objects and tools.

- Attend your biological needs.

While we all have personal preferences and diverse approaches to space management, there is only one configuration that optimizes your cognitive resource use during a focused session: the one that provides everything you need to complete your task and nothing else. Once you have created your private environment, you need to learn how to protect it from external perturbations.

Control the Surrounding Environment

Your surrounding environment needs to protect your state of concentration during your focused sessions; it needs to be a tranquil space that is unlikely to be perturbated by external events.[xli] Creating an efficient private environment is not enough if the persons around you invade it, redirecting your attention and resetting your working memory. Depending on the complexity of the problem that you are trying to solve, one interruption can be sufficient to compromise an entire session.[xlii] Do not hesitate to isolate yourself during your focused sessions that do not require collaboration if it's necessary to protect your state of concentration because, as Pablo Picasso emphasized a century ago, without great solitude, no serious work is possible.[xliii] If you work for a company that provides protected areas for focused thinking, use those spaces. If you instead work for an organization that solely operates in an open-office configuration, consider moving to a vacant meeting room, a quiet corner, or if you are allowed to do it, even search for an alternative place outside the building.

Develop Your Ability to Concentrate

Entering a deep state of concentration requires developing two interconnected aspects of your conscious executive function: manipulating relevant *items of information* using attention and working memory and blocking irrelevant ones utilizing the inhibitory control.[xliv] As illustrated in Figure 4-18, when the intensity of your concentration descends below the level of perturbation of your environment, you become susceptible to distraction.[xlv]

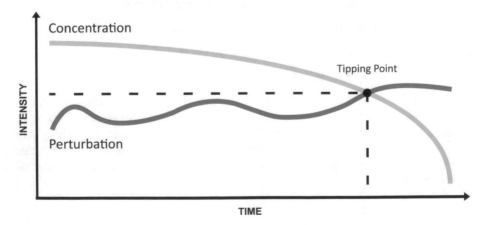

Figure 4-18. The tipping point that defines your distraction threshold

If you relied predominantly on multitasking activities in the past, finding your focus can be challenging initially, but, comparably to other abilities, these functionalities of your prefrontal cortex behave comparably to a muscle: the more you train them, the more they develop and remain in shape. If you blocked time for your top three priorities on your calendar, you have enough focused sessions per week at your disposal to produce rapid improvements.

Optimize Cognitive Disengagement

Exiting a deep state of concentration requires facilitating the process of resetting your cognitive resource configuration from the current task to the subsequent one.[xlvi] This process allows you to be cognitively available again.[xlvii] Every time you attempt to switch activity and disengage your cognitive resources, you inevitably originate a certain amount of residual switching costs that can degrade your performance.[xlviii] Residual switching costs develop from endogenous and exogenous factors.[xlix] Internally, this phenomenon is due to the unconscious transfer of inhibition and stimulus-response bindings from the previous activity to the new one.[l] Externally instead, it is influenced by your ability to create cognitive closure and the presence of a reliable schedule.[li] While you cannot avoid paying the endogenous residual switching costs, you can mitigate the exogenous ones by creating cognitive closure at the end of your sessions.

Before closing a stream of work, take a moment to reflect on what you produced and download your memory using your note application of choice, creating a series of action points to recall these items of information during the next instance of those sessions. The following questions can guide your thinking during this process:

Regarding the daily goal and weekly objectives

What did I complete in this session?

What do I need to complete in the subsequent sessions?

Regarding the current session

What do I need to remember in the subsequent session?

Regarding the next session

What do I need to do in my subsequent session?

Completing this additional step at the end of each focused session will reduce the need to maintain items of information in your memory, minimizing the generation of exogenous residual switching costs and facilitating your process of cognitive disengagement.[lii] Additionally, operating using the predictable schedule that you crafted in this chapter will further enhance your ability to disengage cognitively, nurturing your confidence in your ability to resume your work effectively in one of the subsequential preallocated sessions in your calendar. This condition also contributes to reducing the overall level of unnecessary stress associated with your role.[liii]

Find Career Support

Among all the persons that you can encounter during your professional journey, two figures have the ability to profoundly impact your career trajectory: a design sponsor and a design mentor. A design sponsor is an influencing person inside your organization that voluntarily supports you, placing their reputation on the line, taking responsibility for your promotion.[liv] On the other hand, a design mentor is a person inside or outside your organization who voluntarily supports you, offering psychological support for your personal and professional development.[lv] In some companies, sponsorship and mentorship are part of official programs, but if you work for an organization that does not offer this type of career support, you must be proactive and take full accountability for these initiatives. In Table 4-1, you can examine the distinct aspects of these two figures.[lvi]

Table 4-1. The peculiar aspects of a design sponsor and a design mentor

Design Sponsor	Design Mentor
Must be an influential manager at a senior level in the hierarchy.	Can be anyone at any level in the hierarchy.
Focuses on your professional advancement.	Focuses on your personal development.
Strives to increase your professional reputation.	Strives to increase your self-efficacy and self-worth.
Provides you with exposure to other executives who may help in career professional progression.	Provides you with emotional support, feedback, and advice to improve personal development.
Offers you access to social power, political abilities, and influence.	Offers you support to develop social power, political abilities, and influence.
Offers you protection from negative publicity or damaging contact with senior executives.	Offers you guidance on avoiding negative publicity and damaging contact with senior executives.
Makes sure you are considered for promising opportunities and challenging assignments.	Makes sure you qualify for promising opportunities and challenging assignments.
Serves as a social gateway.	Serves as a role model.

From the perspective of a sponsor or a mentor, having a protégé or a mentee represents a demanding commitment typically reserved for individuals considered "worthy." In the following sections, you will learn how to position yourself to be supported by one of these two figures and the differences between the sponsor-protégé and mentor-mentee relationships.

Position Yourself to Be Supported

Establishing a connection with a design sponsor or a design mentor is not something that happens organically. Analogously to other meaningful relationships in your social system, it requires a purposeful and proactive effort to demonstrate your worth and establish the necessary trust and respect to exchange tangible and mutual value. While sponsors and mentors typically commit to these types of relationships motivated by a feeling of social contribution,[lvii] they nevertheless dedicate their time, attention, and relationship capital only to individuals who perform exceptionally well, providing a tangible return on investment.

In some cases, supporting the rise of a promising designer can be one of the most significant achievements of someone's career. That was certainly true for Robert Brunner when he brought Jony Ive at Apple in 1992.[lviii] Brunner collaborated with Ive during his time at Tangerine, the design consultancy that the former CDO at Apple co-founded in London with Martin Darbyshire and the current Head of Service Design at the Royal College of Art Professor Clive Grinyer, who also wrote the foreword of this book.

Positioning yourself to be sponsored or mentored is about demonstrating that you represent a good investment by being successful in your current role and clearly on an ascending trajectory. The following three strategies can increase your opportunities to be noticed by a potential sponsor or mentor.

Define your professional trajectory. Establishing a productive relationship with your sponsor and mentor necessitates clarity of expectations. Articulating the steps underpinning your aspirations will increase your ability to identify the appropriate sponsor or mentor and spark their interest in your story.

Expand your social system. Demonstrating your worth requires you to be noticed. Expanding your social system will increase your social reach and, consequently, the quantity of sponsorship and mentorship opportunities available to you. In Chapter 5, you will learn how to build your social system.

Achieve results. Being successful in your role is the most important thing you can do to find career support. This concept may sound counterintuitive, but your sponsor and mentor can only help you elevate your career to the next level, leveraging your profound *strengths*. Occasionally, they can also improve your *weaknesses*, but they never help you fix your *fatal flaws*. If the investigation that you conducted in Chapter 3 suggested that you still have fatal flaws in your ability profile that prevent you from leveraging your strengths, you must fix them first using coaching sessions before actively searching for a sponsor or a mentor.

The Sponsor-Protégé Relationship

The relationship between the design sponsor and the protégé provides a strategic approach to the protégé's career advancement. This type of social connection is characterized by periodic structured conversations distributed on different time scales. Examples of these types of communications are occasional email, biweekly or monthly meetings, and quarterly discussions. While personal necessities and preferences uniquely define your relationship with your design sponsor, there are some aspects connected to your role as a protégé, which is essential to observe.

The Role of the Protégé

You can only have one relationship with one sponsor at a time, and your primary responsibility is to demonstrate loyalty. Your sponsor places their reputation on the line for you, and this commitment necessitates devotion in return to be sustained. This type of relationship is typically framed around the period of a given promotion, usually a few years, but in some cases, it can continue after that point if your new hierarchical position allows it or as a natural consequence of the trust, respect, and goodwill built during that time. In this context, you must concentrate on nurturing the social connection with your sole design sponsor, solidifying that person's professional reputation, and supporting the achievement of their objectives by providing complementary strategic perspectives and approaches.

Select Your Design Sponsor

Every person that appreciates your work, willing to endorse it at a more senior level within your organization, and advocating for your role advancement represents a promising candidate for a design sponsor. Depending on the pool of potential sponsors available in your company, you can assess these persons using two diametrically opposite but nonmutual exclusive approaches.

Receive direct career support. This approach focuses on increasing your professional reputation. In this scenario, you target a senior manager who esteems your work by initiating a relationship with that person and creating the social conditions necessary to begin advocating for a role advancement in your favor.

Receive indirect career support. This approach concentrates on helping a senior manager in achieving their objectives. In this scenario, you target a person that needs competencies that you can offer by initiating a relationship that supports their professional journey. Having a crucial role in the success and promotion of a senior manager often translates into role advancement.

Receiving direct support represents the preferable way to fast-track your career progression, but, unfortunately, it's an option that is not always available. In those cases, using an indirect approach can represent a valid alternative.

Make the Ask

While you may feel that you already connected with someone open to advocating for your promotion, you cannot rely on that until you make an official request to formalize the social contract. The nature of your relationship with the potential sponsor inside your organization will affect the formality and circumstances of your in-person ask. Assuming that you already introduced your career trajectory, you can make the formal ask using an approach similar to the following example:

> *... this is the professional trajectory I envisioned.*
>
> *Would you support me in that as my sponsor?*

Given that your potential sponsor works with you in the same company, if you established a good relationship with that person prior to the day of the ask, discussing your career ambitions and delineating what sponsoring you will bring in return for them, the form of the ask is not as important as it can be when instead you make it to one of your potential design mentors, which typically is an external connection in your social system. If your potential design sponsor accepts your offer, express gratitude, and formalize the social contract agreeing on the roles and responsibilities of the relationship. If the person declines your offer instead, demonstrate understanding, appreciation, and express gratitude for the time that they dedicated to you.

The Mentor-Mentee Relationship

The relationship between the mentor and the mentee provides a strategic approach to the mentee's personal development. This type of social connection is characterized by periodic structured and unstructured conversations distributed on different time scales. Examples of these interactions are occasional email, monthly phone calls, and quarterly face-to-face discussions.

While personal necessities and preferences uniquely define your relationship with your design mentor, there are some aspects connected to your role as a mentee, which is essential to observe.

The Role of the Mentee

You can and you must have multiple mentors at the same time, and your primary responsibility is to demonstrate dedication. Your mentor shares their knowledge with you, and this commitment necessitates dedication in return to be sustained. This type of relationship is typically framed around a period of a few years, but it's not uncommon to see this type of relationship evolve beyond that point as a natural consequence of the trust, respect, and goodwill built during that time. In this context, you must concentrate on expanding the overarching benefits of your mentorship, creating your personal *board of mentors*, where each director has the potential of being a design mentor in a specific area of expertise.

Select Your Mentors

Nobody can be considered the perfect mentor, and it's improbable for you to meet someone who possesses precisely the knowledge base that you need to develop all your strengths and improve all your weaknesses. The concept of your board of mentors exists to address this specific condition and aims to expand your opportunities to fulfill your developmental needs and career aspirations. A mentor working for your company can typically provide you an in-depth perspective on your organization dynamics, while mentors from other companies can usually offer you inspiring novel viewpoints. Irrespective of their working context, your board of mentors needs to span across five categories.

Connector

Main competency: Ability to form and maintain relationships. A connector is an individual with an extensive social network that can teach you how to establish and nurture relationships and solve social-oriented problems. This type of mentor can also act as a *superconnector* and introduce you to new individuals belonging to other social systems.

Politician

Main competency: Ability to make sagacious decisions. A politician is an individual sensible and judicious in the circumstances that can teach you the nuances of who supports whom and why. This type of mentor can help you understand how to express your opinion productively, characterize specific actions to protect your collaborators, and interpret the organizational chart identifying invisible relationships based on the influence beyond mere job titles.

Expert

Main competency: Ability to solve complex problems. An expert is an individual with a deep understanding of a narrow design domain that can teach you how to improve your and your team's performance. This type of mentor can also typically connect you with other domain experts to explore different areas of the five pillars underpinning the design Leadership Tent model.

Challenger

Main competency: Ability to push you beyond your perceived limits. A challenger is an individual committed to continuously improving their capabilities that can teach you how to push yourself outside your area of comfort, increasing your zone of proximal development; you will learn more about this dynamic in Chapter 9. This type of mentor can help you distinguish between healthy and toxic forms of stress and find opportunities to increase and reduce them accordingly to promote growth in your current environment.

Sage

Main competency: Ability to see and celebrate things in an empowering way. A sage is an individual that knows how to relate to events positively and constructively. This type of mentor can support you during challenging moments and teach you how to remain motivated and maintain positive momentum by learning from your mistakes effectively and focusing on the positive aspect of the events punctuating your professional journey.

Having your board of mentors composed of subject matter experts on critical areas for your professional development can profoundly impact your career trajectory, augmenting your abilities during challenging moments and propelling your growth to a level otherwise unattainable. Creating a board of mentors requires time, effort, and opportunities and therefore should be considered a priority early in your career.

Make the Ask

While your sponsor always resides inside your company, your potential mentors are more likely to be an external connection in your social system. Regardless of their working context, as you did with your sponsor, you need to make a request to formalize the social contract. In this scenario, the conversation typically begins with an informal discussion to explore the mentorship opportunity. You can approach the desired person using an approach similar to the following example:

> *… I deeply admire your [work/career/leadership/…] and wondered if I could ask your advice on my [work/career/…]?*

During that conversation, you can understand if a possible mentor-mentee relationship could or could not work and eventually prepare the ground for a formal ask. If you feel that your connection is robust enough to advance an offer, you can express your interest using an approach similar to the following example:

> ... *I would like you to consider mentoring me, and I would like to get on your calendar to meet you in person and formally ask you.*

This preparatory step will also provide an opportunity for the counterpart to decline the offer with dignity. If the person accepts the invite, you need to take full ownership of the event, organizing the day, time, location, and agenda of the conversation that typically can have a duration of a few hours, depending on whether you decide to structure the conversation around a coffee, a lunch, or a dinner. The following example provides you with some critical points that your agenda typically needs to capture.

Part 1: Introduce your professional story (1/4 of the time)

- Significant evolutionary milestones
- Significant projects and problems solved
- Significant achievements and awards

Part 2: Articulate your professional trajectory (2/4 of the time)

- Introduce your purpose.
- Introduce your 5, 10, or 15 years of envisioned trajectory.
- Describe your strategic objectives.
- Describe your developmental program.

Part 3: Frame the relationship (1/4 of the time)

- Contextualize the contribution of the mentor.
- Describe the contribution of the mentor.
- Agree on role and responsibilities.

The first part of the conversation is dedicated to introducing yourself and demonstrating that you represent a good investment by articulating the antecedent milestones of your professional story. Before the meeting, capture these milestones in a *storyfolio* and attach it to the calendar invite as a reference.

The second part of the conversation is dedicated to introducing your purpose and connecting it to your envisioned professional trajectory. Break down your vision into strategic objectives and connect them with the goals of your developmental program.

The third part of the conversation is dedicated to contextualizing the mentor's contribution to your professional story and trajectory by describing the type of support that you would like to receive from that person. If that person agrees with your view, you can advance your offer using an approach similar to the following example:

> *... will you mentor me for the next few years?*

If your potential design mentor accepts your offer, express gratitude, and formalize the social contract agreeing on the roles and responsibilities of the relationship. If the person declines your offer instead, demonstrate understanding, appreciation, and express gratitude for the time that they dedicated to you.

Invite Reflection

Reflection is the activity supported by your conscious executive function that unlocks a peculiar characteristic of the creative problem-solving process and broadly human beings: experiential learning. Experiential learning is the process of learning through reflection on doing and relies predominantly on the monitor ability of your prefrontal cortex to evaluate the effect of your actions against a given objective or goal.[lvix] As defined originally by John Dewey, the reflective practice represents the active, persistent, and careful consideration of any belief or supposed form of knowledge in the light of the ground that supports it.[lx] This method implies a questioning approach that challenges your mental models[2] by juxtaposing their underpinning *reflective assumptions* to how they might be with the intention of uncovering new possibilities in the form of experiential learning.

At the beginning of that inquisitory process, the acquisition of new information perturbates the equilibrium of your cognitive state, challenging the reflective assumptions sustaining your mental models.[lxi] This initial reaction generates a condition of *cognitive disequilibrium* that increases the levels of perceived ambiguity.[lxii] In this situation, reflection aims to revise your mental models by restructuring the cognitive associations that connect actions and responses to a given context, assigning new behavioral meanings to those associations. This process redefines your perceptual history and, consequently, your reflective assumptions that the brain uses to interpret the world, restoring a state of *cognitive equilibrium*[lxiii] (Figure 4-19).

---

[2] A mental model is an internal representation of the relations between a set of elements.

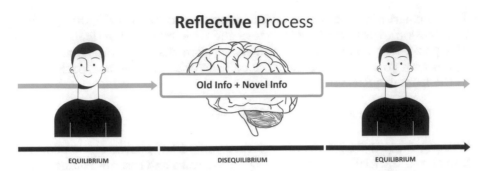

Figure 4-19. The reflective process

One of the implied intentions of your core practices is to set the preconditions required to promote experiential learning. Having time to invite reflection allows you to *learn by doing*, evaluating the effectiveness of your behavioral configuration against your priorities, objectives, and goals.[lxiv] Moreover, the act of reflecting promotes a sense of exploration and discovery that represents a prerequisite to success.[lxv] The reflective process can be classified into two distinct but interconnected practices: reflection-in-action and reflection-on-action.[lxvi]

Reflection-in-action. The practice of reflecting in-action is driven by an unconscious and unintentional process outside your control; it can be promoted but not voluntarily initiated.[lxvii] Reflection-in-action happens when you behave with an inquisitive attitude characterized by the engagement of your conscious executive function in concurrently executing a scheme of action and gauging its effect against goal-directed expectations.[lxviii] Reflecting in-action, technically referred to as *cognitive reappraisal*, is also known as *reframing* and represents one of the most peculiar skills of a designer and a foundational element of the creative problem-solving process.

Reflection-on-action. The practice of reflecting on-action is driven by a conscious and intentional process inside your control; it can be promoted and voluntarily initiated.[lxix] Reflection-on-action happens when you behave with an investigating attitude characterized by the engagement of your conscious executive function in using critical thinking to analyze a past scheme of action and gauge its effect against goal-directed expectations.[lxx]

Reflective Cycle

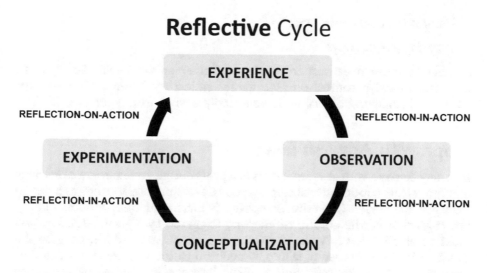

Figure 4-20. The reflective cycle

Figure 4-20 shows the reflective cycle underpinning the *experiential process learning*.[lxxi] Moreover, the practice of reflection creates the necessary conditions to promote and sustain intrinsic motivation.[lxxii] Intrinsic motivation refers to the spontaneous tendency to seek novelty and challenges, exercising and extending the capacity to explore and learn.[lxxiii] Inviting reflection-in-action during the problem-solving process fosters a sense of control that engenders intrinsic motivation, one of the strongest predictors of success.[lxxiv] Inviting reflection-on-action after the problem-solving process, experiencing positive emotions from the attainment of a valued outcome, represents a vital source of intrinsic motivation, which constitutes the foundational enabler of high performance.[lxxv]

Reserve Time for Reflection

If you want to learn and grow continuously, you need to have daily and weekly moments of individual reflection. Allocate time at the end of each day and week to evaluate your behavior against your daily goal and weekly objectives, answering the following set of questions. Enrich your feedback with a glyph, for instance, an emoji, to capture your general sentiment and increase glanceability in future reviews.

Did I achieve my daily goal/weekly objective?

What impeded me?

What have I learned today?

How can I use this learning to improve my future actions?

What was my sentiment today?

Why do I feel this way?

Answering these questions and acting per their answers will allow you to exercise proactive control on your career trajectory by ensuring that your behavioral configuration will not involuntarily deviate from your priorities.

Cope with Adverse Events

Learning to cope with adverse events is a prerequisite to experiential learning and represents a foundational step on your developmental journey as a design manager and leader. Cognitive objectivity is unattainable for a human being, but regardless of the emotional intensity triggered by a given situation, you must be able to detach yourself from the consequences and investigate the events effectively. In case of failure, in addition to the questions dedicated to your daily and weekly reflection, you can interrogate yourself by answering the following ones. You can also explore these questions with your coach or mentor to investigate the situations from different perspectives:

What was my objective/goal?

What was my strategy/behavior?

What went wrong?

What is the lesson to learn?

What can I change?

What are the repercussions on my strategy?

What is my next action?

You may have noticed that no questions are asking "why." This approach is intentional because understanding *why* something happened is not a prerequisite to creative and critical thinking and promotes self-deprecating evaluations.[lxxvi] Your attitude in situations where your behavior doesn't produce the effect that you envisioned defines your ability to learn from your experience and dictate how much knowledge you can extrapolate from it and, consequently, how much control you can exercise on your future. Developing this attitude requires the comprehension of a few foundational concepts.

Learning is omnipresent. Every experience can offer you something to learn if you ask yourself the appropriate questions.

Learning is a skill. The more you navigate adverse situations, the more you develop an affinity with the experiential process learning.

Failure is not a necessary and sufficient condition for learning. The more you concentrate on learning fast instead of failing fast, the more you accelerate your progress.

Failure can motivate you. Disappointment can fuel your determination to succeed if framed as a learning experience.

Avoiding failure is impossible. Life is a wicked problem that cannot be solved up front, and learning by doing, often characterized by unsuccess and disappointment, is an integral part of this journey.

Failure does not define you. There is an essential difference between accepting, processing, and learning from failure and seeing yourself as a failure.

Practice reflection can profoundly affect the trajectory of your career and, more broadly, your life. Based on a conservative estimation of 250 working days per year, holistically speaking, if you improve yourself by 1% every working day, you will have generated a 12X growth at the end of the year.[3]

Endnotes

i. Fuster, Joaquín M. The Prefrontal Cortex. Elsevier, AP, 2011.

ii. Fuster, Joaquín M. The Prefrontal Cortex. Elsevier, AP, 2011.

iii. Fuster, Joaquín M. The Prefrontal Cortex. Elsevier, AP, 2011.

iv. Fuster, Joaquín M. The Prefrontal Cortex. Elsevier, AP, 2011.

v. Fuster, Joaquín M. The Prefrontal Cortex. Elsevier, AP, 2011.

vi. Fuster, Joaquín M. The Prefrontal Cortex. Elsevier, AP, 2011.

vii. Fuster, Joaquín M. The Prefrontal Cortex. Elsevier, AP, 2011.

viii. Fuster, Joaquín M. The Prefrontal Cortex. Elsevier, AP, 2011.

ix. Fuster, Joaquín M. The Prefrontal Cortex. Elsevier, AP, 2011.

x. Fuster, Joaquín M. The Prefrontal Cortex. Elsevier, AP, 2011.

xi. Fuster, Joaquín M. The Prefrontal Cortex. Elsevier, AP, 2011.

xii. Fuster, Joaquín M. The Prefrontal Cortex. Elsevier, AP, 2011.

xiii. Collins, Anne, and Etienne Koechlin. "Reasoning, Learning, and Creativity: Frontal Lobe Function and Human Decision-Making." PLoS Biology, vol. 10, no. 3, 2012, doi:10.1371/journal.pbio.1001293.

[3] $1.01^{250} = 12.0321557683$

xiv. Judge, Timothy A., and Ronald F. Piccolo. "Transformational and Transactional Leadership: A Meta-Analytic Test of Their Relative Validity." Journal of Applied Psychology, vol. 89, no. 5, 2004, pp. 755–768., doi:10.1037/0021-9010.89.5.755.

xv. Grawe, Klaus. Neuropsychotherapy: How the Neurosciences Inform Effective Psychotherapy. Earlbaum, 2007.

xvi. Oyserman, Daphna. "Values, Psychology Of." International Encyclopedia of the Social & Behavioral Sciences, 2015, pp. 36–40., doi:10.1016/b978-0-08-097086-8.24030-0.

xvii. Parks-Leduc, Laura, et al. "Personality Traits and Personal Values." Personality and Social Psychology Review, vol. 19, no. 1, 2014, pp. 3–29., doi:10.1177/1088868314538548.

xviii. Fuster, Joaquín M. The Prefrontal Cortex. Elsevier, AP, 2011.

xix. Fuster, Joaquín M. The Prefrontal Cortex. Elsevier, AP, 2011.

xx. Fuster, Joaquín M. The Prefrontal Cortex. Elsevier, AP, 2011.

xxi. Mehrabian, Albert. Nonverbal Communication. Aldine Transaction, 2010.

xxii. Drucker, Peter. The Effective Executive. Routledge, 2018.

xxiii. Dalton, Amy N., and Stephen A. Spiller. "Too Much of a Good Thing: The Benefits of Implementation Intentions Depend on the Number of Goals." Journal of Consumer Research, vol. 39, no. 3, 2012, pp. 600–614., doi:10.1086/664500.

xxiv. Dalton, Amy N., and Stephen A. Spiller. "Too Much of a Good Thing: The Benefits of Implementation Intentions Depend on the Number of Goals." Journal of Consumer Research, vol. 39, no. 3, 2012, pp. 600–614., doi:10.1086/664500.

xxv. Gazzaniga, Michael S., et al. Cognitive Neuroscience: The Biology of the Mind. Norton, 2014.

xxvi. Gazzaniga, Michael S., et al. Cognitive Neuroscience: The Biology of the Mind. Norton, 2014.

xxvii. Gazzaniga, Michael S., et al. Cognitive Neuroscience: The Biology of the Mind. Norton, 2014.

xxviii. Newport, Cal. Deep Work: Rules for Focused Success in a Distracted World. Piatkus, 2016.

xxix. Kushlev, Kostadin, and Elizabeth W. Dunn. "Checking Email Less Frequently Reduces Stress." Computers in Human Behavior, vol. 43, 2015, pp. 220–228., doi:10.1016/j.chb.2014.11.005.

xxx. Kushlev, Kostadin, and Elizabeth W. Dunn. "Checking Email Less Frequently Reduces Stress." Computers in Human Behavior, vol. 43, 2015, pp. 220–228., doi:10.1016/j.chb.2014.11.005.

xxxi. Baddeley, Alan. "Working Memory." Current Biology, vol. 20, no. 4, 2010, doi:10.1016/j.cub.2009.12.014.

xxxii. Poeppel, David, et al. The Cognitive Neurosciences. The MIT Press, 2020.

xxxiii. Nieuwenhuis, Sander, and Stephen Monsell. "Residual Costs in Task Switching." Psychonomic Bulletin & Review, vol. 9, no. 1, 2002, pp. 86–92., doi:10.3758/bf03196259.

xxxiv. Leroy, Sophie. "Why Is It so Hard to Do My Work? The Challenge of Attention Residue When Switching between Work Tasks." Organizational Behavior and Human Decision Processes, vol. 109, no. 2, 2009, pp. 168–181., doi:10.1016/j.obhdp.2009.04.002.

xxxv. Newport, Cal. Deep Work: Rules for Focused Success in a Distracted World. Piatkus, 2016.

xxxvi. Vanity Fair. "Apple's Jony Ive on the Lessons He Learned from Steve Jobs." YouTube, 9 Oct. 2014, www.youtube.com/watch?v=2oksetv3i90.

xxxvii. Newport, Cal. Deep Work: Rules for Focused Success in a Distracted World. Piatkus, 2016.

xxxviii. Newport, Cal. Deep Work: Rules for Focused Success in a Distracted World. Piatkus, 2016.

xxxix. Fuster, Joaquín M. The Prefrontal Cortex. Elsevier, AP, 2011.

xl. Fuster, Joaquín M. The Prefrontal Cortex. Elsevier, AP, 2011.

xli. Newport, Cal. Deep Work: Rules for Focused Success in a Distracted World. Piatkus, 2016.

xlii. Monsell, Stephen, and Jon Driver. Control of Cognitive Processes. MIT, 2000.

xliii. Telephone Conversation with Pablo Picasso. ABC Diario Ilustrado De Madrid, 15 June 1932, pp. 35–36.

xliv.   Baddeley, Alan. "Working Memory." Current Biology, vol. 20, no. 4, 2010, doi:10.1016/j.cub.2009.12.014.

xlv.   Gazzaniga, Michael S., et al. Cognitive Neuroscience: The Biology of the Mind. Norton, 2014.

xlvi.   Leroy, Sophie, and Aaron M. Schmidt. "The Effect of Regulatory Focus on Attention Residue and Performance During Interruptions." Organizational Behavior and Human Decision Processes, vol. 137, 2016, pp. 218–235., doi:10.1016/j.obhdp.2016.07.006.

xlvii.   Leroy, Sophie, and Aaron M. Schmidt. "The Effect of Regulatory Focus on Attention Residue and Performance During Interruptions." Organizational Behavior and Human Decision Processes, vol. 137, 2016, pp. 218–235., doi:10.1016/j.obhdp.2016.07.006.

xlviii.   Nieuwenhuis, Sander, and Stephen Monsell. "Residual Costs in Task Switching." Psychonomic Bulletin & Review, vol. 9, no. 1, 2002, pp. 86–92., doi:10.3758/bf03196259.

xlix.   Leroy, Sophie, and Aaron M. Schmidt. "The Effect of Regulatory Focus on Attention Residue and Performance During Interruptions." Organizational Behavior and Human Decision Processes, vol. 137, 2016, pp. 218–235., doi:10.1016/j.obhdp.2016.07.006.

l.   Monsell, Stephen, and Jon Driver. Control of Cognitive Processes. MIT, 2000.

li.   Monsell, Stephen, and Jon Driver. Control of Cognitive Processes. MIT, 2000.

lii.   Leroy, Sophie, and Aaron M. Schmidt. "The Effect of Regulatory Focus on Attention Residue and Performance During Interruptions." Organizational Behavior and Human Decision Processes, vol. 137, 2016, pp. 218–235., doi:10.1016/j.obhdp.2016.07.006.

liii.   Peters, Achim, et al. "Uncertainty and Stress: Why It Causes Diseases and How It Is Mastered by the Brain." Progress in Neurobiology, vol. 156, 2017, pp. 164–188., doi:10.1016/j.pneurobio.2017.05.004.

liv.   HBR Guide to Getting the Mentoring You Need. Harvard Business Review Press, 2014.

lv.   HBR Guide to Getting the Mentoring You Need. Harvard Business Review Press, 2014.

lvi.   HBR Guide to Getting the Mentoring You Need. Harvard Business Review Press, 2014.

lvii. Hewlett, Sylvia Ann, et al. "The Relationship You Need to Get Right." Harvard Business Review, 1 Aug. 2014, hbr. org/2011/10/the-relationship-you-need-to-get-right.

lviii. Kahney, Leander. Jony Ive: The Genius behind Apple's Greatest Products. Portfolio Penguin, 2014.

lix.   Fuster, Joaquín M. The Prefrontal Cortex. Elsevier, AP, 2011.

lx.   Dewey, John. How We Think. Martino Publishing, 1910.

lxi.   Berger, Peter Ludwig, and Thomas Luckmann. The Social Construction of Reality a Treatise in the Sociology of Knowledge. Penguin, 1991.

lxii. Berger, Peter Ludwig, and Thomas Luckmann. The Social Construction of Reality a Treatise in the Sociology of Knowledge. Penguin, 1991.

lxiii. Berger, Peter Ludwig, and Thomas Luckmann. The Social Construction of Reality a Treatise in the Sociology of Knowledge. Penguin, 1991.

lxiv. Kolb, David A. Experimental Learning: Experience as the Source of Learning and Development. Prentice-Hall, 1984.

lxv.   Fogg, Brian Jeffrey. Tiny Habits: The Small Changes That Change Everything. Virgin Books, 2020.

lxvi. Schön, Donald. The Reflective Practitioner: How Professionals Think In Action. Basic Books, 1983.

lxvii. Munby, Hugh. "Reflection-in-Action and Reflection-on-Action." Education and Culture, vol. 9, no. 1, 1989, docs. lib.purdue.edu/eandc/vol09/iss1/art4/.

lxviii. Munby, Hugh. "Reflection-in-Action and Reflection-on-Action." Education and Culture, vol. 9, no. 1, 1989, docs. lib.purdue.edu/eandc/vol09/iss1/art4/.

lxix. Munby, Hugh. "Reflection-in-Action and Reflection-on-Action." Education and Culture, vol. 9, no. 1, 1989, docs. lib.purdue.edu/eandc/vol09/iss1/art4/.

lxx.   Munby, Hugh. "Reflection-in-Action and Reflection-on-Action." Education and Culture, vol. 9, no. 1, 1989, docs. lib.purdue.edu/eandc/vol09/iss1/art4/.

lxxi. Kolb, David A. Experimental Learning: Experience as the Source of Learning and Development. Prentice-Hall, 1984.

lxxii. Sansone, Carol, and Judith M. Harackiewicz, editors. Intrinsic and Extrinsic Motivation: The Search for Optimal Motivation and Performance. Academic Press, 2000.

lxxiii. Ryan, Richard M., and Edward L. Deci. "Self-Determination Theory and the Facilitation of Intrinsic Motivation, Social Development, and Well-Being." American Psychologist, vol. 55, no. 1, 2000, pp. 68–78., doi:10.1037/000 3-066x.55.1.68.

lxxiv. Ryan, Richard M., and Edward L. Deci. "Self-Determination Theory and the Facilitation of Intrinsic Motivation, Social Development, and Well-Being." American Psychologist, vol. 55, no. 1, 2000, pp. 68–78., doi:10.1037/000 3-066x.55.1.68.

lxxv. Di Domenico, Stefano I., and Richard M. Ryan. "The Emerging Neuroscience of Intrinsic Motivation: A New Frontier in Self-Determination Research." Frontiers in Human Neuroscience, vol. 11, 2017, doi:10.3389/ fnhum.2017.00145.

lxxvi. Sansone, Carol, and Judith M. Harackiewicz, editors. Intrinsic and Extrinsic Motivation: The Search for Optimal Motivation and Performance. Academic Press, 2000.

Build and Project Influence

Management, leadership, and influence are not the same entities, but they present deep intertwined elements. Coping with complexity and coping with change are the core activities of management and leadership, but it's the ability to influence individuals that enables a person to instantiate these intentions. Knowing is not enough; you must actualize the expertise, and your chances of success are connected directly with your propensity to project influence to deploy effective leadership and managerial competencies concurrently.[i] In the management age of empathy, influence is a central element of your behavioral profile that enables your capacity to achieve via abilities such as networking, collaboration, negotiation, and persuasion.[ii]

Influence is multifactorial in nature, and its constituents and dynamics, directly and indirectly, are present in every subject discussed in this book. In this chapter, you will learn how to build and project influence. You will explore the essence of influence, going beyond the widespread misunderstandings of concepts like social power and politics. You will analyze how to create the intention to perform a behavior and convert that intention into action. You will also examine how to create and maintain your social system to generate and project healthy and sustainable levels of influence.

© Andrea Picchi 2022
A. Picchi, *Design Management*, https://doi.org/10.1007/978-1-4842-6954-1_5

Comprehend Influence

Influence can be described as the power to have an effect on something or someone in a given context,[iii] and, as Cade Massey reminds in his MBA lectures at the Wharton Business School, it represents the power to move others to take actions that they would not otherwise.[iv] Social influence materializes via an active or passive act of persuasion[v] and instantiates the quintessential ability of a manager with leadership capabilities. Without this capacity, an individual in charge of a group that contributes via others is powerless.[vi] In the management age of empathy, power, and therefore influence, is granted by interlocutors, not acquired against the will of the counterparts in a Machiavellian manner.[vii] Under this condition, influence, in all its forms, represents all you possess to propel your team and organization to success.

The Nature of Influence

The nature of influence behind an act of persuasion is vastly misinterpreted and often demonized; nonetheless, it represents a foundational event of our lives as social beings.[viii] Understanding or modifying another person's perspective is one of the most vital abilities that human beings developed. After more than 2000 years of social evolution, the nature of influence with its underpinning concepts of ethos, pathos, logos, topos, and kairos, articulated by Aristotle in his treatise *Rhetoric*,[ix] is still unchanged. Within a modern organizational context, this type of social interaction permeates six fundamental dimensions.[x]

Sources: Where does influence originate?

Channels: Where can influence be deployed?

Scope: What is the influence range of compliance?

Efficacy: How much resistance can influence overcome?

Immediacy: How fast can influence overcome resistance?

Longevity: How long can influence be successfully exercised?

In opposition to interactions such as manipulation and coercion, a healthy and sustainable act of persuasion is characterized by specific attributes spanning across five domains.[xi]

Intention: Influence implies a behavior driven by a deliberate effort. It's an act that the influencer executes with a specific objective in mind as a means to an end.

Change: Influence implies the modification of a mental state. It's an act that the influencer executes to change the mental state of the influencee as a precursor to a change in behavior.

Freedom: Influence implies the presence of a degree of freedom. It's an act that the influencer executes in respect of the free will of the influencee.

Success: Influence implies the attainment of an objective. It's an act that the influencer executes to cause a specific behavior on the side of the influencee.

Communication: Influence implies the sole use of exchange of information. It's an act that the influencer executes predominantly via the medium of language.

Given a specific context, your ability to build and project influence is determined by your political skills and social power. In the following sections, you will explore the role of these two crucial factors.

Political Abilities

The level of influence projected is a function of your political abilities. A politician is defined as an individual who has the ability to make the right decisions,[xii] someone sensible and judicious in the circumstances.[xiii] Gerald Ferris, Professor of Management and Psychology at Florida State University, continuing the work on social influence in organizations initiated by Jeffrey Pfeffer, Professor of Organizational Behavior at the Graduate School of Business at Stanford University, elaborated and validated the concept of the *Political Skill Inventory*[xiv] founded on four abilities.

Social astuteness: Leverage self-awareness and social awareness to interpret the environment's intricacies and others' emotional and behavioral substrate.[xv]

Interpersonal power: Leverage self-management and social management to adapt to the environment's intricacies and others' emotional and behavioral substrate.[xvi]

Networking proficiency: Forge strong bonds for friendships, alliances, and coalitions, leveraging vast networks to achieve objectives.[xvii]

Apparent sincerity: Interact with other individuals projecting trustworthiness and credibility, avoiding being perceived as manipulative or coercive.[xviii]

As you may have noticed, political skills leverage all the four domains of emotional intelligence to connect with other individuals. The questionnaire in Table 5-1 can help you to self-assess your political skills and develop an understanding of your current proficiency.[xix]

Table 5-1. Survey questionnaire to assess the four domains of the Political Skill Inventory

Social astuteness$^\alpha$
- I sense the intention and motivations of others.
- I pay attention to the nonverbal communicative cues of others.
- I listen to understand the perspective of others.
- I assess a situation before presenting an idea to others.
- I am conscious of how I am perceived by others.

Interpersonal power$^\alpha$
- I communicate effortlessly with others.
- I develop rapport with others.
- I find common ground with others.
- I collaborate well with others.
- I solve problems leveraging the perspective of others.

Networking proficiency$^\alpha$
- I build relationships with influential persons.
- I have developed an extensive network of influential persons.
- I make myself visible with influential persons in my organization.
- I can rely on my connections when I need support.
- I help my connections to build their network.

Apparent sincerity$^\alpha$
- I discuss my intentions with others.
- I do not hide my emotions.
- I align my behavior with my feelings.
- I align my actions with my communication.
- I demonstrate an interest in the needs of others.

α 7-Point scale: Strongly disagree, Disagree, Slightly disagree, Neutral, Slightly agree, Agree, Strongly agree.

Politics is what you practice when you cultivate creative collaboration, optimize the design operations, and illuminate the business impact of design across your organization. As you will learn in Chapter 9, your political skills are a determinant of team cohesiveness, which in turn represents a precondition to leverage diversity and enhance creative collaboration.[xx]

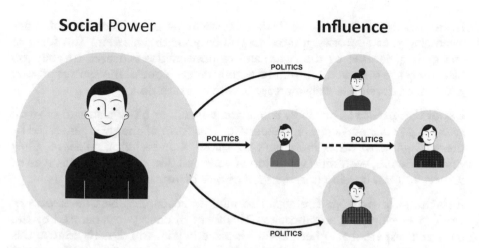

Figure 5-1. The interdependence between social power, politics, and influence

Managing like a leader requires influence, projecting influence necessitates political skills, and political skills require social power (Figure 5-1). Politics, and as you will learn in the next section, social power, is not inherently good or bad; it's a tool defined by its use.

Social Power

Social power is the capacity to initiate, sustain, or block a behavior[xxi] or, from an operational perspective, a stream of work. Within a given context, a person's potential level of projectable influence is a function of their social power. From a collective perspective, social power is vital to maintain a functional organization.

Today, it is not uncommon to encounter an aversion toward the use of the word "power," which frequently identifies a socially negative connotation. Influence, and by extension social power, can be undoubtedly acquired and exercised for malicious purposes, but this doesn't inherently corrupt it. Power must be approached as a tool, like the post-its, the whiteboards, or the laptop that you use every day at work, and the way you use these instruments defines more yourself than the tools themselves.

The Three Types of Social Power

The influence that you project on something or someone is originated from three foundational types of social power.[xxii] While these three forces are intrinsically interconnected, and to some degree, simultaneously processed by your interlocutor's mind, the resultant degree of influence is always driven by one primary social driver: the one that generates the vaster amount of energy. The following three types of social power define the "source" of influence, the dimension that identifies the *origin of compliance*.

Role power: The ability to initiate, sustain, or block a stream of work leveraging your authority granted to you by your organization. This form of energy is generated by the rights and responsibilities allocated to your job title. Only the organization, with its authority, can control this type of power; you cannot develop it, only increase it with a promotion.

Expertise power: The ability to initiate, sustain, or block a stream of work leveraging your knowledge and expertise. This form of energy is generated by the willingness of other individuals to follow your technical guidance. Only your colleagues, with their validation of your abilities, can control this type of power; you can develop it in the span of one or more decades.

Relationship power: The ability to initiate, sustain, or block a stream of work leveraging your social capital. This form of energy is generated by the trust and respect that other individuals place in you. You directly control this type of energy with your behavior; typically, you can develop it in the span of months or years.

While relationship power represents undoubtedly the healthier generator of influence, all the enduring examples of leadership come from individuals who understood how to draw from all three different sources continuously and often concurrently. All three types of social power can be decisive at some point in your career. Utilizing them consciously to achieve your objectives demands you an understanding of the effects deriving from their usage. As you will learn in the following sections, the three different social powers affect the *perceived relationship quality* between the influencer and the influencee in distinct manners.[xxiii]

Effects of Role Power

If you drive the design conversation leveraging your authority conferred by your role, you are using your role power. The predominant effect of using role power is a decrease in the levels of *perceived group cohesiveness*[xxiv] (Figure 5-2). When you leverage your authority to achieve a goal, you erode the group's perception of their "collective efficacy" and, consequently, their ability to collaborate creatively.

Role Power   **Influence**

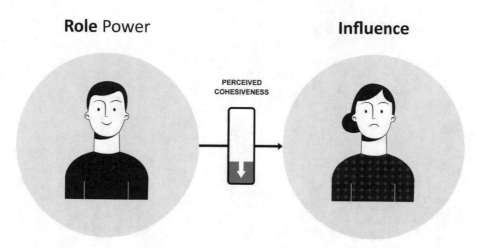

PERCEIVED
COHESIVENESS

Figure 5-2. The primary effect of the use of role power

Role power is the form of social energy required in certain moments to maintain operational momentum unblocking design conversation that reached an impasse. The perception of role power is based on the position of your job title inside the organizational hierarchy. You must consider the use of role power only in specific situations such as decision-making, promoting, and hiring and when you need to "lead behind" through culture creating rituals and managing creative tension.

Effects of Expertise Power

If you drive the design conversation using your technical abilities, you are using your expertise power. The predominant effect of using expertise power is an increase in the levels of perceived proficiency[xxv] (Figure 5-3) attributed to you by your team. When you leverage your technical abilities to achieve a goal, you increase the group's perception of your efficacy and, consequently, their ability to collaborate creatively.

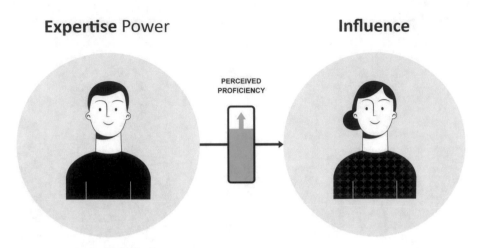

Expertise Power

PERCEIVED
PROFICIENCY

Influence

Figure 5-3. The primary effect of the use of expertise power

Expertise power is the form of social energy required to elevate behavioral standards and grow the three foundational pillars of an organization: persons, processes, and products. The perception of expertise power is relative to the context, and it's a function of the difference between your level of abilities and the average level of competence present in your environment. You must consider the use of expertise power when you need to "lead alongside" your team in direct contact with their activities.

Effects of Relationship Power

If you drive the design conversation leveraging your social capital, you are using your relationship power. The predominant effect of using relationship power is an increase in the level of *perceived trustworthiness*[xxvi] (Figure 5-4) attributed to you by your team. Leveraging your social capital to achieve a goal increases the group's perception of their cohesiveness and, consequently, their ability to collaborate creatively.

Relationship Power   Influence

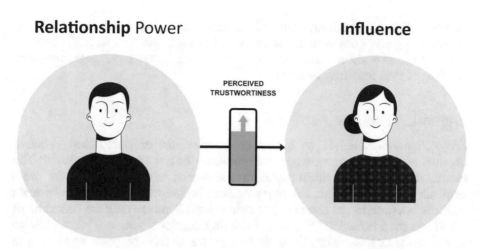

Figure 5-4. The primary effect of the use of relationship power

Relationship power is the form of energy required to support the behavioral dynamics of creativity and create a healthy, sustainable, and productive environment for diversity. The perception of relationship power is built on a feeling of mutual trust and respect developed from experiencing moments characterized by shared passions, challenges, and activities. Once trust and respect are established, relationship power is predominantly propelled by the principle of *liking* and *reciprocity.*[xxvii] You must consider relationship power as your primary form of cooperation and when you need to "lead ahead" with a strong viewpoint.

Project Influence

Projecting influence requires an active or passive act of persuasion. Ethically persuading another individual, technically referred to as *extracting compliance*, is a complex, context-dependent, cognitive process underpinned by two main phases: creating the intention to perform a given behavior and converting that specific intention into action.[xxviii]

Create the Intention to Perform a Behavior

Generating the willingness to perform a behavior in your interlocutor's mind is a bespoke event characterized by three complementary activities.[xxix] Fulfilling the requirements of these activities represents a necessary and sufficient condition for creating the intention to perform any behavior:[xxx]

- **Activity 1:** Engender positive attitude.

- **Activity 2:** Manage social considerations.

- **Activity 3:** Promote perceived ability.

In the following sections, you will explore the foundation of influence, investigating the constituents of an act of persuasion and learning how to approach these three activities and, when it's necessary, overcome the challenges that they may present.

Engender Positive Attitude

Projecting influence via an active or passive act of persuasion requires producing a mental state change as a precursor to a change in behavior.[xxxi] The mental state that behavioral psychology research has identified as the most centrally implicated in the act of persuasion is the *attitude*.[xxxii] An individual's attitude is a function of the person's salient beliefs concerning the outcome of the demanded behavior.[xxxiii] In the following sections, you will learn how to engender a positive attitude articulating the details of your request and address eventual counterarguments efficiently.

Leverage Consequence-Based Arguments

Using a consequence-based argument to extract compliance requires you to articulate the implications derived from the details of your request using a conditional approach under a *Gain-Loss Framing*: if this happens, then that happens. The strongest is considered the causal relationship between your request and its consequences, the more the message will result to be persuasive.[xxxiv] Your request's details can be phrased using a *gain-framed* message that emphasizes the advantages of undertaking the advocated course of actions or using a *loss-framed* message that instead emphasizes the disadvantages of not engaging in the urged behavior.[xxxv] The most appropriate method to use is the one that can be perceived as having the most substantial causal relationship from your interlocutor's perspective; the decisional context and the interlocutor's personality are discriminants in this situation.[xxxvi]

Unequivocal evidence demonstrates that human beings tend to manifest an asymmetry in the way that positive vs. negative information is perceived and interpreted during the judgmental and decision-making process.[xxxvii] As a consequence of the correlations between emotions and perceptions, knowledge with a negative connotation tends to influence evaluation and decision-making processes more strongly than comparably positive information.[xxxviii] This tendency is called *negativity bias*[1] and serves a critical evolutionarily adaptive function. Neuroscience demonstrates how our brain uses three times more neurons to investigate the surrounding environment looking for potential adverse events and harmful consequences than positive

---

[1] The negativity bias refers to our proclivity to attend to, learn from, and use negative feedback far more than positive information.

ones.[xxxix] In Chapter 9, you will explore the interplay between emotions, judgment, and decision-making.

Imagine your team or workgroup is not in a position to test the design properly, and you want to request additional funds to create a usability lab. A possible gain-framed approach can be based on the fact that producing a continuous stream of user feedback can allow the team to make user-centered decisions that precisely fulfill specific user needs and, consequently, increase the annual revenues. Instead, a possible loss-framed approach can be based on the fact that not producing a continuous stream of user feedback will force the team to make assumption-based decisions that will increase the annual expenditure due to the trial-and-error approach.

Leverage Consequence Desirability

Using the desirability of a consequence to extract compliance requires you to articulate the implications of your request under the priorities of your interlocutor. The more a consequence is perceived as desirable, the more the message will result to be persuasive.[xl] The most desirable implication is the one that your interlocutor considers to be so and can be identified by answering the question: What is the main priority of my interlocutor concerning this request?

Imagine you want to promote human-centered design during a business strategy conversation. If you know that your interlocutor's priority is to explore new streamlines of revenues, you can observe that adopting a human-centered approach will allow the team to explore unmet needs and develop multiple concepts safely and inexpensively.

Leverage Consequence Likelihood

Using the likelihood of a consequence to extract compliance requires you to articulate the implications of your request, emphasizing the positive probability that these consequences will occur. The more a consequence is considered probable, the more the message will result to be persuasive.[xli] The perception of the likelihood of a consequence is affected by an individual bias captured by a construct called *Consideration of Future Consequences* (CFC), which refers to the differences in the degree to which individuals consider shorter-term as opposed to longer-term behavioral effects.[xlii]

A message that emphasizes short-term consequences results in being more persuasive to individuals with a bias toward tactical decisions. In contrast, a communication that emphasizes long-term effects results in being more persuasive to individuals with a tendency toward strategic commitments.[xliii] Irrespective of these two opposite mindsets, you can use two connected strategies to illustrate the likelihood of a consequence.

Invoke parallel cases: Describe similar scenarios in which the claimed consequence happened.

Illuminate the underlying mechanisms: Describe the details of how the advocated action can produce the claimed consequence.

Imagine you want to promote a human-centered approach to business within your organization during a conversation with your company's CEO. Using the first strategy, you can invoke parallel cases describing that other companies in a similar situation embraced human-centered practices and increased, for instance, their revenues, brand reputation, and employee satisfaction. Individuals will find it easier to believe that an outcome will occur if they see that it has already happened in the past in similar circumstances.[xliv] Using the second strategy, you can illuminate the underlying mechanism describing how these other companies developed a human-centered approach to business, presenting a detailed representation of their journey. Individuals will find it easier to believe that an outcome will occur if they can understand how that outcome has already been achieved in the past.[xlv]

Regardless of the strategy deployed, invoking parallel cases, or illuminating the underlying mechanisms of those given stories, you must always leverage a combination of qualitative and quantitative content. The quantitative data will help you set the rational foundation of your argument, while the qualitative details will allow you to engage with the interlocutor emotionally. In Chapter 8, you will learn how to balance these two essential elements to craft compelling stories.

Address Counterarguments

A counterargument represents the contradiction of one or more specific points in a thesis during a persuasive message.[xlvi] Engendering a positive attitude in your interlocutor's mind requires you to be prepared to handle possible disagreements and counterarguments during the conversation, leveraging three distinct communicative strategies.[xlvii]

One-sided message: This strategy does not mention or acknowledge counterarguments and does not attempt to refute them. This type of communicative approach concentrates on offering constructive and supporting arguments only.

Nonrefutational two-sided message: This strategy acknowledges and attempts to refute opposing arguments indirectly. This type of communicative approach concentrates on overwhelming a counterargument with supportive ones without challenging its underlying claim and relevance.

Refutational two-sided message: This strategy acknowledges and attempts to refute opposing arguments directly. This type of communicative approach concentrates on challenging the relevance of a counterargument criticizing its reasoning and offering evidence to undermine its underlying claim.

Multiple studies demonstrated that the *refutational two-sided* approach possesses a dramatically superior persuasive effect compared to the *nonrefutational two-sided* one.[xlviii] The *nonrefutational two-sided* strategy also results in being counterintuitively less effective than a *one-sided* message.[xlix] When your goal is to be persuasive, unless the counterargument is out of context and utterly irrelevant, you cannot ignore it, and you must acknowledge and refute it directly.[l] Figure 5-5 illustrates this relationship.

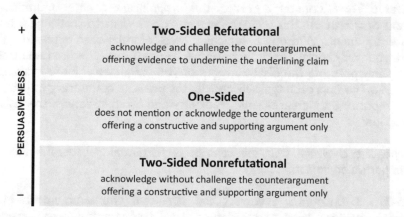

Figure 5-5. The relationship in persuasiveness between the three communicative strategies

In the management age of empathy, a refutational two-sided approach to counterarguments, while effectual, on some occasions, can originate unnecessary confrontational and potentially socially endangering moments in environments where the preconditions for candor are not present, and the act of challenging a person cannot be associated with care.[li] Striving to protect the healthiness of your relationships, you can leverage the two-sided communicative approach in a more nuanced manner. The following three scenarios delineate this counterargument management strategy.

Scenario 1: Encounter an incorrect counterargument. In this scenario, you have to challenge the invalid counterargument to prevent the conversation from following a fallacious premise. This situation forces you to acknowledge and refute the counterargument directly using a two-sided refutational message and leveraging your expertise power. The following example demonstrates a possible implementation of this strategy:

That is an understandable concern, however ... [refutation].

Scenario 2a: Encounter a correct but insignificant counterargument in a context of trust. In this scenario, you don't want to challenge the valid counterargument. Leveraging your relationship power, you can deploy a less confrontational approach using a two-sided nonrefutational message aimed to overwhelm the opposite argument with supportive ones and indirectly remarking its insignificance. The relationship power and trust that you built in the past with your audience will enhance the persuasiveness of an otherwise less effective message:

> While I agree with your point, it is crucial to consider that ... [supportive argument].

Scenario 2b: Encounter a correct but insignificant counterargument in a context of distrust. In this scenario, you still don't want to challenge the valid counterargument. Without a sufficient level of relationship power in the room, your only socially healthy[2] option is to use a two-sided candid message that openly presents the disadvantages of your argument before introducing your idea. This approach typically anticipates possible counterarguments, but, more importantly, increases your credibility and often initiates a trustworthy interaction:

> A possible challenge with this concept is [disadvantage], despite that ... [supportive argument].

Scenario 3: Encounter a correct and significant counterargument. In this scenario, you don't want to challenge the valid counterargument. Leveraging your relationship power, you can make yourself vulnerable and openly acknowledge the relevance of the counterargument, reflect on it, and factor it into your message modifying your point of view accordingly. Changing your perspective and incorporating significant counterarguments is a foundational principle to refine your message's effectiveness. Furthermore, modeling this behavior also engenders psychological safety and cultivates creative collaboration:[lii]

> This is a valid observation, and I believe we can ... [perspective shift].

Figure 5-6 captures and illustrates the logical ramifications of this counterargument management strategy. The two-sided refutational approach is the message with the vaster degree of potential persuasiveness. Nevertheless, when feasible, you can decide to leverage relationship power and its underpinning prebuilt trust to deploy a less confrontational approach without necessarily compromising the effectiveness of your message.

---

[2] A two-sided refutational strategy is always an option if you can afford to decrease your relationship power. In some critical situations, it can be considered an undesired but acceptable trade-off.

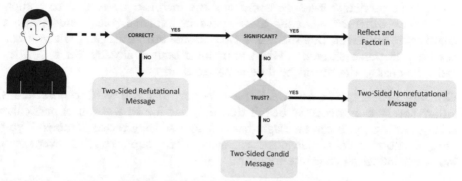

Figure 5-6. The counterargument management strategy

Manage Social Considerations

Projecting influence via an active or passive act of persuasion, in some circumstances, involves exercising a certain degree of control over social forces.[liii] Even if an individual has developed a positive attitude toward a request that you advanced, the person can be influenced by opposing social norms and still refrain from undertaking the advocated course of action.

Social norms are unwritten rules shared by members of the same group or society.[liv] These norms are a central part of our social regulations that govern the socially accepted ways of acting in specific contexts and situations.[lv] These unwritten rules discern what we perceive as socially accepted by our salient social circles, informing our thinking and action-oriented decision-making process.[lvi] There are various social considerations that an individual consciously and unconsciously factors during the decisional process before taking action; in the following sections, you will explore two of the most influential ones: prescriptive and descriptive norms.

Prescriptive Norms

Prescriptive norms, also called injunctive norms, are defined by our perception of whether persons, which we consider important to us, desire the performance or nonperformance of a given behavior.[lvii] They signal social approval or disproval and define what we believe other individuals think we should or shouldn't do in a given context. Knowing that you have to test the design before releasing it or not releasing it without validation represents an example of a prescriptive norm.

A prescriptive norm is a function of the normative beliefs that an individual ascribes to particular salient persons and the intrinsic motivation to comply with those others.[lviii] Affecting these types of informal rules tends to be a challenging endeavor, but there are two connected strategies that you can consider in these situations when your interlocutor already has a positive attitude: change the norms or de-emphasize them.

Change prescriptive norms. If the individual that you want to persuade is influenced by the normative beliefs that they ascribe to a group of particular salient persons, you can leverage that fact by involving those persons. If you engage those salient individuals receiving their support, this event can indirectly influence your interlocutor.[lix]

Attempting to change your interlocutor's prescriptive norms requires investigating external sources of influence that can affect the perspective of that person. For instance, trying to evangelize the benefits of a human-centered approach to business with a senior stakeholder in your organization, you can ask yourself the following questions:

Is there a salient person I can leverage?

Is there someone whose views will influence that person?

Is there someone whose views matter to that person?

Changing prescriptive norms represents an arduous task, but in some situations, it can be the only option you have to manage the social considerations of your interlocutor.

De-emphasize prescriptive norms. If the individual that you want to persuade is influenced by the normative beliefs that they ascribe to a group of particular salient persons, but you cannot leverage those norms or engage these individuals, you can try to redirect the attention of your interlocutor. If you shift the attention of that person from the salient persons' desire to their positive attitude toward the requested behavior, this event can directly influence your interlocutor.[lx]

Attempting to de-emphasize your interlocutor's prescriptive norms requires reminding that person that we should all be in control of our decisions and focus more on what we know is the best decision and less on what we believe the others want from us. For instance, trying to persuade a colleague that they shouldn't consider a managerial role solely because the line manager recommended it, you can emphasize to your interlocutor the following concepts:

This choice is something that affects you more than anybody else, and it should be your decision.

This decision is the kind of choice that other persons can't take for you; you have to make it yourself.

You should do what you think is right.

De-emphasizing prescriptive norms represents an easier task than changing them, especially if you developed a significant level of expertise and relationship power with that person.

Descriptive Norms

Descriptive norms are defined by our perception of whether persons, which we consider important to us, already perform or not perform a given behavior.[lxi] They signal social prevalence and indicate what we believe other persons do in a given context. Knowing that a designer we respect always presents the design to senior stakeholders using a business-oriented vocabulary represents an example of a descriptive norm.

A descriptive norm is not a function of normative beliefs and tends to arise from perceptions that parallel those determining prescriptive norms.[lxii] Typically, an individual tends to have a set of salient descriptive norm referents whose behavior is perceived as a source of guidance.[lxiii] Affecting these types of informal rules tends to be a delicate endeavor because it can produce adverse effects and unwanted repercussions, and for reasons that researchers have still not identified, it is not always effective in creating a change in behavior. With this fact in mind, in contexts where your interlocutor has already developed a positive attitude toward a given behavior, there are two connected strategies that you can consider: enhance the norms and/or alter them.

Enhance descriptive norms. If the individual that you want to persuade is influenced by a set of salient descriptive norm referents whose behavior might be seen as a source of guidance, you can leverage that fact by providing descriptive norm information in relation to that given set of referents. If you illuminate the connection between those salient referents and their behavior, this event can directly influence your interlocutor.[lxiv]

Attempting to enhance the descriptive norms of your interlocutor requires presenting factual information aimed to remind that person what other influential individuals or groups are doing. For instance, if you want to advocate the use of research with a senior stakeholder, you can emphasize to your interlocutor the following connections:

The design team at [company name] steadily increased conversion last year by testing their design at the end of every iteration.

The team in [department name] increased the creative output last year by regularly interviewing customers.

While enhancing descriptive norms does not always produce a change in behavior, it undoubtedly represents an easy strategy worth deploying.

Alter descriptive norms. If the individual that you want to persuade is influenced by a set of salient descriptive norm referents whose behavior is seen as a source of guidance, but they have a misperception of those norms, you can leverage that fact providing correct descriptive norm information in relation to that given set of referents.

Attempting to alter the descriptive norms of your interlocutor requires presenting factual information aimed to demonstrate to that person what other influential individuals or groups are doing in opposition to what they believed they were doing. For instance, if you want to advance a request to one of your connections, you can ask yourself the following questions:

- What are the erroneous perceptions that can impact some of the priorities of my interlocutor?

- What are the facts related to the priorities of my interlocutor that can elucidate these misperceptions?

- How can the actual social norm support the priorities of my interlocutor?

While altering descriptive norms does not always create a change in behavior, it always represents an altruistic act toward your connection, which in return can fortify your relationship with that person.

Promote Perceived Ability

Projecting influence via an active or passive act of persuasion, in some other cases, involves exercising a certain degree of control over potential context-dependent barriers to action.[lxv] An individual's perceived ability and control over a given behavior is a function of that person's beliefs in regard to the resources and obstacles relevant to the demanded actions. Specifically, it is a function that evaluates the person's perception of the likelihood or frequency that a given control factor will occur and the power of the control factor to inhibit or facilitate the behavior.[lxvi]

Promoting the perceived ability, technically referred to as self-efficacy,[3] in your interlocutor requires instilling in that person the conviction that what you are urging is mentally and physically doable. In the following section, you will learn how to overcome four types of obstacles that commonly prevent individuals from taking action.

Remove Obstacles

Creating the intention to perform a behavior frequently requires removing different types of barriers,[lxvii] in some situations, multiple ones concurrently (Figure 5-7). These obstacles are not always obvious, and without a prudent observation, a lack of behavioral compliance can erroneously be imputed to incompetence. There are four main types of barriers that can prevent the performance of a given behavior, and for each of them, you can use a precise strategy to influence your interlocutor's perceived ability.

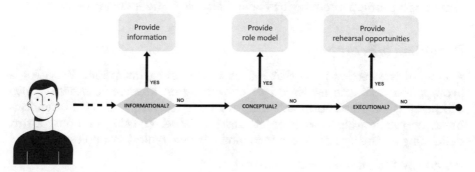

Figure 5-7. The obstacle management strategy

Informational obstacle: The person lacks the information required to perform the requested behavior. For instance, a designer hesitates to make a design decision because of data limitations. Removing this type of obstacle requires you to provide the information necessary to perform the requested behavior.

Conceptual obstacle: The person possesses the information but lacks the concept required to perform the requested behavior. For instance, a designer hesitates to run a workshop because of conceptual limitations. Removing this type of obstacle requires you to provide a role model capable of promoting a mental model and unlock the requested behavior.

---

[3] Self-efficacy is the individual's subjective perception of their capability to perform in a given setting or to attain desired results.

Executional obstacle: The person possesses the information and the concept but lacks the practice required to perform the requested behavior. For instance, a designer hesitates to present the work because of practice limitations. Removing this type of obstacle requires you to provide opportunities for rehearsal to promote the requested behavior.

Convert the Intention to Perform a Behavior into Action

In some contexts, creating the intention to perform a behavior is not sufficient to unlock that specific course of action. Despite engendering a positive attitude, managing the social considerations, and promoting the perceived ability of your interlocutor, that person may still fail to engage in the requested behavior.[lxviii] When an *intender*[4] still fails to engage in the requested action, you face the challenge of converting an intention into action.[lxix] In this situation, you have at your disposal three strategies to deliberately guide attention and memory recall to increase the likelihood of the intention-behavior relationship: providing a prompt, encouraging explicit planning, and inducing hypocrisy.[lxx]

Providing a Prompt

A prompt is a retrieval cue that makes a given behavior salient. Providing a prompt increases the likelihood of converting an intender's intention into action by guiding attention and memory recall toward a specific intention-behavior relationship.[lxxi] A prompt can be delivered using different forms depending on the situation; the following are four typical examples:

As an invitation: *Have you considered …*

As a suggestion: *Don't forget to …*

As a reflection: *How might we …*

As a rationale: *Suspending judgment can promote creative thinking …*

Depending on the scenario, the message containing the prompt can be delivered using different communication channels, including but not limited to

Analogic: Via in-person communication

Physical: Via signs and posters

Digital: Via instant messages and emails

---

[4] An *intender* is an individual that developed the intention to perform a specific behavior in opposition to a *nonintender* that did not develop that given intention.

As you learned in Chapter 1, discussing the behavioral needs of the physical environment, you can use prompts to trigger behaviors in line with a design-oriented mindset; the following are some instantiations of this practice:

- Create signs and labels for different spaces to trigger specific emotional states and mindsets.

- Create separate spaces for divergent and convergent thinking to trigger a precise approach to idea management.

- Create separate spaces for individual and collective thinking to trigger a precise approach to production.

- Provide movable elements to trigger space adaptation.

- Expose raw materials to trigger a bias toward actions.

You can deploy this strategy based on prompts in almost every scenario, and it can represent an invaluable tool to potentially unlock intenders that, for some reason, still failed to engage in the requested behavior. In Chapter 7, you will explore the use of prompts within the context of coaching.

Encouraging Explicit Planning

An explicit plan is a detailed description that instantiates a given behavior. Encouraging explicit planning increases the likelihood of converting an intender's intention into action by increasing the person's perceived ability and providing direction to a behavioral scheme.[lxxii] This type of plan defines the *when*, *what*, and *how* of a given behavior; the following is its typical structure:[lxxiii]

> When [trigger], I will [response] by [action].

> When a stakeholder questions a design, I will address the comment by presenting the research finding that informed that specific solution.

This practice stimulates the intender to move from abstract thinking to concrete thinking, from a conceptual intention to an implementation intention.[lxxiv]

Inducing Hypocrisy

In moral psychology, hypocrisy captures the misalignment between a set of beliefs, abstract thinking, and a given behavior.[lxxv] Inducing hypocrisy increases the likelihood of converting an intender's intention into action by indicating the details of a counterattitudinal behavior.[lxxvi] This practice is based on encouraging attitude-behavior consistency by generating cognitive dissonance and motivating the intender to realign the behavior with the intention.

Encouraging attitude-behavior consistency requires two key phases; when solely one of them is completed, hypocrisy induction tends to be weaker or nonexistent:[lxxvii]

- **Ensuring the salience of the relevant intention:** Via proattitudinal advocacy or prompts

- **Ensuring the salience of past failures to act consistently with that given intention:** Via feedback or storytelling

Suppose a stakeholder evangelizes customer centricity within the organization, but that person tends to advance feature requests based on untested assumptions with the team. In that case, you can counterargument emphasizing the importance of validating new feature requests with real users, reminding that person how not following that approach already created negative repercussions in the past. Inducing hypocrisy is an effective strategy that must be utilized with caution and sensibility because under certain conditions it presents potential risks and can produce two significant undesired effects.

Boomerang effect. Inducing hypocrisy under certain conditions can generate an opposite result. Indicating the details of a counterattitudinal behavior to a person characterized by low perceived ability and, consequently, behavioral control can push the intender to reframe the intention to perform the intended behavior rather than the inconsistent action.[lxxviii] If the intender perceives that the behavioral change is unattainable, that person may unconsciously decide to change their intention to reduce cognitive dissonance.[lxxix] A possible counterstrategy in these situations consists in arousing cognitive dissonance around the requested behavior while eliminating all the other possible modes of dissonance reduction except the one linked with the desired action.

Psychological discomfort. Inducing hypocrisy under certain conditions can arouse a significant level of cognitive distress that can undermine trust, erode psychological safety, and decrease relationship power.[lxxx] Indicating the details of a counterattitudinal behavior to a person characterized by low tolerance to emotionally based extremely aversive feelings can push the intender to reframe the intention to perform the intended behavior rather than the inconsistent action. In a diverse design team, the probability is that different individuals have different levels of tolerance regarding aversive feelings. A possible counterstrategy in these situations consists in limiting its use to contexts permeated by reciprocal trust and respect that can help the intender to interpret your communication positively.

Build Your Social System

Careers are built on two foundational elements: results and relationships. Both components are indispensable if you want to express your full potential and, if this is one of your objectives, possibly reach the pinnacle of your profession. Contrary to popular belief, producing results is a necessary but not sufficient condition to ensure a continuous career progression. The omnipresent need for collaboration demanded in the management age of empathy and the systemic problems that a contemporary designer navigates are the predominant motives behind the need to build a network that can support the achievement of your objectives. Architect what sociology describes as a social system[5] that you can leverage to develop social power, project influence, and form healthy and productive relationships.

Building a resourceful social system represents a time-demanding activity to preemptively initiate during the beginning of your career when you traverse Stages 1 and 2 of contribution. A social system becomes a contributory element of your success at Stage 3 and a blocker to enter Stage 4: a strong network of relationships is a nonwritten must-have requirement for an executive-level role. In the subsequent sections, you will learn how to build your social system by establishing and maintaining healthy connections.

Define the Social System Structure

An effective social system must be able to support you in several different situations during your career. The following four criteria can facilitate your effort to define and build this social structure accordingly.

Focus. This criterion captures the primary domain of contribution of a connection. It answers the question: *In which professional circumstances can that person potentially support me?* The following are the three areas of focus:

- **Craft:** The area that identifies your abilities, supporting your initiatives and providing access to resources that you can leverage to achieve the objectives defined by your role

- **Career:** The area that identifies your profession, providing access to individuals that you can leverage to advance your work and achievements inside and outside your organization

---

[5] In sociology, a social system is the patterned network of relationships constituting a coherent whole that exists between individuals, groups, and institutions.

- **Reach:** The area that identifies your social range, providing access to new individuals that you can leverage to expand your network and perspective on significant subjects

Location. This criterion captures the place of a connection. It answers the question: *Is that person working in my organization?* The following are the two types of locations:

- **Internal:** The location that identifies connections working inside your organization

- **External:** The location that identifies connections working outside your organization

Proximity. This criterion captures the frequency of communication established with that person. It answers the question: *What is the cadence of our interactions?* The following are the three levels of proximity:

- **Private:** The proximity level that characterizes a connection in your team or workgroup that you can refer to as "friend"

- **Social:** The proximity level that characterizes a connection outside your team that you can refer to as "acquaintance"

- **Public:** The proximity level that characterizes a connection outside your team or organization that you can refer to as "stranger"

Tie. This criterion captures the nature of the emotional investment that you established with that person. It answers the question: *How close is my relationship with that person?*

- **Strong:** Characterized by a high emotional investment that tends to produce strong bonds, typically associated with a private level of proximity. A friend or close colleague in your team that genuinely shares intimate personal life details with you represents a typical example of a strong tie.

- **Weak:** Characterized by a low emotional investment that tends to produce weak bonds, typically associated with a social level of proximity. A mentor that meets you a few times a year represents a typical example of a weak tie.

- **Invisible:** Characterized by the absence of emotional investment, typically associated with a public level of proximity. A prominent figure in the design community with whom you feel connected despite not having a formal relationship represents a typical example of an invisible tie.

Social System

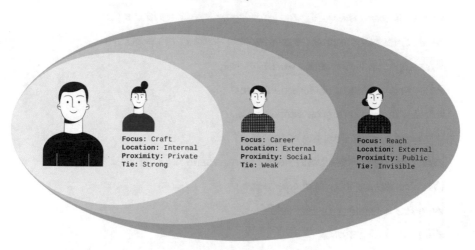

Figure 5-8. The three types of relationships in your social system

Relationships (Figure 5-8) are a fascinating and incredibly nuanced social element of our life. With these criteria, you can abstract some of its constituents and use them to strategize your interactions, concentrate the time invested, and optimize your return on investment.

Populate the Social System

When you begin to populate your social system, your objective must be to build a substantial foundation composed of internal connections belonging to your organization and external ones belonging more broadly to the design community. These connections need to span across the three areas of focus: craft, career, and social reach. Browse social networks, event websites, or any other resource that can help you create an initial list of potential candidates. Reflect on current relationships worth nurturing, old connections worth reviving, and potential ones worth pursuing. The following questions can help you to frame this initial exploration:

// Identify current connections

Who is supporting my professional growth?

Who is supporting me during the challenging moments?

Who is representing a positive role model for me?

// Identify connections worth reviving

Who supported my professional growth in the past?

Who supported me during the challenging moments in the past?

Who represented a positive role model for me in the past?

// Identify new connections worth pursuing

Whom can I help to achieve objectives?

Whom can I help to grow professionally?

Who can benefit from my support?

Who has the abilities that I need to progress in my career?

Who has access to critical information?

Who has a resourceful network?

Who has a high status?

Who can help me to achieve my objectives?

Create a list of possible connections balancing relationships with a different area of focus, location, level of proximity, and social tie. This initial list captures your current potential levels of social capital intended as the value of these social resources, both tangible and intangible, and the impact that these relationships can have on your professional life.

Capital Network Matrix

The most practical and efficient way to organize your social system's population under the four criteria previously introduced is to use a Capital Network Matrix (CNM) (Figure 5-9). You can implement this structure with your spreadsheet software of choice.

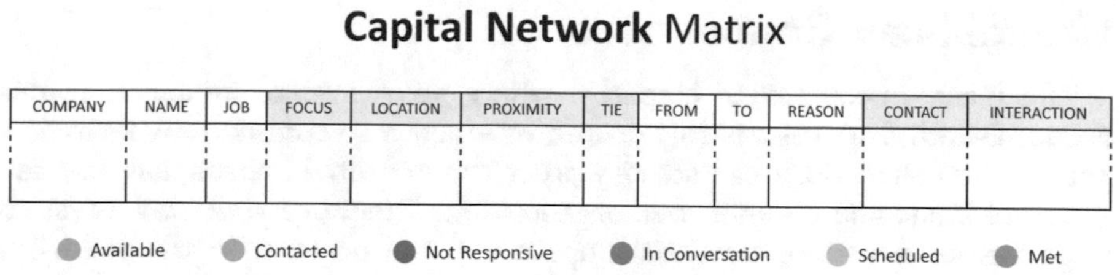

Capital Network Matrix

COMPANY	NAME	JOB	FOCUS	LOCATION	PROXIMITY	TIE	FROM	TO	REASON	CONTACT	INTERACTION

● Available ● Contacted ● Not Responsive ● In Conversation ● Scheduled ● Met

Figure 5-9. The structure of the Capital Network Matrix

The Capital Network Matrix, from left to right, is structured as follows. Three columns are allocated to the company, name, and job title of the connection. Four columns are assigned to the focus, location, level of proximity, and social tie that characterize that person. Three columns are reserved for who introduced you to that person, also known as *superconnector*, eventually the person that you presented to that individual, and the reason for that interaction. Two columns are allocated to the day of the last contact and the last interaction with that person. Each connection in your social system can be in one of the following six states; each condition can be associated with a specific color to enhance its glanceability:

- Available (for a meeting)
- Contacted
- Not responsive ($\geq$ 2 weeks)
- In conversation
- Scheduled
- Met

A Capital Network Matrix, and its population, is a live document that must evolve alongside the necessities of your professional life. During the beginning of your career, at Stages 1 and 2 of contribution, you need to concentrate on creating more connections that can develop your craft by growing your technical abilities. Instead, when you operate at Stages 3 and 4 of contribution, you need to focus predominantly on creating more relationships that can build your managerial and leadership competencies while increasing your social reach, promoting your work, and introducing you to new connections with the capacity to raise your social capital and, therefore, potential relationship power.

Create New Connections

In the management age of empathy, where your level of influence comes predominantly from relationship power, your ability to continuously form new productive connections can steadily grow the potential impact that you can generate inside and outside your organization. The more your social system is pervasive, the more potential impact you can achieve. In the following section, you will explore how to maximize the opportunity at your disposal to extend your internal and external network.

Establish Internal Connections

Developing an internal network represents an instrumental activity to your success regardless of your stage of contribution; your internal connections represent the foundation of everything you can achieve in your organization. This activity requires you to interact mainly with individuals with a private and social level of proximity.

Internal Network Matrix

Irrespective of your team or workgroup nature, whether collocated, distributed, physical, or virtual, your ability to collaborate and receive support in your organization needs to span vertically and horizontally across your company's hierarchy structure (Figure 5-10). The vertical dimension denotes the *height* of your network and captures the layers defined by your and your colleagues' level of seniority. The horizontal dimension denotes the *width* of your network and captures the layers defined by other departments and their structure.

The domain determined by these two dimensions visualizes your potential *scope* and *channels* of influence within your organization and can be captured using an Internal Network Matrix (INM). You can implement this structure, creating a matrix with your spreadsheet software of choice, adding as many columns and rows as necessary to represent the key actors inside your organizational hierarchy.

Internal Network Matrix

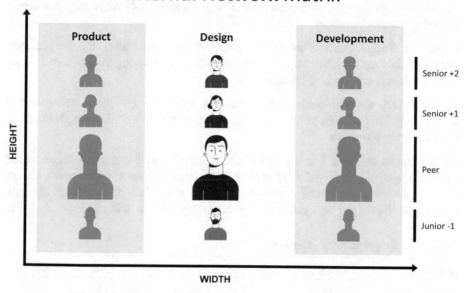

Figure 5-10. The structure of the Internal Network Matrix

During your career, you will need to overcome a diverse set of challenges that will require you to collaborate across the different areas of your internal network. Maintaining an updated Internal Network Matrix allows you to build productive collaborations by providing a structure that you can utilize to assess and manage two critically related factors: the internal levels of social power and design maturity.

Assess the Internal Levels of Design Maturity

The different levels of design maturity involved affect the behavioral nuances of a given creative collaboration. Individuals with a high degree of design maturity tend to be more comfortable with some of the inevitable aspects of human-centered problem-solving, such as the level of ambiguity and psychological discomfort generated during the process.[lxxxi]

Under that circumstance, to adapt your behavioral configuration accordingly, you need to identify individuals with whom you must collaborate that are characterized by low levels of design maturity. Different models have tried to represent the different levels of design maturity within an organization, but while these structures use various labels, they all depict comparable progressions from different perspectives. The following list represents a structure that can easily capture this cultural development from a human-centered problem-solving viewpoint.

Level 1: Executor. The design team does not frame challenges into problems and implements predefined solutions. The group typically manifests low levels of creative collaboration.

Level 2: Supporter. The design team does not frame challenges into problems but defines and implements solutions. The group typically manifests medium levels of creative collaboration.

Level 3: Driver. The design team does frame challenges into problems and defines and implements solutions. The group typically manifests high levels of creative collaboration.

As illustrated in Table 5-2, if you feel the necessity, you can use a spreadsheet to represent the Internal Network Matrix and utilize a three-point Likert scale to assess the level of design maturity of your internal connections and, consequently, their perception of the design team's contribution.

Table 5-2. Design maturity assessment

Name	Department	Role	Design Maturity
Name Surname	Product	Head of Product	Level 2
Name Surname	Development	Development Lead	Level 1
Name Surname	Marketing	Marketing Manager	Level 3

Congruently to other aspects of your organizational culture, the global level of design maturity is a function of several "pockets" identified by subcultures.[lxxxii] This document can help you map this social element of your company, and by increasing your self-awareness regarding how the design team is perceived within the organization, you can develop a more nuanced and effective communicative strategy.

Assess the Internal Levels of Social Power

The different levels of social power involved affect the behavioral nuances of a given creative collaboration.[lxxxiii] Individuals with a high degree of social power tend to perceive other persons more stereotypically and judge them more negatively.[lxxxiv]

Under that circumstance, to adapt your behavioral configuration accordingly, you need to identify individuals with whom you must collaborate that are characterized by high levels of role, expertise, and relationship power. In Chapter 6, you will learn more about the job title nomenclature used in this example by exploring a dual-track career framework for managers and individual contributors.

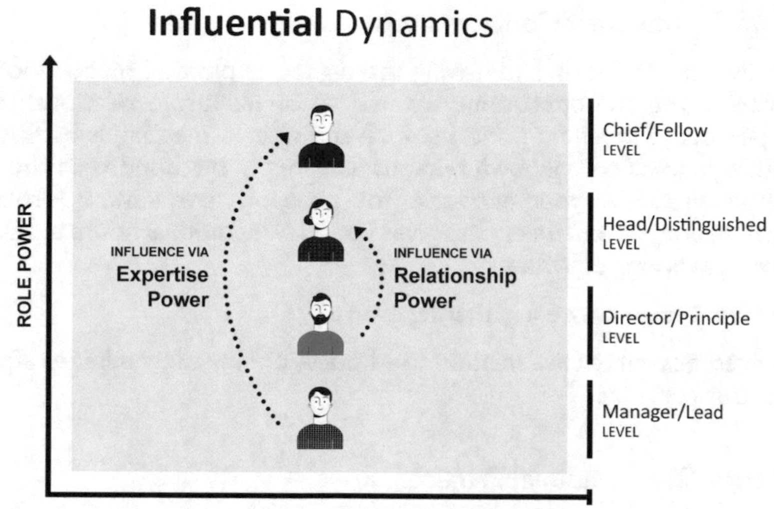

Influential Dynamics

Figure 5-11. A possible example of influential dynamics within a team

In Figure 5-11, you can see a possible scenario that depicts the influential dynamics of a group derived from the projection of expertise and relationship power. These forces can often traverse and overcome formal hierarchies and represent decisive decision-making factors. In Chapter 9, you will learn how to minimize these undesired influential dynamics and regulate the decisional process by separating the act of preparing and making a given decision. Actively observing the social synergies within a group can allow you to assess the internal levels of social power beyond the apparent presence of the role power defined by a person's job title. Besides actively observing individuals, spending time talking with them represents an opportunity to reveal their beliefs, refine your assessment, and prove or disprove your initial assumptions.

Invite for Breakfast and Lunch

Sharing a meal with someone represents one of the most potent ways to initiate and nurture a relationship with another person. Sharing the fulfillment of this deep, biological need for food can promote the establishment of profound bonds. You have at least five breakfast and five lunch events at your disposal during the week, which translates to circa ten hours a week that you can dedicate to your networking activities.

Objective: Never eat alone.

Goal: Invite at least three persons a week outside your design team for breakfast or lunch.

Pre- and Postmeeting Conversations

Presenting yourself early and leaving the digital or physical meeting location late provide you the opportunity to initiate or nurture a relationship with other persons. The short period before and after a meeting is sufficient to introduce yourself to unknown persons and fortify the bond with the other individuals already in your network. You probably have several formal and informal meetings every week; this gives you a few additional hours to dedicate to your networking activities.

Objective: Never waste a gathering event.

Goal: Present yourself five minutes in advance, or stay a few minutes after the end of your meetings.

Join Cross-Organizational Projects

Contributing to interdepartmental projects provides a unique opportunity to initiate a relationship with other persons with whom you would not otherwise get in contact. If you work for a small company, you may not have that opportunity, but if you work for a large organization, and your manager offers you to join this type of project, accept it without hesitation, even if you feel that your agenda is already busy. Cross-organizational projects will expose you to an unprecedented number of new potential connections that you can quickly formalize over a meal or before and after a meeting.

Objective: Never avoid unfamiliar contexts.

Goal: No goal as the cadence of this scenario is not under your control.

Show Appreciation for Project Achievements and Promotions

Demonstrating authentic appreciation when an individual in your organization achieves a significant objective represents an excellent opportunity to initiate or nurture a relationship with that person. Inform yourself about the company initiatives, be aware of other individuals' challenges, and be prepared to demonstrate authentic appreciation when one of your connections achieves significant results. Face-to-face communication is always, and by far, the most effective way to display any genuine and altruistic interest because it implies a time and effort investment made to celebrate these persons without asking anything in return.

Objective: Never miss an opportunity to celebrate the success of others.

Goal: No goal as the cadence of this scenario is not under your control.

Establish External Connections

Developing your external network represents a crucial activity for your long-term success that, once you have established a critical mass of internal connections, must be elevated to an integral part of your professional life by reserving it a place in your calendar. This activity requires you to interact mainly with individuals with a social and public level of proximity.

Invite for Breakfast and Lunch

Inviting other persons for breakfast and lunch must remain your primary way to initiate or nurture your relationships. In this scenario, you can increase the probability of acceptance, making yourself available to meet your connections in a location close to their office.

Objective: Never eat alone.

Goal: Invite at least one person a week outside your organization for breakfast or lunch.

Invite for Virtual Coffee

Sharing a coffee virtually with a connection allows you to initiate or nurture a relationship when in-person meetings are not a viable option. Virtual coffee breaks represent a fantastic tool when you manage a virtual or distributed team, implement social distancing, or when you or someone in your group spends a significant amount of time working from home. A virtual coffee can happen any day at any time during the day, and it's typically a 15-minute meeting via video call during one of the potential breaks during the day.

Objective: Never waste a break from work.

Goal: Invite at least one person a week outside your organization for a virtual coffee.

Send Updates

Sharing information via email represents an efficient way to nurture the relationship with persons that are not immediately involved in your short- and mid-term design operations, with whom you want to remain in touch, waiting for the right opportunity to spend quality time with them. You can decide to send a direct update via email or indirectly tagging your connection in a post on a social network like LinkedIn.

Objective: Never miss an opportunity to share updates with others.

Goal: Send one round of update emails every three months.

Attend Design Events

Participating in industry events represents one of the most prolific opportunities to initiate a relationship with another person. Sharing passions, challenges, and achievements can establish deep bonds between individuals. Before attending the event, whether a large conference or a small meetup, create a list with the persons that you want to meet and approach them one month before the event investigating their availability for a meeting. Divide the list into two categories:

- **Persons that you know:** Connections with a social level of proximity

- **Persons that you do not know:** Connections with a public level of proximity or individuals not currently part of your social system

An even better way to leverage the social potential of an event is to be the organizer. Organizing an event will provide you with even greater exposure to potential new relationships. Being the organizer of an event allows you to play the role of the *superconnector*: an individual that connects two or more individuals and their social systems, helping them to achieve their objectives. In that role, you will be able to accumulate a vast amount of social credit that you can subsequently transmute into influence if necessary.

Objective: Never miss an opportunity to share your passions with others.

Goal: Attend at least one major design event a year.

Create Travel Margins

Organizing travels represents an additional opportunity to work on your social system. Leaving some temporal margins around your journey schedule allows you to create the opportunity to initiate or nurture relationships beyond your primary working location. If you manage a distributed team, you can use those margins to spend unstructured time with the group. If you speak at a conference, you can instead use that time to meet external connections that are usually distant for a face-to-face meeting.

Objective: Never waste time when you are out of the office.

Goal: No goal as the cadence of this scenario is not under your control.

Promote a Healthy Social System

Developing healthy and longevous relationships requires wisdom and tactfulness. The healthiness characterizing a connection is a function of the trust perceived by the individuals engaged in that given relationship. Healthy relationships represent a necessary precondition to developing influence longevity.

Engender Trust

As we introduced in Chapter 1, trust is defined as the expectation that others' future actions will be favorable to one's interests.[lxxxv] In this context, this feeling represents a crucial social element because it has the ability to lower the transactional costs of interpersonal interactions.[lxxxvi] The principal currency of trust is the willingness to do things for one another, which represents the foundation of relationship power. The following are three critical behaviors that you can model to promote that desire.[lxxxvii]

Be present: Stay in touch and make yourself available by allocating time for it in your calendar. Block at least one hour a week to manage your network and elevate this activity to the same level as other events in your diary.

Be candid: Manifest openness and honesty in your conversations by providing sincere and constructive feedback when common sense suggests otherwise. Ask for permission to provide feedback, acknowledge your biased perspective, and be always specific and never personal.

Be altruistic: Demonstrate a genuine interest in the issues and passions of your connections, understanding their needs and helping them to achieve their goals. Make two favors before asking for one: give, give, then ask.

In Chapter 6, you will explore how to build and assess trust within the context of creating your team.

Ponderate the Use of Communication

Communication underpins the process of building trust with other individuals, and using the most appropriate communicative tool represents a decisive factor in the relationship equation.[lxxxviii] The exchange of information produced during an interaction can be classified using three variables: frequency, reach, and depth. Modulating three variables allows you to adapt your communicative strategy based on the needs of a given situation.[lxxxix] For instance, using a digital channel, you can achieve a high frequency and reach of communication, while using an in-person interaction, you can obtain an increased depth of communication creating bonds and promoting trust. Unfortunately, you cannot optimize these two variables simultaneously, and different situations demand different strategies to maximize the degree of influence projected.

When the objective is to build trust and project influence, it's important to remember that while technology creates novel scenarios where *shallow trust* can be promoted, it's unable to establish *deep trust*; only face-to-face in-person interaction can promote that feeling.[xc] The richness of nonverbal cues that permeate in-person communication and the chemical response that only physical proximity can trigger represent the primary drivers for empathy and a precondition for deep trust and, consequently, any form of influence.[xci] As a corollary to this dynamic, consider organizing an in-person meeting when you need to advance a significant request to one of your connections. The nonverbal cues conveyed during your in-person interaction will promote empathy, increase the veracity of your message, and enhance the perceived legitimacy of your request.[xcii]

While technology extended the potential "reach" and "frequency" of our digital communication, it still represents an inadequate tool to develop the "depth" required in certain circumstances.[xciii] Developing the ability to discern which communication variable you need to maximize under the requirements of a given conversation will profoundly impact your effectiveness as a design manager and leader.

Maintain a Healthy Social System

The healthiness of a network is a function of the amount of value exchanged in the form of experiences. The social exchange shaping those experiences can be positive or negative when tensions and moments of irritation occur. Equivalent interactions validate one another, reinforcing the perception of the positive or negative connotation associated with that given experience.[xciv] Conversely, positive and negative elements from past interactions don't nullify one another and remain simultaneously present in a relationship originating a perpetual ambivalent state.[xcv]

Maintaining a healthy social system is not synonymous with eradicating negative experiences; tensions and moments of irritation are an intrinsic part of life. As you will explore in Chapter 8, this form of energy can represent an essential catalyst of creative collaboration when managed adequately.

Tension is defined as the feeling of psychological strain and uneasiness derived from a situation characterized by antagonist forces such as ideas, attitudes, behaviors, and emotions.[xcvi] As you will learn in the next section, maintaining a healthy social system means preserving predominantly positive relationships by preventing that tensions and moments of irritation escalate into conflicts.

Recognize the Moments of Irritation

At the beginning of this chapter, we introduced that social environments are permeated by unwritten norms that inform our action-oriented decisional processes.[xcvii] These norms are cross-disciplinary and based on multiple criteria, including but not limited to the context of the interaction and the culture of the individuals involved.[xcviii] Driven by survival intentions, social norms fulfill the onerous task of facilitating our social exchanges by providing us with a behavioral model to follow.[xcix] However, when persons with diverse behavioral models related to a given context interact, someone can, voluntarily or involuntarily, violate another individual's norms generating tension and a moment of irritation. In this scenario, the parties must discuss and clarify the dynamics that originated the strain and that irritating event and, when necessary, advance an apology and renegotiate the violated norms of engagement. If the parties do not have that conversation, more moments of irritation can accumulate, and the situation can escalate into a conflict (Figure 5-12) that can jeopardize and potentially terminate the relationship.

Conflict Development

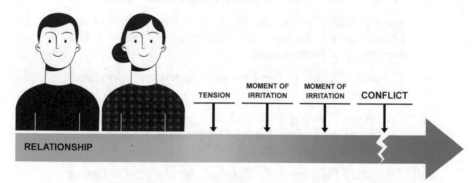

Figure 5-12. The escalation of tensions and moments of irritation into a conflict

When you interact with your social system, you must be aware of these potential moments of irritation and proactively manage them before they escalate into something perilous. The approach that you consciously or unconsciously deploy during these defining moments will define the healthiness of your relationships, the resourcefulness of your network, and the longevity of your relationship power.

Endnotes

i.   Rodgers, Mark. Persuasion Equation: The Subtle Science of Getting Your Way. AMACOM, 2015.

ii. Johnson, Lauren Keller. "Exerting Influence Without Authority." Harvard Business Review, 2 Aug. 2016, hbr. org/2008/02/exerting-influence-without-aut.

iii. Gardner, John W. On Leadership. The Free Press, 1993.

iv. Massey, Cade. "Influence." MBA. Wharton School of the University of Pennsylvania, Wharton School of the University of Pennsylvania.

v. Rodgers, Mark. Persuasion Equation: The Subtle Science of Getting Your Way. AMACOM, 2015.

vi. Aristotle. The Art of Rhetoric. Penguin Books, 2004.

vii. Keltner, Dacher. The Power Paradox: How We Gain and Lose Influence. Penguin Books, 2017.

viii. Aristotle. The Art of Rhetoric. Penguin Books, 2004.

ix. Aristotle. The Art of Rhetoric. Penguin Books, 2004.

x. Massey, Cade. "Influence." MBA. Wharton School of the University of Pennsylvania, Wharton School of the University of Pennsylvania.

xi. O'Keefe, Daniel J. Persuasion: Theory and Research. SAGE, 2016.

xii. "Politics: Meaning of Politics by Lexico." Lexico Dictionaries | English, Lexico Dictionaries, www.lexico.com/definition/politics.

xiii. "Politics: Meaning of Politics by Lexico." Lexico Dictionaries | English, Lexico Dictionaries, www.lexico.com/definition/politics.

xiv. Ferris, Gerald R., et al. "Development and Validation of the Political Skill Inventory." Journal of Management, vol. 31, no. 1, 2005, pp. 126–152., doi:10.1177/0149206304271386.

xv. Ferris, G.r. "Political Skill in Organizations." Development and Learning in Organizations: An International Journal, vol. 22, no. 1, 2007, doi:10.1108/dlo.2008.08122aad.009.

xvi. Ferris, G.r. "Political Skill in Organizations." Development and Learning in Organizations: An International Journal, vol. 22, no. 1, 2007, doi:10.1108/dlo.2008.08122aad.009.

xvii. Ferris, G.r. "Political Skill in Organizations." Development and Learning in Organizations: An International Journal, vol. 22, no. 1, 2007, doi:10.1108/dlo.2008.08122aad.009.

xviii. Ferris, G.r. "Political Skill in Organizations." Development and Learning in Organizations: An International Journal, vol. 22, no. 1, 2007, doi:10.1108/dlo.2008.08122aad.009.

xix. Ferris, Gerald R., et al. "Development and Validation of the Political Skill Inventory." Journal of Management, vol. 31, no. 1, 2005, pp. 126–152., doi:10.1177/0149206304271386.

xx. Lvina, Elena, et al. "Team Political Skill Composition as a Determinant of Team Cohesiveness and Performance." Journal of Management, vol. 44, no. 3, 2015, pp. 1001–1028., doi:10.1177/0149206315598371.

xxi. Hook, Derek, et al. The Social Psychology of Communication. Palgrave Macmillan, 2011.

xxii. Horstman, Mark, and Michael Auzenne. "Manager Tools Podcast." Podcasts | Manager Tools, Jan. 2005, www.manager-tools.com/podcasts.

xxiii. Pfeffer, Jeffrey. Managing with Power: Politics and Influence in Organization. Harvard Business Review Press, 1994.

xxiv. Pfeffer, Jeffrey. Power: Why Some People Have It and Others Don't. Harper Business, 2010.

xxv. Pfeffer, Jeffrey. Power: Why Some People Have It and Others Don't. Harper Business, 2010.

xxvi. Pfeffer, Jeffrey. Power: Why Some People Have It and Others Don't. Harper Business, 2010.

xxvii. Cialdini, Robert B. Influence: the Psychology of Persuasion: Robert B. Cialdini. Collins, 2007.

xxviii. O'Keefe, Daniel J. Persuasion: Theory and Research. SAGE, 2016.

xxix. O'Keefe, Daniel J. Persuasion: Theory and Research. SAGE, 2016.

xxx. Ajzen, Icek. "The Theory of Planned Behavior." Organizational Behavior and Human Decision Processes, vol. 50, no. 2, 1991, pp. 179–211., doi:10.1016/0749-5978 (91)90020-t.

xxxi. O'Keefe, Daniel J. Persuasion: Theory and Research. SAGE, 2016.

xxxii. O'Keefe, Daniel J. Persuasion: Theory and Research. SAGE, 2016.

xxxiii. Ajzen, Icek. "The Theory of Planned Behavior." Organizational Behavior and Human Decision Processes, vol. 50, no. 2, 1991, pp. 179–211., doi:10.1016/0749-5978 (91)90020-t.

xxxiv. Levin, Irwin P., et al. "All Frames Are Not Created Equal: A Typology and Critical Analysis of Framing Effects." Organizational Behavior and Human Decision Processes, vol. 76, no. 2, 1998, pp. 149–188., doi:10.1006/obhd.1998.2804.

xxxv. Levin, Irwin P., et al. "All Frames Are Not Created Equal: A Typology and Critical Analysis of Framing Effects." Organizational Behavior and Human Decision Processes, vol. 76, no. 2, 1998, pp. 149–188., doi:10.1006/obhd.1998.2804.

xxxvi. Levin, Irwin P., et al. "All Frames Are Not Created Equal: A Typology and Critical Analysis of Framing Effects." Organizational Behavior and Human Decision Processes, vol. 76, no. 2, 1998, pp. 149–188., doi:10.1006/obhd.1998.2804.

xxxvii. Vaish, Amrisha, et al. "Not All Emotions Are Created Equal: The Negativity Bias in Social-Emotional Development." Psychological Bulletin, vol. 134, no. 3, 2008, pp. 383–403., doi:10.1037/0033-2909.134.3.383.

xxxviii. Ito, Tiffany A., et al. "Negative Information Weighs More Heavily on the Brain: The Negativity Bias in Evaluative Categorizations." Journal of Personality and Social Psychology, vol. 75, no. 4, 1998, pp. 887–900., doi:10.1037/0022-3514.75.4.887.

xxxix. Peeters, Guido, and Janusz Czapinski. "Positive-Negative Asymmetry in Evaluations: The Distinction Between Affective and Informational Negativity Effects." European Review of Social Psychology, vol. 1, no. 1, 1990, pp. 33–60., doi:10.1080/14792779108401856.

xl. O'Keefe, Daniel J. Persuasion: Theory and Research. SAGE, 2016.

xli. O'Keefe, Daniel J. Persuasion: Theory and Research. SAGE, 2016.

xlii. Strathman, Alan, et al. "The Consideration of Future Consequences: Weighing Immediate and Distant Outcomes of Behavior." Journal of Personality and Social Psychology, vol. 66, no. 4, 1994, pp. 742–752., doi:10.1037/0022-3514.66.4.742.

xliii. Orbell, Sheina, and Maria Kyriakaki. "Temporal Framing and Persuasion to Adopt Preventive Health Behavior: Moderating Effects of Individual Differences in Consideration of Future Consequences on Sunscreen Use." Health Psychology, vol. 27, no. 6, 2008, pp. 770–779., doi:10.1037/0278-6133.27.6.770.

xliv. O'Keefe, Daniel J. Persuasion: Theory and Research. SAGE, 2016.

xlv. O'Keefe, Daniel J. Persuasion: Theory and Research. SAGE, 2016.

xlvi. VandenBos, Gary R. APA Dictionary of Psychology. American Psychological Association, 2015.

xlvii. O'Keefe, Daniel J. Persuasion: Theory and Research. SAGE, 2016.

xlviii. O'Keefe, Daniel J. Persuasion: Theory and Research. SAGE, 2016.

xlix. O'Keefe, Daniel J. Persuasion: Theory and Research. SAGE, 2016.

l. Haaften, T. van. Bending Opinion Essays on Persuasion in the Public Domain. Leiden University Press, 2011.

li. Scott, Kim. Radical Candor. Pan Books, 2019.

lii. Edmondson, Amy C. The Fearless Organization: Creating Psychological Safety in the Workplace for Learning, Innovation, and Growth. Wiley, 2018.

liii. O'Keefe, Daniel J. Persuasion: Theory and Research. SAGE, 2016.

liv. Hechter, Michael, and Karl-Dieter Opp. Social Norms. Russell Sage Foundation, 2001.

lv. Yamin, Paulius, et al. "Using Social Norms to Change Behavior and Increase Sustainability in the Real World: A Systematic Review of the Literature." Sustainability, vol. 11, no. 20, 2019, p. 5847., doi:10.3390/su11205847.

lvi. Legros, Sophie, and Beniamino Cislaghi. "Mapping the Social-Norms Literature: An Overview of Reviews." Perspectives on Psychological Science, vol. 15, no. 1, 2019, pp. 62–80., doi:10.1177/1745691619866455.

lvii. O'Keefe, Daniel J. Persuasion: Theory and Research. SAGE, 2016.

lviii. O'Keefe, Daniel J. Persuasion: Theory and Research. SAGE, 2016.

lix. O'Keefe, Daniel J. Persuasion: Theory and Research. SAGE, 2016.

lx. O'Keefe, Daniel J. Persuasion: Theory and Research. SAGE, 2016.

lxi. O'Keefe, Daniel J. Persuasion: Theory and Research. SAGE, 2016.

lxii. Fishbein, Martin, and Icek Ajzen. Predicting and Changing Behavior: The Reasoned Action Approach. Routledge, 2015.

lxiii. O'Keefe, Daniel J. Persuasion: Theory and Research. SAGE, 2016.

lxiv. O'Keefe, Daniel J. Persuasion: Theory and Research. SAGE, 2016.

lxv. O'Keefe, Daniel J. Persuasion: Theory and Research. SAGE, 2016.

lxvi. Ajzen, Icek. "The Theory of Planned Behavior." Organizational Behavior and Human Decision Processes, vol. 50, no. 2, 1991, pp. 179–211., doi:10.1016/0749-5978(91)90020-t.

lxvii. O'Keefe, Daniel J. Persuasion: Theory and Research. SAGE, 2016.

lxviii. O'Keefe, Daniel J. Persuasion: Theory and Research. SAGE, 2016.

lxix. O'Keefe, Daniel J. Persuasion: Theory and Research. SAGE, 2016.

lxx. O'Keefe, Daniel J. Persuasion: Theory and Research. SAGE, 2016.

lxxi. Ajzen, Icek. "The Theory of Planned Behavior." Organizational Behavior and Human Decision Processes, vol. 50, no. 2, 1991, pp. 179–211., doi:10.1016/0749-5978(91)90020-t.

lxxii. O'Keefe, Daniel J. Persuasion: Theory and Research. SAGE, 2016.

lxxiii. Gollwitzer, Peter M., and Paschal Sheeran. "Implementation Intentions and Goal Achievement: A Meta-Analysis of Effects and Processes." Advances in Experimental Social Psychology Advances in Experimental Social Psychology Volume 38, 2006, pp. 69–119., doi:10.1016/s0065-2601(06)38002-1.

lxxiv. Gollwitzer, Peter M., and Paschal Sheeran. "Implementation Intentions and Goal Achievement: A Meta-Analysis of Effects and Processes." Advances in Experimental Social Psychology Advances in Experimental Social Psychology Volume 38, 2006, pp. 69–119., doi:10.1016/s0065-2601(06)38002-1.

lxxv. Ellemers, Naomi, et al. "The Psychology of Morality: A Review and Analysis of Empirical Studies Published From 1940 Through 2017." Personality and Social Psychology Review, vol. 23, no. 4, 2019, pp. 332–366., doi:10.1177/1088868318811759.

lxxvi. Stone, Jeff, and Nicholas C. Fernandez. "To Practice What We Preach: The Use of Hypocrisy and Cognitive Dissonance to Motivate Behavior Change." Social and Personality Psychology Compass, vol. 2, no. 2, 2008, pp. 1024–1051., doi:10.1111/j.1751-9004.2008.00088.x.

lxxvii. Stone, Jeff, and Nicholas C. Fernandez. "To Practice What We Preach: The Use of Hypocrisy and Cognitive Dissonance to Motivate Behavior Change." Social and Personality Psychology Compass, vol. 2, no. 2, 2008, pp. 1024–1051., doi:10.1111/j.1751-9004.2008.00088.x.

lxxviii. Fried, Carrie B. "Hypocrisy and Identification With Transgressions: A Case of Undetected Dissonance." Basic and Applied Social Psychology, vol. 20, no. 2, 1998, pp. 145–154., doi:10.1207/15324839851036769.

lxxix. Chen, Mei-Fang, and Pei-Ju Tung. "The Moderating Effect of Perceived Lack of Facilities on Consumers' Recycling Intentions." Environment and Behavior, vol. 42, no. 6, 2009, pp. 824–844., doi:10.1177/0013916509352833.

lxxx. Priolo, Daniel, et al. "Three Decades of Research on Induced Hypocrisy: A Meta-Analysis." Personality and Social Psychology Bulletin, vol. 45, no. 12, 2019, pp. 1681–1701., doi:10.1177/0146167219841621.

lxxxi. Feist, Gregory J. "A Meta-Analysis of Personality in Scientific and Artistic Creativity." Personality and Social Psychology Review, vol. 2, no. 4, 1998, pp. 290–309., doi:10.1207/s15327957pspr0204_5.

lxxxii. Alvard, Michael S. "The Adaptive Nature of Culture." Evolutionary Anthropology: Issues, News, and Reviews, vol. 12, no. 3, 2003, pp. 136–149., doi:10.1002/evan.10109.

lxxxiii. O'Keefe, Daniel J. Persuasion: Theory and Research. SAGE, 2016.

lxxxiv. Brauer, Markus, and Richard Y. Bourhis. "Social Power." European Journal of Social Psychology, vol. 36, no. 4, 2006, pp. 601–616., doi:10.1002/ejsp.355.

lxxxv. Robinson, Sandra L. "Trust and Breach of the Psychological Contract." Administrative Science Quarterly, vol. 41, no. 4, 1996, p. 574., doi:10.2307/2393868.

lxxxvi. Edmondson, Amy C. "Psychological Safety, Trust, and Learning in Organizations: A Group-Level Lens." Trust and Distrust in Organizations: Dilemmas and Approaches, by Roderick M. Kramer, Russell Sage Foundation, 2004, pp. 239–272.

lxxxvii. Cook, Karen S., editor. Trust in Society. Russell Sage Foundation, 2003.

lxxxviii. Cook, Karen S., editor. Trust in Society. Russell Sage Foundation, 2003.

lxxxix. Cook, Karen S., editor. Trust in Society. Russell Sage Foundation, 2003.

xc. Thompson, Leigh L. The Truth About Negotiations. FT Press, 2013.

xci. Nooteboom, Bart. "The Dynamics of Trust: Communication, Action and Third Parties." Comparative Sociology, vol. 10, no. 2, 1 Jan. 2011, pp. 166–185., doi:10.1163/156913311X566553.

xcii. Bohns, Vanessa K. "A Face-to-Face Request Is 34 Times More Successful than an Email." Harvard Business Review, 11 Apr. 2017, hbr.org/2017/04/a-face-to-face-request-is-34-times-more-successful-than-an-email.

xciii.   Feng, Jinjuan et al. "Empathy And Online Interpersonal Trust: A Fragile Relationship". Behaviour & Information Technology, vol 23, no. 2, 2004, pp. 97-106. Informa UK Limited, doi:10.1080/0144929031000l659240.

xciv.   Dutton, Jane E., and Belle Rose Ragins. Exploring Positive Relationships at Work: Building a Theoretical and Research Foundation. Psychology Press, 2017.

xcv.   Dutton, Jane E., and Belle Rose Ragins. Exploring Positive Relationships at Work: Building a Theoretical and Research Foundation. Psychology Press, 2017.

xcvi.   VandenBos, Gary R. APA Dictionary of Psychology. American Psychological Association, 2015.

xcvii.   Legros, Sophie, and Beniamino Cislaghi. "Mapping the Social-Norms Literature: An Overview of Reviews." Perspectives on Psychological Science, vol. 15, no. 1, 2019, pp. 62–80., doi:10.1177/1745691619866455.

xcviii.   Bicchieri, Cristina. Norms in the Wild. How to Diagnose, Measure, and Change Social Norms. Oxford University Press, 2017.

xcix.   Hechter, Michael, and Karl-Dieter Opp. Social Norms. Russell Sage Foundation, 2001.

Managing Designers

Create the Team

A team is a stable bounded group of individuals interdependent in achieving a shared goal.[i] In the management age of empathy, creating a team goes beyond merely hiring talented designers and requires establishing the precondition for these individuals to collaborate and attack ambiguous and often business-critical problems. The most differentiating attitude of this small autonomous network of interconnected individuals derives from their willingness to take interpersonal risks to unlock learning during the problem-solving process.[ii] This peculiar behavioral configuration that distinguishes these groups originates from the presence of two social climates that affect their mindset and working practices:[iii] trust at the individual level and psychological safety at the collective level.

Establishing trust as a precursor of physiological safety represents one of the most crucial objectives for a design manager who manages like a leader. Ideally, the C-level management in your company expresses their top-down support to these two essential social conditions openly, providing you sustain in accomplishing this foundational objective. If you instead operate in an unsupported context, you must be conscious of the fact that you can still create *pockets* of trust and psychological safety within the organization where a learning-oriented climate can be sustained.

In this chapter, you will learn how to create a team. You will learn how to build trust and psychological safety as preconditions for the fulfillment of the other four social needs of the working environment: dependability, structure and clarity, meaning of work, and impact of work. You will analyze the nature of fear, how to discern between which type of failure you must and must not reward, and how to clarify your role and contribution to the team. You will also examine how to institute a hiring practice founded on high hiring standards

© Andrea Picchi 2022
A. Picchi, *Design Management*, https://doi.org/10.1007/978-1-4842-6954-1_6

and a career framework that can support both individual contributors' and managers' professional progression. In the end, you will explore how to onboard new team members and how to plan your first meeting with a new hire.

Join a New Team

While your ultimate objective as a design manager and leader is always to use your abilities to fulfill the needs of your team or workgroup and enable the business success of your company, when you join a new team, there are essential social preconditions connected to the development of trust, and by extension respect, that you have to consider. When you join a new team, depending on the context of the role, generally, you can be in one of the following four scenarios.

Apprentice: You manage a part of your team under the direct request and supervision of your line manager. This scenario represents the safest way to progress from an individual contributor to a managerial role because you receive assistance from your manager within a social environment that is already familiar to you. In this context, a defining challenge is to clearly articulate your accountability and responsibility with your line manager to align expectations and promote a trustworthy collaboration.

Successor: You replace your line manager that received a promotion or decided to leave the company. This scenario presents some commonalities with the "apprentice" one because you operate in a social environment that is already familiar to you. In this context, a defining challenge is to internalize the fact that you are now accountable for the outcome of your team and reframe the social dynamics with your former individual contributor colleagues to promote trustworthy collaborations.

New hire: You join an already established team either within your organization or at a new one. This scenario presents social and operational challenges because it combines the need to understand the intricacies of a new business and the necessity to model your behavior under a new set of social norms. In this context, a defining challenge is to map the relationship power ramifications within your company to initiate new meaningful relationships and promote trustworthy collaborations.

Founder: You join a company as the first designer with the primary objective to build a new team. This scenario offers ample freedom of action because it allows you to craft the social dynamics of your team or workgroup. In this context, a defining challenge is to articulate your vision and align stakeholders' expectations to promote trustworthy collaborations inside and outside the group.

In the following sections, considering the contextual differences delineated by these four typical scenarios, you will learn how to build trust and create psychological safety within your team or workgroup.

Build Trust

Trust is an essential element in managing individuals and building high-performing teams.[iv] It constitutes the substrate upon which all relationships are built and therefore represents a precursor of psychological safety and the primary conduit of influence and collaboration.[v] As a complement to their initial research, Zenger Folkman demonstrated that trust has a positive impact on all the five clusters of leadership capabilities: personal character (Pch), technical skills (Tsk), results-oriented skills (Ros), social-oriented skills (Sos), and organizational change skills (Ocs).

When you develop a high level of trust, you increase the efficacy of your 16 differentiating design leadership abilities and, consequently, the levels of engagement with your collaborators.[vi] From a leadership development perspective, trust represents a universal *competency companion* that forms a *powerful combination* with each of the 16 differentiating design leadership abilities producing a nonlinear increase in performance. Figure 6-1 illustrates the positive correlation between trust and leadership effectiveness.[vii]

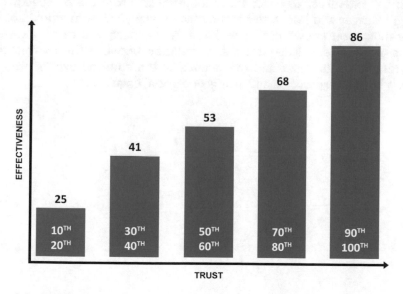

Figure 6-1. The positive correlation between trust and leadership effectiveness

Building trust proactively requires you to frame this social element from a broader perspective. Trust is a complex attitude that involves affective filiations and normative expectations that fulfills a crucial evolutionary purpose at the foundation of human *ultrasociality*.[1] Trust is defined as the expectation that others' future actions will be favorable to one's interests.[viii] This definition emphasizes the two intertwined components at the foundation of trust and, by extension, respect: warmth and competence. Warmth can also be identified as *societal trust*, while competence as *technical trust*. The cognitive expectation derived from the assessment of the behavioral intention and capability of an individual and the consequent willingness to be vulnerable in response to that evaluation is conceived as an antecedent and a consequence of warmth and competence, trust and respect.[ix] These two interpersonal states promote a condition of confidence that shapes emotions and feelings governing social judgment and decision-making and, consequently, the behavior of individual and collective entities.[x] The level of confidence associated with a given individual is a function of the degree of trust and respect attributed to that person:

Trust + Respect = Confidence

From a survival perspective, in encounters with conspecifics, human beings must immediately determine whether a person represents a potential antagonist or a collaborator by deciphering that individual's behavioral intention[2] and ability to enact those intentions.[xi] From the perspective of a design manager and leader, this is the reason why your team would not care "how much you know" until they know "how much you care." Figure 6-2 shows a model that illustrates the correlation between the perception of warmth and competence and the valence of the emotion experienced that models both interpersonal and intergroup social cognition.[xii]

---

[1] The social organization of species that have become highly adaptive at the group level.
[2] An intention is a form of pattern recognition that extends across space and time.

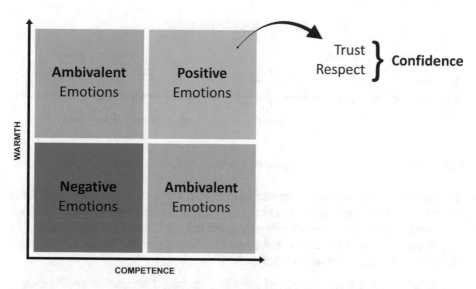

Figure 6-2. The correlation between the perception of warmth (societal trust) and competence (technical trust) and the valence of the emotion experienced

Individuals or groups perceived as warm and competent elicit uniformly positive emotions, feelings, and behaviors, whereas those who lack warmth and competence evoke constant negativity.[xiii] Entities assessed as high on competence and low on warmth, or vice versa, trigger ambivalent emotions, feelings, and behavioral reactions.[xiv]

Build Trust Across Cultures

While trust and respect represent a universal substrate upon which all relationships are built across cultures, the degree of perceived relevance of these two elements diversifies among different social groups.[xv] Some cultures developed a predominant focus on warmth and, by extension, trust, while others on competence and, by extension, respect.[xvi] Both warmth and competence need to be cognitively present to elicit trust and respect and consequently promote a condition of confidence, but, initially, one interpersonal state must be prioritized based on the cultural fabric of the social group in question[xvii] (Figure 6-3).

Trust Development

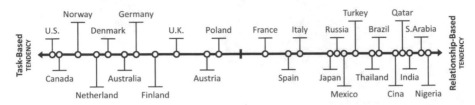

Figure 6-3. The distribution of task-based and relationship-based tendencies across cultures

In cultures with a results-oriented task-based tendency, confidence is built initially via business-oriented activities.[xviii] Typically, task-based relationships are rapidly established and, if necessary, abandoned based on the practicality of the situation. In task-based cultures, a condition of confidence is initiated by demonstrating technical proficiency.[xix]

In cultures with a social-oriented relationship-based tendency, confidence is built initially via convivial-oriented activities.[xx] Typically, social-oriented relationships are slowly established, and they persist beyond the practicality of the situation. In social-based cultures, a condition of confidence is initiated by demonstrating social openness.[xxi]

Within an organization, the level of social confidence derived from the perceived warmth and competence attributed to individuals and groups determines significant professional and organizational outcomes, including, but not limited to, creative collaboration, candidate and employee evaluation, and allocation of tasks and resources.[xxii]

Elicit Warmth and Competence

The correlated relationship between trust and leadership entails all the 16 differentiating design leadership abilities as a factor in eliciting warmth and competence. Among these competencies, three clusters of abilities result to be crucial in promoting the underlying behaviors that contribute to the development of trust.[xxiii] Each cluster engenders a specific determinant of this social dynamic: affinity, proficiency, and consistency.

Affinity. An individual perceives "warmth" when that person senses that you are genuinely interested in their uniqueness, demonstrating the following abilities:[xxiv]

- (Sos) Ability to communicate powerfully and prolifically

- (Sos) Ability to inspire and motivate others to high performance

- (Sos) Ability to build relationships
- (Sos) Ability to develop individual and collective skills
- (Sos) Ability to collaborate

The following are some rituals that can complement and substantiate these abilities:

- Design a ritual to articulate strategic directions.
- Design a ritual to coach for performance.
- Design a ritual to know individuals personally and professionally.
- Design a ritual to coach for development.
- Design a ritual to articulate decision-preparation and decision-making processes.

Proficiency. An individual perceives "competence" when that person senses that you are informed and knowledgeable, demonstrating the following abilities:[xxv]

- (Tsk) Ability to understand persons, business, and technological needs and implications
- (Tsk) Ability to solve problems and analyze issues
- (Ocs) Ability to articulate visions and strategies

The following are some rituals that can complement and substantiate these abilities:

- Design a ritual to translate the strategic direction into personal objectives for the group.
- Design a ritual to contextualize problems and issues holistically.
- Design a ritual to connect the contribution of your group to the organization's strategy.

Consistency. An individual validates the perception of "warmth" and "competence" when that person senses that your actions confirm your words, demonstrating the following abilities:[xxvi]

- (Pch) Ability to lead by example demonstrating high integrity and credibility
- (Ros) Ability to establish challenging objectives
- (Ros) Ability to take initiative

The following are some rituals that can complement and substantiate these abilities:

- Design a ritual to demonstrate trustworthiness and commitments.

- Design a ritual to articulate the meaning and impact of objectives and goals.

- Design a ritual to initiate programs, projects, and processes.

The perceptual system of human beings is designed to identify risks detecting and processing differences in stimuli from the environment.[xxvii] As a direct consequence of that nature, eliciting warmth and competence inevitably includes an element of consistency, without which trust and respect cannot be established or conserved when present.[xxviii]

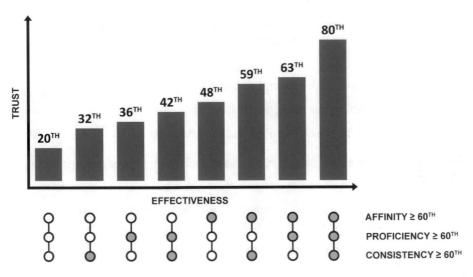

Figure 6-4. The correlation between the development of trust and the three clusters of abilities

When a person scores at or above the 60th percentile on all three clusters of abilities, the overall trust score is at the 80th percentile[xxix] (Figure 6-4). Moreover, the "affinity" cluster alone is able to place a person at the 48th percentile above the 42nd percentile, where individuals demonstrate abilities connected simultaneously to "judgment" and "consistency."[xxx] This insight reemphasizes the importance of achieving results and nurturing relationships equally for a design manager and leader. The individual willingness to accept the vulnerability that defines trust, and by extension respect, represents a precondition for the development of a collective climate of psychological safety.[xxxi] Trust represents the individual construct necessary to allow each team member to be socially predisposed to the interactions required to develop a collective construct like psychological safety.[xxxii] Under that condition, trust can be framed as an enabler that induces a psychologically safe climate.

Create Psychological Safety

Psychological safety is broadly defined as a climate in which individuals are comfortable expressing and being themselves.[xxxiii] More specifically, in a design context, psychological safety is a collective construct that promotes candid discussions without reciprocal negative emotions where conflicts of ideas are leveraged to encourage creativity without allowing them to escalate into conflicts of relationship. It's a climate that requires trust and respect to form and constitute a crucial element in environments founded on collaboration and knowledge exchange.

The need for psychological safety is based on the premise that we do not see reality or perceive the world as it is.[xxxiv] This condition implies that for a human being it is impossible to solve complex problems in isolation and perform "perfectly" in every situation. Establishing a psychologically safe climate within a design team ensures the manifestation of several desired behaviors, including, but not limited to, the following ones.[xxxv]

Enables candid conversations: Psychological safety mitigates the social tensions allowing the exploration of conflicting ideas without permitting them to escalate into conflicts of relationship; this dynamic increases the efficacy of divergent and convergent thinking.

Enhances creative collaboration: Psychological safety relieves the interpersonal fear related to cooperation; this dynamic increases the cognitive resources available for creative collaboration.

Amplifies bias toward action: Psychological safety alleviates the concerns regarding other individuals' reactions to behaviors or activities that have the potential for embarrassment; this dynamic increases the propensity to take action.

Promotes ownership: Psychological safety attenuates the aversion to taking risks; this dynamic increases the ability to take the initiative and work autonomously.

Enables continuous learning: Psychological safety reduces the negative perception of unsuccessful initiatives; this dynamic increases the ability to reflect in-action and on-action.

Magnifies individual and collective performances: Psychological safety promotes and sustains self-expression; this dynamic increases the cognitive resources devoted to achieving goals rather than self-protection.

Furthermore, psychological safety also positively influences group cohesiveness,[xxxvi] which, as you will learn in Chapter 9, represents a precondition for a healthy integration of diversity. It is impossible to comprehend psychological safety without understanding its antagonist: fear. The following section will explore the nature of fear and the typical fear-inducing stimuli present within a design team.

The Nature of Fear

Neuroeconomist Gregory Berns, Professor at Emory University, utilizing data from functional magnetic resonance imaging (fMRI), demonstrated that when the *neural circuit* (Figure 6-5) involved in the process of fear activates, it absorbs a significant amount of cognitive resources and leaves the brain with "less" neural processing power available for exploratory activities.[xxxvii] From a survival perspective, once your brain identifies a potential threat, everything else becomes momentarily irrelevant.

Neuronal Circuit of Fear

Figure 6-5. Two foundational elements of the neuronal circuit of fear: the prefrontal cortex and the basal ganglia

Fear is characterized by an adaptable nature in terms of cognition and behavioral response, where the relationship between stimuli and behaviors mediated by fear is flexible and context dependent.[xxxviii] A design manager and leader who attempts to minimize *interpersonal fear* must comprehend the context-dependent implications and the temporally extended dynamic nature of a fear state. The following model presents five dimensions that can help you to frame a state of fear and investigate a fear-inducing stimulus.[xxxix]

Type: What is the nature of the threat?

Circumstance: Is the threat avoidable?

Vicinity: Is the threat physically near?

Imminence: Is the threat temporally near (in the future)?

Separation: Is the threat temporally near (in the past)?

Fear-inducing stimuli are a function of the organizational culture and its underlying social norms.[xl] These types of events can be associated with persons, events, or locations and can be divided into four nonmutual exclusive categories.[xli]

Fear of being seen as ignorant: When an individual asks questions or seeks information expressing concerns and requesting clarifications. For instance, asking repeated inquiries on a design solution during a review meeting when the silence in the room implies that you are expected to know that information.

Fear of being seen as incompetent: When an individual admits a mistake requesting assistance. For instance, confessing during a team meeting that your design hypothesis concerning a problem that you are expected to solve was incorrect.

Fear of being seen as antagonistic: When an individual critically evaluates activities and performances. For instance, expressing an opinion during a retrospective meeting on individual or collective unsatisfying outcomes.

Fear of being seen as disruptive: When an individual seeks collaboration asking other persons' time and contribution. For instance, requesting feedback and support on a design solution during a one-on-one conversation with a colleague.

A conscious or unconscious state of fear has access to a vast number of cognitive and behavioral processes with the result of modulating attention, memory, perception, and decision-making.[xlii] Fear induces a survival modality that, among other functionalities, inhibits creative thinking and impairs bias toward action, two essential dynamics of the design operations.[xliii]

Understand Your Role and Contribution

From the perspective of a design manager and leader, one of the aspects necessary to cultivate creative collaboration is to adopt a behavioral configuration capable of minimizing interpersonal fear and maximizing social participation.[xliv] This conduct can be broadly delineated by the following three steps:

1. Set the direction for the work.
2. Invite relevant input to clarify the direction.
3. Reflect on and factor in the appropriate information.

In contrast to an environment where the person in charge of the team possesses all the answers and assesses the execution of directives, in this context, the team members actively contribute to the operations beyond the mere implementation, providing crucial knowledge and insights. When implemented effectively, this behavioral strategy unlocks a feeling of ownership that allows you to engender candor and participation, two conditions necessary to develop a psychologically safe climate. The following section will introduce how these three steps can be instantiated to cultivate a psychologically safe environment.

Cultivate a Psychologically Safe Environment

Based on the research conducted in the past two decades by Amy Edmondson, Professor of Leadership and Management at Harvard Business School, Figure 6-6 illustrates the three interrelated practices required to cultivate psychological safety.[xlv]

Cultivate Psychological Safety

Establish Preconditions	Invite Participation	Respond Productively
Emphasize the Purpose	Demonstrate Situational Humility	Express Appreciation
Reframe Failure	Support Candid Communication	Embrace Failure
Clarify the Need for Voice	Practice Proactive Inquire	Sanction Clear Violations

Figure 6-6. The three interrelated practices required to cultivate psychological safety

In an environment permeated by psychological safety, interpersonal fear is minimized, and individuals are more concerned about repressing their full participation than sharing potentially sensitive, challenging, or inaccurate ideas.[xlvi] This attitude is essential to cultivate creative collaboration and optimize design operations.

Establish the Preconditions

Creating the preconditions for psychological safety allows you to set shared expectations and define meanings that encourage candid conversations that promote experiential learning via reflection-in-action and reflection-on-action.[xlvii]

Emphasize the purpose: Articulating a compelling purpose for the team allows you to generate a source of intrinsic motivational energy that helps the team to reduce the perception of stress associated with intelligent failure and the asymmetry of voice and silence during challenging moments.[xlviii] Moreover, when you connect your team's and company's purpose, substantiating their *why*, *what*, and *how*, you also contribute to the fulfillment of two of your team's social needs: the "meaning of work" and "impact of work."

Reframe failure: Presenting failure as a natural constituent of experiential learning allows you to integrate this inescapable event into the design process productively. The fear of reporting failure and consequently the avoidance to elaborate it[3] represent the strongest indicators of an environment characterized by low levels of psychological safety.[xlix] Failure is never the objective of the problem-solving process, but promoting a cognitive frame that processes it as a valuable source of insights permits the team to maximize the learning opportunities.

A cognitive frame consists of a set of assumptions or beliefs in regard to a given situation.[l] Unless you proactively provide a perspective that explicitly presents failure as an accepted practice, your team will consciously or unconsciously seek to avoid it.[li] For instance, David Kelley, co-founder of IDEO, reframed failure saying, "fail faster to succeed sooner." Reframing failure requires you to comprehend the basic failure archetypes (Figure 6-7) because not every mistake unlocks knowledge, and not every unsuccessful event deserves to be celebrated.

Failure Archetypes

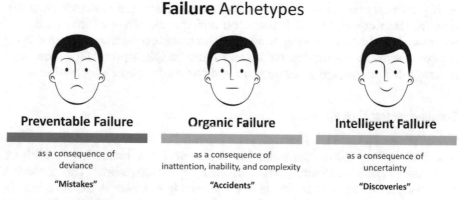

Preventable Failure	Organic Failure	Intelligent Failure
as a consequence of deviance	as a consequence of inattention, inability, and complexity	as a consequence of uncertainty
"Mistakes"	"Accidents"	"Discoveries"

Figure 6-7. The three basic archetypes of failure

"Preventable failure" is a consequence of deviance, where the divergence from pre-agreed social norms, typically in the form of a process or practice, produces unwanted outcomes.[lii] This type of failure must be framed as a "mistake" that does not unlock learning; it must not be celebrated and require intervention. Decreasing the incidence of "preventable failure" is directly correlated to an increase in the stability of the team's cohesiveness.[liii]

"Organic failure" is a consequence of inattention, inability, and complexity, where novel permutations of the problem domain elements generate unwanted outcomes.[liv] This type of failure must be framed as an "accident"

---

[3] The inability of a person to use critical thinking to elaborate adverse events is also correlated to low levels of self-esteem and self-efficacy.

that does not unlock learning; it must not be celebrated and require support in the form of coaching or collaboration. Decreasing the incidence of "organic failure" is directly correlated to an increase in the team's productivity.[lv]

"Intelligent failure" is a consequence of uncertainty, where novel forays into new areas of the problem domain lead to unwanted outcomes.[lvi] This type of failure must be framed as a "discovery" that unlocks learning; it must be celebrated openly. In opposition to the other forms, intelligent failure represents the by-product of a structured and deliberate exploration that cannot prescind from taking exploratory risks to understand a given problem domain. Decreasing the sensibility of the team to intelligent failure is directly correlated to an increase in the team's creative potential.[lvii]

Clarify the need for voice: Introducing the necessity for candid conversations allows you to sustain experiential learning via an effective analysis of exploratory initiatives and a continuous acquisition of knowledge. You can emphasize the need for voice and help the group overcome the inherent asymmetry of voice and silence by reminding them of the ambiguous and interdependence nature of creative collaboration. If you remind that the ambiguity that characterizes the problem-solving process demands curiosity, that message can promote exploration and the exchange of information. If you elucidate that the interdependence that characterizes the problem-solving process requires ownership to sustain multidisciplinary collaboration, that message can encourage proactiveness and the exchange of information.

Strategize for Failure

When your objective is to establish the preconditions for psychological safety, you have to conciliate the need for voice with the necessity to mitigate "mistakes" and "accidents" while promoting "discoveries." This situation requires you to strategize for failure by setting behavioral expectations in regard to these three forms of undesired outcomes with the intent to minimize interpersonal fear and, as you will learn in Chapter 7, the likeliness of triggering a threat response and social pain.

Creating a psychologically safe climate does not translate into "being nice" or, more technically, prioritizing relationships over results accepting and celebrating behaviors and outcomes even when unproductive. As you will learn in the next section, in order to create a healthy learning environment, your reaction to "mistakes," "accidents," and "discoveries" must be diversified.

Your responsibility as a design manager and leader is to clarify this crucial distinction to the team and provide them with the mental frame necessary to interpret your behavioral response accordingly. Beyond the definitions of the three archetypes of unsuccessful outcomes introduced in the previous section, you also must discuss with the team the five core elements that characterize intelligent failure and constitute the preconditions for a continuous learning practice.

The opportunity is significant. The initiative needs to be aligned with the purpose and vision of the organization.

The outcome is informative. The initiative needs to unlock learning and inform the problem-solving process.

The scope is minimized. The initiative needs to focus on a specific question to answer.

The assumption is formalized. The initiative is founded on a hypothesis that can be proved or disproved.

The assumption is investigated. The initiative is conducted to prove or disprove the hypothesis.

Once you have established the behavioral expectations in regard to the three archetypes of failure, you can concentrate on concurrently mitigating "mistakes" and "accidents" while promoting "discoveries" by leveraging three specific modalities of creative leadership.

Reduce preventable failure. You achieve this objective by leading through culture to indirectly influence the beliefs, behaviors, and abilities of the team.

Anticipate organic failure. You achieve this objective by leading in direct contact with your team's activities to remain in touch with the needs and challenges of the group.

Promote intelligent failure. You achieve this objective by leading with a strong viewpoint articulating a destination that motivates the team to explore and learn.

In Chapter 8, you will explore these three modalities of creative leadership in detail, learning how to deploy them based on the problem to solve and the specific needs of the team.

Invite Participation

Inviting participation allows you to engender the confidence that voice is welcomed, lowering the social threshold related to candid conversations and what is considered an appropriate contribution.

Demonstrate situational humility: Displaying a learning mindset allows you to invite participation by substantiating the beliefs that no one can "know everything" and, consequently, lowering the social threshold for interpersonal risks in candid conversations. Situational humility is a learning-oriented mindset characterized by *curiosity* and substantiated demonstrating *selective vulnerability*. When you display selective vulnerability[4] in a given situation, you

---

[4] Before demonstrating vulnerability, you must elicit warmth and competence to build or conserve trust and respect.

connect with your team members at an interpersonal level without oversharing at the detriment of trust and respect. You candidly open up to the group while still prioritizing their boundaries, as well as your limits. This comportment implicitly promotes the team's ownership and participation.[lviii]

Support candid communication: Providing guidelines for discussions and forums for input allows you to invite participation by institutionalizing a candid exchange of information and, consequently, experiential learning.

Practice proactive inquiry: Posing purposefully crafted questions allows you to invite participation by encouraging your team to learn more about a given problem domain. Practicing proactive inquiry and manifesting a genuine interest in the answers demonstrates respect for the interlocutor and represents a precondition for creative collaboration. Questions are the *atomic unit of exploration*; they must be generative, inspiring, and generate reflection. Typically, you should use *why* to form queries aimed to identify the root cause of a problem domain and *what (if)* and *how (might we)* to respectively diverge and converge the exploratory thinking.

Respond Productively

Responding productively allows you to reinforce a psychologically safe climate oriented toward continuous learning providing support to activities that can potentially present interpersonal risks.

Express appreciation: Expressing genuine gratitude without judgment allows you to respond productively by reinforcing a behavior that, overcoming interpersonal risks, initiates candid conversations and supports experiential learning. Stanford Professor Carol Dweck demonstrated with her research on the human mindset that to establish a resilient learning-oriented environment, it is crucial to praise individuals for efforts, regardless of the outcome.[lvix] When a member of your team advances a suggestion or poses a question, irrespective of its relevance, your initial response must be welcoming and appreciative. Subsequently, if necessary, you can elaborate on your reply by providing feedback or technical details. The ambiguity, uncertainty, and absence of direct cause-effect characterizing the design process require you to manifest an effective response to all kinds of contributions and outcomes.

Embrace intelligent failure: Reminding the distinction between failing and being a failure allows you to respond productively by creating the cognitive preconditions for the integration of intelligent failure into the problem-solving process. In doing that, you must convey that intelligent failure is never the objective but solely a means to an end, experiential learning. Remind the team that exploratory and experimental failure in a condition of uncertainty is a by-product of the design process and that top performers produce, exploit, and share lessons from intelligent failure. In order to complement your behavioral response, create events where the team can discuss experiences, present lessons learned, and you can provide support and contribution to the design direction.

Sanction clear violations: Taking decisive actions to sanction a clear, repeated violation of the pre-agreed social norms that enable psychological safety allows you to reinforce this mental construct[5] and your integrity as a leader.[ix] Psychological safety requires supportive boundaries to prosper, because, similarly to other mental constructs, it's subject to be influenced by antagonist behaviors. Despite that, a design manager and leader who responds equally to every form of failure does not create a healthy learning environment. Table 6-1 shows the appropriate responses in relation to the three basic archetypes of failure.

Table 6-1. Appropriate responses to the three basic archetypes of failure

Preventable	Organic			Intelligent
Mistakes	**Accidents**			**Discoveries**
Deviance	**Inattention**	**Inability**	**Complexity**	**Uncertainty**
1. Analysis	1. Analysis	1. Analysis	1. Analysis	1. Celebration
2. Reflection	2. Reflection	2. Reflection	2. Reflection	2. Analysis
3. Sanction	3. Support	3. Coaching	3. Support	3. Reflection
4. Coaching				4. Ideation

A sophisticated understanding of the appropriate responses to the three basic archetypes of failure allows you to institute an effective strategy for learning from failure discerning the events that deserve celebration from the ones that require intervention.

Assess Psychological Safety

Psychological safety is a climate that originates from a mental construct that develops and regresses as its preconditions are fulfilled or neglected. As the person in charge of the group, you must always be aware of the team's current psychological climate. You can do it qualitatively by monitoring its behavioral indicators during your daily interaction with your team members and quantitatively by asking them to complete a survey. Table 6-2 shows a questionnaire[lxi] that your team can complete anonymously, typically annually or semiannually.

---

[5] A mental construct is formed by a set of ideas and beliefs.

Table 6-2. Psychological safety and team learning climate questionnaire

Advocacy[α]

  1. I would recommend working in this team to others.

Psychological safety[β]

  1. In this team, if I make a mistake, it is often held against me.[R]

  2. In this team, I can bring up problems and severe issues.

  3. In this team, I get rejected for being different.[R]

  4. In this team, it is safe to take a risk.

  5. In this team, it is difficult to ask other members for help.[R]

  6. In this team, no one would deliberately act in a way that undermines my efforts.

  7. In this team, my unique skills and talents are valued and utilized.

Team learning behavior (internal)[β]

  1. In this team, we always communicate problems and errors to the relevant persons to take appropriate action.

  2. In this team, we often take time to figure out ways to improve our team's work processes.

  3. In this team, we talk about mistakes and ways to prevent and learn from them.

  4. In this team, we tend to handle conflicts and differences of opinion privately or offline rather than addressing them directly as a group.[R]

  5. In this team, we frequently obtain new information that leads us to make essential changes in our plans or work processes.

  6. In this team, we often raise concerns about team plans or decisions.

  7. In this team, we continuously encounter unexpected hurdles and get stuck.[R]

  8. In this team, we try to discover assumptions or beliefs about issues under discussion.

  9. In this team, we help others to understand our unique area of expertise.

  10. In this team, we gain a significant understanding of other areas of expertise.

Team learning behavior (external)[β]

  1. In this team, we frequently coordinate with other teams to meet organizational objectives.

  2. In this team, we cooperate effectively with other teams or shifts to meet corporate objectives or satisfy customer needs.

  3. In this team, we are not great at keeping everyone informed who needs to buy in to what the team is planning and accomplishing.[R]

  4. In this team, we go out and get all the information we possibly can from a lot of different sources.

  5. In this team, we do not have time to communicate information about our team's work to others outside the team.[R]

  6. In this team, we invite individuals from outside to present information or have discussions with us.

(continued)

Table 6-2. (continued)

Recommendationsχ

- What would you recommend the team keep doing?
- What would you have the team change?

R Reverse score: Value 1 is converted to a value of 5, and 2 to a value of 4.

α Binary answer: Yes and No.

β 5-Point scale: Never, Rarely, Sometimes, Often, Always.

χ Open answer: Text field.

The questionnaire is structured into three main sections. The first section assesses the essence of psychological safety, capturing the extent to which an individual perceives the social climate as conducive to interpersonal risk; it is a measure of the willingness to trust others not to attempt to obtain a personal advantage at their expense.[lxii] The second section assesses the learning behavior internal to the team, capturing the extent to which an individual engages in actions conceived to monitor progress and performance against goals and the tendency of that person to participate in activities designed to test assumptions and create new possibilities.[lxiii] The third section assesses the learning behavior external to the team, capturing the extent to which an individual engages in actions intended to obtain information and feedback from the customers or collaborators within the organization.[lxiv]

At some point in your role, you will need to grow your team and hire new talents. In the second part of this chapter, you will learn how to establish an efficient hiring practice characterized by high behavioral standards.

Hire Talents

When you operate at Stages 3 and 4 of the Four Stages of Contribution model contributing via teams and workgroups, hiring represents the most effective instrument for ensuring your success and the one of your team and organization. There are always brilliant individuals behind a prolific creative environment, efficient design operations, and successful outcomes. If human beings are the source of success of an organization, the decisions made to determine which individuals join the company are crucial. In this section, you will explore the principles of effective hiring and learn how to implement these principles to establish an effective and efficient hiring practice.

The Principles of Effective Hiring

An effective hiring practice is able to recurrently discern candidates capable of being successful in a given role with a high degree of confidence. This practice is founded on four essential principles.

Principle 1: Define high behavioral standards for the roles. Creating demanding behavioral requirements to join the team allows you to positively affect performance and retention.[lxv] Having a high behavioral standard enables you to hire talented designers, which increases the collective ability and, consequently, performance.[lxvi] At the same time, setting demanding requirements allows you to preserve the collective ability, which reinforces the sense of belonging within the team and, consequently, retention.[lxvii] Moreover, defining high behavioral standards and creating a group of talented designers contributes to developing an appealing reputation in the industry that attracts accomplished individuals and facilitates your sourcing effort.[lxviii]

Principle 2: Define abilities but investigate behaviors. Capabilities do not exist in the abstract; they are always instantiated and substantiated in the form of a concrete comportment in a given context. Knowledge per se is necessary but not sufficient to engage in a given set of actions required to complete a task successfully. Therefore, investigating the behavioral requirements of a given role represents the most reliable strategy to interview a candidate.

Principle 3: Evaluate candidates using a default to "not hire" mindset. The purpose of evaluating candidates is to find a reason to say "no" by looking for weaknesses, trying to eliminate all the candidates who would not match the behavioral standards of a given role. We do not perceive the world as it is, and our perception is biased.[lxix] If we look for strengths in a candidate, we will probably find them, but that insight does not prevent us from hiring a false positive. Counterintuitively, the primary risk of hiring is not to fail to fill the vacancy but to add false positive to the team that does not respect the behavioral standards of the role. Figure 6-8 illustrates the four possible outcomes of the hiring process.

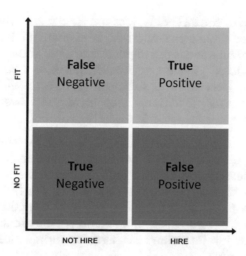

Figure 6-8. The four possible outcomes of the hiring process

If you concentrate on protecting the team from hiring someone who would not match the hiring criteria, you remain with two possible outcomes: hiring a good fit, or true positive, and not hiring a good fit, or false negative. The latter represents an undesired but acceptable consequence of achieving this objective.

Principle 4: Provide a reason to join the team. An effective hiring process cannot prescind from providing a compelling articulation of why an individual should invest the most limited and precious resource, time, on "you" being part of "your" team. During the process, you must introduce the company's purpose, how that aspiration informs the team's purpose, and leverage the contribution of a specific role. This message also clarifies the expectation related to the expected level of fulfillment associated with the "meaning of work" and the "impact of work." Lastly, this part of the communication must also mention the support for personal development that a team member will receive via personalized coaching activities.

Establish a Structured Process

Frank Schmidt and John Hunter, two psychologists and university professors that received the Distinguished Scientific Award for their contribution to Applied Psychology, published a meta-analysis of 85 years of research on 19 different assessment techniques finding that structured evaluation practices possess a high *predictive validity*.[6] A structured method is defined in terms of

---

[6] Predictive validity is the capacity of an assessment method to predict how a candidate would perform once hired.

the standardization of the interview question format and the response scoring framework adopted where any variation in assessment is a function of the candidate's performance and is not imputable to a variation in the interviewer's standards or questions.[lxx] Evaluating a candidate using a structured approach composed of four interviews is sufficient to hire an individual with 86% confidence, where every additional step increments averagely by 1% the predictive validity.[lxxi]

The evaluation process constitutes a significant moderator of the interview validity, where the predictive factor increases proportionally with the increase in structure.[lxxii] Assessing a candidate using a structured method, complemented by the integration of multiple perspectives during the hiring process, is the most effective way to mitigate the inherently and inevitably biased nature of human perception.[lxxiii] Structured investigations represent the standard de facto not solely for job interviews but also for survey interviews and other rigorous research forms.

Define the Behavioral Requirements

As the person in charge of the team, you are responsible for determining the exact behavioral requirements of the group. The behavioral requirement of a role is defined as a function of the abilities necessary to succeed in that given position. Your objective is to formalize these abilities using the same structure adopted by the Leadership Tent model. If you have a large team, you may need to create two separate models to distinguish the requirements defined for individual contributors and managers. Given a role, specify the desired behavior in terms of the abilities necessary and structure them under the five clusters of capabilities of the Leadership Tent framework:

- Personal character
- Technical skills
- Results-oriented skills
- Social-oriented skills
- Organizational change skills

You can use the following four critical areas of design expertise that typically characterize prolific problem-solvers to stimulate your initial thinking.

Navigate ambiguity: Being able to explore a problem domain that is loosely defined.

Build intentionally: Being able to create as a way to think and explore problem domains.

Socialize knowledge: Being able to leverage a growth mindset to share information and learn from others as a gate to collaboration.

Communicate deliberately: Being able to articulate and narrate stories that paint compelling pictures of the future.

In the case of an individual contributor role, the behavioral requirement of a position does not have to span across the entire spectrum of capabilities. For instance, organizational change skills are not a core requirement for a Junior and a Senior Designer at Stages 1 and 2, but they become instrumental at Stage 3 for a Head of Design and indispensable at Stage 4 for a Chief Design Officer. The following list presents some recurrent abilities that typically characterize the behavior of effectual individual contributors:

- Ability to identify and understand challenges.
- Ability to frame challenges into problems.
- Ability to experiment and explore problems.
- Ability to build to think and make possibilities tangible.
- Ability to reflect and learn.
- Ability to create stories and articulate design decisions.

Once you have specified the abilities behind a given behavior, in order to assess them, you have to decompose them into a set of actions that can be investigated directly using a behavioral question. In the following example, you can see one of the previously introduced abilities decomposed into its elementary activities.

Ability to experiment and explore problems:

- Craft driving questions to inspire and guide the team.
- Conduct data-informed, insight-driven experiments.
- Contribute to a diversity of perspectives.
- Explore, process, and integrate divergent ideas.
- Make decisions leveraging analysis, experience, and judgment.

Specifying the abilities necessary to demonstrate a given behavior allows you to create clear job descriptions, precise behavioral questions, and uniform criteria of evaluation. With a clear understanding of the abilities and actions that drive the behaviors that you deem necessary for the roles in your team, you can now define the behavioral standard that a candidate must respect during the hiring process to be considered for each position.

Define the Behavioral Standards

The behavioral standard of a role is expressed as a quantification of the minimum level of expertise associated with each ability defined for that given position. The behavior is quantified using a five-point Likert scale with a value ranging from 0 associated with "not qualified" and 4 with "world-class." In Figure 6-9, you can see the visual representation of a behavioral standard for a role delineated by a "world-class" (A5), three "extremely qualified" (A2, A3, A4), and a series of "highly qualified" abilities (A5, ..., AN). The profile delineated by the behavioral standard of a given role can be considered the archetype of the desired candidate for that position, and you can save it digitally using your spreadsheet software of choice.

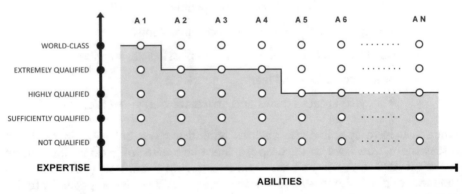

Figure 6-9. An example of a behavioral standard implemented using a five-point Likert scale

Having clear and quantifiable expectations for each ability considered necessary for a job does not necessarily imply that a candidate cannot be considered if that person does not precisely match all the requirements. Hiring, exactly as designing, requires an investigative, open, data-informed approach, not a blind, close, data-driven one. Having a behavioral standard for a role allows you to make direct and effective decisions in the vast majority of cases and deliberately break the rule and place smart, informed bets when necessary.

Define the Hiring Process

The hiring process has two main objectives: maximize the confidence level related to true positive outcomes and minimize the incidence of personal biases during the decisional process, all in strict compliance with the hiring standards. The first objective is achieved by assessing the candidate's compliance with the behavioral standard, investigating the absence of the required abilities instead of searching for signs of their presence. The underlying and counterintuitive intention is to find a reason to "not hire" the candidate. The second objective is achieved by scheduling a sufficient number of conversations with an adequate number of individuals to inform a comprehensive and statistically significant candidate evaluation. The underlying intention is to increase the degree of diversity of the dataset acquired. The following example presents a process that implements the aforementioned intentions:

1. **Resume and portfolio screening**

   Who: Head of Design, human resources, or someone familiar with all the design roles

2. **Phone/video call screening**

   Who: Head of Design or human resources

3. **In-person/video call interviews**

   Who: Peers, subordinates (if any), cross-functional colleagues, Head of Design

4. **Interview evaluation meeting**

   Who: All interviewers

5. **Candidate prioritization meeting**

   Who: All interviewers

6. **Hiring committee review**

   Who: Individuals with a design background who would not directly work with the candidate

7. **Reference check and offer**

   Who: Human resources or Head of Design

Step 1: Resume and portfolio screening. This session needs to be conducted by a person familiar with the different design roles available within the team and, if your organization operates on a global scale, with the cultural differences characterizing educational degrees. During the sourcing and screening phase, it's important to remember that some great designers do not necessarily have an updated portfolio, sometimes because some of their work is under a "nondisclosure agreement" (NDA) or simply because they were occupied creating something remarkable. If you encounter this type of candidate, don't be judgmental on the lack of a proper portfolio and move them to the next step where you can adequately explore their abilities and potential contribution to the team.

Step 2: Phone/video call screening. This session has a duration of 30 minutes and represents the first direct contact with the candidate. A person from the human resources department can make this initial phone or video call; however, when possible, this represents a unique opportunity for the Head of Design to demonstrate a genuine interest in the candidate and increase the possibilities of acceptance if an offer is extended at the end of the process. The following is a possible structure of this screening:

1. Process overview (1 minute)

2. Company and role introduction (4 minutes)

3. Opening question: "Tell me about yourself" ($\leq$ 10 minutes)

4. Follow-up question: "Describe to me how your experience prepares you for this position" ($\leq$ 10 minutes)

5. Questions from the candidate (5 minutes)

The call is typically concluded by informing the candidate that the hiring group will review the answers and provide a response within a week.

Step 3: In-person/video call interviews. These sessions have a duration of 60 minutes each for peers, subordinates (if any), and cross-functional colleagues. The last session is with the Head of Design and has a duration of 90 minutes. Typically, a team can comfortably manage one round of in-person interviews per day. The following is a possible structure of this interview:

1. Introduction, process overview (5 minutes)

2. Five behavioral questions (variable)

3. Questions from the candidate (5 minutes)

The five behavioral questions are predefined and listed in order of importance; every interviewer asks the same questions during the session. The diversity of perspective is not necessarily defined by the answer to the initial behavioral question but more likely by the candidate's responses to the additional and specific follow-up questions that each interviewer asks during the conversation. This approach represents an intentional strategy to compare each candidate against the hiring standards instead of against each other. A system based on comparing candidates can solely identify the local optimum, the optimal candidate within a neighboring set of contenders, and is incapable of intentionally ensuring respect for the behavioral standards defined for the roles.

Despite having a structured interview, the follow-up questions are often unscripted, and every interviewer can decide to probe in whichever direction feels appropriate, adding a unique perspective and contribution to the evaluation process. After the behavioral questions, the interviewer answers questions from the candidate. At the end of the interview, each interviewer must immediately reflect on the assessment and summarize the candidate's evaluation with one word: "hire" or "not hire." Figure 6-10 illustrates the first page of the interview template document used to score the answers and annotate comments during the conversation.

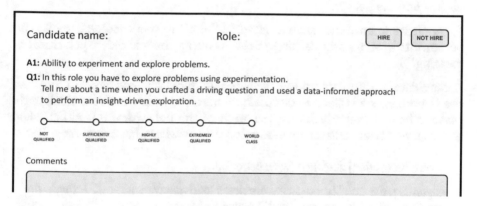

Figure 6-10. The interview template document

Each page of the interview template presents a behavioral question, a five-point Likert scale to quantify the answer, and a comment section to capture additional qualitative details. If the candidate attends an in-person session, the interviewer must use pen and paper to take notes for two crucial reasons. The first one is related to the relationship between the interviewer and the interviewee: taking notes using a laptop creates a physical barrier that divides and obscures the interviewer's activities, impeding the creation of rapport and decreasing the perceived level of psychological safety.[lxxiv] The second one is related to the contribution of the interviewer to the "interview evaluation meeting": compared to typing on a laptop, taking handwritten notes represents a richer cognitive experience that facilitates recall from memory.[lxxv]

Step 4: Interview evaluation meeting. This session has a duration of 30 minutes circa and is held rigorously on the same day as the "in-person/video call interviews." The following is the agenda of this meeting:

1. Process overview (1 minute)

2. Individual report (1 minute)

3. Collective discussion (variable)

After a brief reminder of the process, you introduce the candidate, and then the group silently performs a session of *pollice verso voting* where each interviewer takes a turn and reports the result of the evaluation using a thumb up for "hire" and a thumb down for "not hire." Using a gesture to express the decision silently reduces the perceived stress for *predominantly introverted* individuals and increases the impartiality of the process by removing the potential influence of the tone of voice. Depending on the evaluations, at this point of the process, there are three scenarios.

A candidate is unanimously evaluated as a "not hire." In this case, you disqualified the person.

A candidate is unanimously evaluated as "hire." In this case, you move that person into a list of potential hires to review during the "candidate prioritization meeting."

A candidate is not unanimously evaluated as "hire" or "not hire." In this case, the team begins a collective discussion where the interviewers articulate the reasons behind their decisions. For each of the behavioral questions asked, the interviewer articulates its evaluation using a *what-why* structure:

> [what: perception] because [why: behaviour].
>
> *Regarding the "ability to solve problems and analyze issues", I think the candidate is "extremely qualified" because ...*

After the discussion, you decide if the person can be moved to the list of potential hires reviewed during the "candidate prioritization meeting." If a decision requires a deeper analysis of the quantitative data collected, you can add all the evaluations to your spreadsheet software of choice and apply the standard deviation to visualize the dispersion in evaluation for each ability.

Step 5: Candidate prioritization meeting. This session has a duration of 30 minutes and is held after the team has completed the round of "in-person/video call interviews" for a given role. The objective of this conversation is to decide which candidate must be submitted to the hiring committee. The following is the agenda of this meeting:

1. Process overview (1 minute)

2. Candidate prioritization (10 minutes)

3. Final decision (variable)

After a brief reminder of the process, you introduce the candidate on a wall or a whiteboard, and then the group silently performs a session of "dot voting" where each interviewer uses one dot to express the final preference on the pool of candidates.

If a hiring committee is not present in your organization, at this point, the Head of Design must make the final decision factoring in the dot-voting session and the evaluation reports from the group.

If a hiring committee is instead present, the Head of Design selects the candidate to submit to the commission, preparing a *hiring proposal* that includes the candidate's detail, the role description with the associated behavioral standard, and an interview evaluation report. A hiring committee is often not constituted in a small startup, but it plays a crucial role in ensuring the respect of the hiring standards while scaling a team from a single- to a double- or triple-digit head count.

Step 6: Hiring committee review. This session has a variable duration depending on how many hiring proposals the commission has to review. It is typically held periodically or occasionally, depending on the frequency of the hiring activities. This commission is constituted by an odd number of persons with a design background that will not work directly with the candidate in question. If your organization does not include multiple design teams but has a Chief Design Officer, that person can cover this role individually. If having a hiring committee is not possible in your company, this step is removed from the process.

Step 7: Reference check and offer. The reference check session has an average duration of 15 minutes per reference and is held after the hiring committee or the Head of Design has decided to extend an offer to a given candidate. During the conversation with the referenced person, ask an opinion on the proficiency level of the abilities investigated using the behavioral questions. If the candidate passes the reference check, the Head of Design calls the candidate to personally offer the job specifying a precise deadline for the answer. If and when the candidate accepts the job offer, human resources prepare the job contract.

Assess the Hiring Process

Establishing an efficacious hiring practice requires experience and rigorousness. Interviewers need to develop specific skills to evaluate candidates effectively, and the hiring group requires time to optimize the collaborative dynamics requested by the process. Under these conditions, assessing the effectiveness of the hiring process is crucial because it's complicated to improve something that you don't measure. You can conduct this analysis at an individual and a collective level, typically annually or semiannually, depending on the intensity of your hiring activity.

At an individual level, you focus on measuring the reliability of the evaluation of a given interviewer, investigating their accuracy calculated as the number of candidates evaluated as "hire" at the end of the process against the ones that corroborated the evaluation receiving an offer and potentially passing the probation period, expressed in percentage.

At a collective level, you concentrate on measuring the reliability of the evaluation of a given hiring group, investigating its accuracy calculated as the average score of each ability at the end of the hiring process against the one produced after a reevaluation at the end of the probation period. In this case, you can also apply the standard deviation visualizing the dispersion of the dataset relative to its mean and examine the degree of alignment or misalignment within the group in regard to the evaluation of one candidate or a pool of candidates.

Write Job Descriptions

A compelling job description does not merely inform candidates on the details of a vacancy, but it maximizes the number of applications that fulfill the behavioral requirements of that given role while minimizing the number of irrelevant ones. This type of document must be emotionally engaging, alluding to the social needs of high-performing teams, and rationally attractive, introducing the abilities required for the job. The following are the sections composing the layout of a web page presenting this type of job description.

Section 1: Why this role is important for [company name]

- Introduce the company's purpose.
- Articulate the "impact of work."
- Describe the "meaning of work."

Section 2: What [company name] offers you?

- Specify the line manager and role position within the context of the organization.
- Delineate the social norms alluding to elements of "psychological safety," "dependability," and "structure and clarity."

Section 3: Your mindset and experience

- Outline the abilities required for the role.
- Include additional examples to understand the balance between hands-on and hands-off activities.

Section 4: A question for you

- Ask a behavioral question targeting the primary objective of the role (specify that the answer must be 200 words or less).

Section 5: Apply for the job

- Introduce the hiring process briefly.
- List the required documents and information (if any).
- "Apply Now" button.

Moreover, applying and interviewing for a job represents a challenging experience and a source of stress, reducing the candidate's performance.[lxxvi] Providing a clear job description also reduces the degree of uncertainty associated with the hiring process and expected commitment, which tends to mitigate the initial level of stress related to this type of experience.[lxxvii]

Create the Behavioral Questions

The most reliable strategy to predict the future behavior of a designer is to investigate the past actions of that person or, in the case of a junior designer without antecedent experiences, the mental models behind those actions.[lxxviii] Observing the candidate performing the behaviors requested by a given job within the expected role's environment would be the most accurate method to complete this investigation; this is why a probation period always represents the last phase of the hiring process. In the absence of the opportunity to complete this qualitative observation, behavioral questions represent the most reliable instrument to investigate the candidate's past behavior in relation to a specific situation. A comprehensive behavioral question is composed of three parts:

1. The behavioral context
2. The open-ended request
3. The behavior investigated

The following example presents a behavioral question created to investigate the behavior connected to the "Ability to solve problems and analyze issues":

In this role, you have to solve challenging problems and analyze complex issues. Tell me about a time when you conceptualized a challenge holistically and used a data-informed approach to perform insight-driven explorations.

Part 1: The context. This part of the question describes the context of the desired behavior to the candidate, reducing the opportunity for misunderstanding. This introduction is often omitted as a reaction to the unfounded fear of leading the answer. Providing the context offers an indication solely to the candidates who possess the specific experience necessary to answer the question.

Part 2: The open-ended request. From a linguistic perspective, this part transforms the behavioral question into a demand; the following are some examples that you can use:

Tell me about a time when you ...

Describe a situation where you ...

Give me an example of a time when you ...

Give me an example of a situation where you ...

Share with me an example of you demonstrating ...

Walk me through an experience where you ...

Part 3: The behavior investigated. This part specifies one of the previously identified actions associated with the examined ability.

Probe Behavioral Answers

Candidates answer behavioral questions using different approaches. The most prepared individuals use a predefined structure like the Situation-Task-Action-Result (STAR) model that increases their opportunities to explore the question domain effectively. Despite that, given the degree of openness of behavioral questions, more frequently, a candidate's response requires additional queries to allow the interviewer to score it confidently. Moreover, as mentioned earlier, this practice increases the diversity of perspective by enabling interviewers to explore the candidate's specific experiential domains from their unique perspective and provide a distinct contribution to the evaluation process.

An effective interviewer probes a candidate's answer with empathy and sensibility because it may require interrupting that person's narrative. This approach requires two precautions. Initially, introduce and frame the need for this indispensable act of investigation during the first minutes of the interview when you provide an overview of the process. Subsequently, always initiate probing with an expression of regret. The following are some practical examples that you can use:

Sorry, can you please tell me more about [action]?

Sorry, can you please help me to understand why [action]?

Sorry, can you please go back quickly and elaborate why [action]?

Sorry, can you help me to understand why [action]?

Sorry, what would you have done instead if [scenario]?

You can begin probing every time you feel that the candidate deviated from the intent of the answer or when you think that the response fails to address specific requirements. When not adequately articulated by the candidate, two elements that always require further investigations are the thinking underlying a decisional process and the implied behaviors behind those conclusions. Investigating these two elements allows you to indirectly explore the reflective assumptions that the brain uses to interpret the world and shape an individual's perception and behavior.

Create a Career Framework

The growth represents a foundational human need, and designing the preconditions for career progression constitutes a nonobvious but essential step to achieving results and managing relationships.[lxxix] Fulfilling the need for career progression increases happiness and retention within the team and promotes the perception of the "meaning of work," one of the five social needs of a high-performing team.[lxxx] Irrespective of the nomenclature used to frame design roles in your organization, fulfilling the designers' need for progression requires you to create two interconnected but independent tracks that can support the professional advancement of individual contributors and managers equally (Figure 6-11).

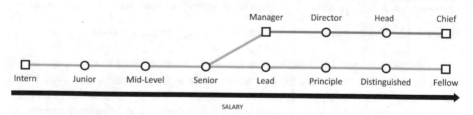

Figure 6-11. A dual-track career framework

This type of approach to career management also avoids forcing talented designers that love being individual contributors to begin a career in management against their professional purpose for the sole reason to pursue a false and ephemeral sense of progression. After you defined the job titles punctuating the framework, you need to specify the set of abilities necessary for each of the roles present in your team. Use the behavioral standards created for the hiring process to inform this work and determine the level of competency required to move from one level to the next.

Onboard New Hires

From a candidate perspective, onboarding defines the process that informs new team members on how to adapt and utilize their results-oriented and social-oriented skills within their new company's social and organizational context. From a managerial and leadership viewpoint instead, onboarding aims to establish the preconditions necessary to initiate the fulfillment of the five social needs that enable and magnify the team's effectiveness: psychological safety, dependability, structure and clarity, meaning of work, and impact of work.

While the early hours, days, and weeks within a new organization are critical, research demonstrates that the process of moving from *organizational outsider* to *organizational insider* typically develops progressively during the first year of the job.[lxxxi] During this time, the behavior of the person in charge of the team has the capacity to promote the success or the failure of new hires.

Onboarding Events

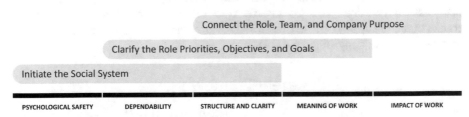

Figure 6-12. The road map that punctuates the primary onboarding events

As a design manager and leader, you have to create a road map (Figure 6-12) that punctuates the onboarding process with events categorized into three main categories: "initiate the social system"; "clarify the role priorities, objectives, and goals"; and "connect the role, team, and company purpose." Some of the necessary actions on this type of checklist are the following:

- Set up the first meeting with the new hire.
- Set up a monthly onboarding check-in.
- Assign a peer buddy.
- Suggest effective strategies to build a social system.
- Encourage candor and participation.
- Introduce the social norms in existence.
- Introduce the new hire to the team.
- Introduce the new hire to the key collaborators.

- Introduce the design strategy direction.

- Introduce the design process.

- Introduce the short- and long-term design priorities.

- Introduce the current design objectives and goals.

- Introduce the role expectations.

- Introduce the performance evaluation system.

Once you have defined the events that delineate the onboarding process, you can add them to an onboarding calendar that can guide the new hire through this experience. The most critical moment during the early hours of the onboarding process is the first meeting with the new hire; you will never have a second opportunity to provide a great first impression.

The First Meeting with the New Hire

For the person in charge of the group, the first meeting with a new hire represents a unique opportunity to show how much you care about that person. One tangible way of demonstrating that sentiment is to prioritize that encounter over everything else on your calendar during that first day. Schedule the meeting as the first official event during the first day at work. For the new hire instead, it is a unique opportunity to receive a cognitive frame that facilitates the interpretation of essential social norms and other vital elements of the working environment. Some examples are connecting the team's *why*, *how*, and *what* to the organization's purpose, explaining how the team makes important decisions, and clarifying the responsibilities and expectations of roles.

During that first meeting, you can also provide a "welcome document" that captures additional information that you consider essential to a successful onboarding process. Some of the typical sections on this type of inventory are the following:

- Technical tasks (computer, calendar, etc.)

- Social tasks (key collaborators)

- User research (studies and lessons learned)

- Competitors (product, strategy, customer details)

- Team structure (role and responsibilities)

- Org structure (design and other interconnected teams)

- Seating map (building/floor map)

- Resources (road map, backlog, wiki, etc.)
- Glossary (common terms, acronyms, etc.)

Fulfilling the need for "structure and clarity" also has a complementary effect on psychological safety[lxxxii] and must be considered a priority during the first month. During this initial phase, the information provided in conjunction with your continuous support initiates the new hire transition from organizational outsider to organizational insider creating the precondition necessary to foster engagement, a prerequisite for a discretionary effort, and high levels of performance.[lxxxiii] Close the session deciding the day and time of the weekly one-on-one meeting and a session dedicated to introducing the probationary objectives.

The Probationary Objective Meeting

After the first meeting, where you introduced the new team member to the working environment, you need to clarify your behavioral expectations regarding the probation period. This conversation represents an opportunity for you to present and agree on specific objectives and goals and an occasion for the new hire to advance eventual requests necessary to fulfill those expectations. The following shows a list of questions that, after you have introduced your behavioral expectations, can help you to initiate this type of discussion:

Are there any potential additional objectives?

What do you want to achieve in your role?

What do you need to learn to be successful?

What support will you need?

How can I help you?

At the end of the meeting, you must have an agreed list of objectives and goals that define the terms of the probation period. If necessary, you can provide to the new team member some additional time for reflection and finalize the list during the next weekly one-on-one.

The First Meeting with the Team

The first meeting of the new hire with the team is equally crucial to ensure a successful onboarding. As a design manager and leader, it is your responsibility to provide an opportunity to every new designer to be perceived as warm and competent to elicit uniformly positive emotions, feelings, and behaviors among the team members.[lxxxiv] In preparation for the first weekly team meeting, ask the new hire to prepare a 15-minute presentation answering the following five questions using five slides:

What is your story?

Why did you decide to be a designer?

What is your favorite project, and why?

Why did you decide to join [your company]?

What do you want to learn in the next 5 days, 5 weeks, and 5 months?

When crafted with intention, this experience can set the foundation for a trustworthy relationship facilitating the new hire's transition from an *organizational outsider* to an *organizational insider*. The session must be informal and interactive, providing an occasion to share meaningful stories intended to promote social connection. Collaboration is an ability that requires empathy to comprehend other individuals and establish with them productive social dynamics. Introducing the answers to those five questions in front of the team typically initiates a conversation that promotes empathy and facilitates an organic and gradual integration into the psychologically safe climate of the group.[lxxxv] In Chapter 9, you will continue to explore the challenge of aligning your team around a shared practice, learning how to craft team agreements and social norms to optimize the design operations.

Endnotes

i.   Northouse, Peter Guy. Leadership: Theory and Practice. Sage, 2016.

ii.   Edmondson, Amy. Teaming: How Organizations Learn, Innovate, and Compete in the Knowledge Economy. Wiley, J., 2012.

iii. Edmondson, Amy. Teaming: How Organizations Learn, Innovate, and Compete in the Knowledge Economy. Wiley, J., 2012.

iv.   Cook, Karen S., editor. Trust in Society. Russell Sage Foundation, 2003.

v.   Cook, Karen S., editor. Trust in Society. Russell Sage Foundation, 2003.

vi. Zenger, John, and Joseph Folkman. "Understanding Trust: The Salt Of Leadership." Zenger | Folkman, 8 Aug. 2020, zengerfolkman.com/articles/understanding-trust-the-salt-of-leadership/.

vii. Zenger, John, and Joseph Folkman. "The Trifecta of Trust." Zenger | Folkman, 26 Sept. 2019, zengerfolkman.com/webinars/trifecta-of-trust/.

viii. Robinson, Sandra L. "Trust and Breach of the Psychological Contract." Administrative Science Quarterly, vol. 41, no. 4, 1996, p. 574., doi:10.2307/2393868.

ix. Li, Fuan, and Stephen C. Betts. "Trust: What It Is And What It Is Not." International Business; Economics Research Journal, vol. 2, no. 7, 2011, doi:10.19030/iber.v2i7.3825.

x. Cuddy, Amy, et al. "The Dynamics of Warmth and Competence Judgments, and Their Outcomes in Organizations." Research in Organizational Behavior, vol. 31, 2011, pp. 73–98., doi:10.1016/j.riob.2011.10.004.

xi. Fiske, Susan T., et al. "Universal Dimensions of Social Cognition: Warmth and Competence." Trends in Cognitive Sciences, vol. 11, no. 2, 2007, pp. 77–83., doi:10.1016/j.tics.2006.11.005.

xii. Fiske, Susan T., et al. "Universal Dimensions of Social Cognition: Warmth and Competence." Trends in Cognitive Sciences, vol. 11, no. 2, 2007, pp. 77–83., doi:10.1016/j.tics.2006.11.005.

xiii. Fiske, Susan T., et al. "Universal Dimensions of Social Cognition: Warmth and Competence." Trends in Cognitive Sciences, vol. 11, no. 2, 2007, pp. 77–83., doi:10.1016/j.tics.2006.11.005.

xiv. Fiske, Susan T., et al. "Universal Dimensions of Social Cognition: Warmth and Competence." Trends in Cognitive Sciences, vol. 11, no. 2, 2007, pp. 77–83., doi:10.1016/j.tics.2006.11.005.

xv. REF: Meyer, Erin. The Culture Map: Decoding How People Think, Lead, and Get Things Done Across Cultures. PublicAffairs, 2014.

xvi. REF: Meyer, Erin. The Culture Map: Decoding How People Think, Lead, and Get Things Done Across Cultures. PublicAffairs, 2014.

xvii. Cuddy, Amy, et al. "The Dynamics of Warmth and Competence Judgments, and Their Outcomes in Organizations." Research in Organizational Behavior, vol. 31, 2011, pp. 73–98., doi:10.1016/j.riob.2011.10.004.

xviii. REF: Meyer, Erin. The Culture Map: Decoding How People Think, Lead, and Get Things Done Across Cultures. PublicAffairs, 2014.

xix. REF: Meyer, Erin. The Culture Map: Decoding How People Think, Lead, and Get Things Done Across Cultures. PublicAffairs, 2014.

xx. REF: Meyer, Erin. The Culture Map: Decoding How People Think, Lead, and Get Things Done Across Cultures. PublicAffairs, 2014.

xxi. REF: Meyer, Erin. The Culture Map: Decoding How People Think, Lead, and Get Things Done Across Cultures. PublicAffairs, 2014.

xxii. Cuddy, Amy, et al. "The Dynamics of Warmth and Competence Judgments, and Their Outcomes in Organizations." Research in Organizational Behavior, vol. 31, 2011, pp. 73–98., doi:10.1016/j.riob.2011.10.004.

xxiii. Zenger, Jack, and Joseph Folkman. "The 3 Elements of Trust." Harvard Business Review, 23 June 2019, hbr.org/2019/02/the-3-elements-of-trust.

xxiv. Cuddy, Amy. Presence: Bringing Your Boldest Self to Your Biggest Challenges. Orion, 2016.

xxv. Cuddy, Amy. Presence: Bringing Your Boldest Self to Your Biggest Challenges. Orion, 2016.

xxvi. Cuddy, Amy. Presence: Bringing Your Boldest Self to Your Biggest Challenges. Orion, 2016.

xxvii. Trimpop Rüdiger. The Psychology of Risk Taking Behavior. North-Holland, 1994.

xxviii. Zenger, John, and Joseph Folkman. "The Trifecta of Trust." Zenger | Folkman, 29 Apr. 2019, zengerfolkman.com/webinars/the-trifecta-of-trust/.

xxix. Zenger, Jack, and Joseph Folkman. "The 3 Elements of Trust." Harvard Business Review, 23 June 2019, hbr.org/2019/02/the-3-elements-of-trust.

xxx. Zenger, Jack, and Joseph Folkman. "The 3 Elements of Trust." Harvard Business Review, 23 June 2019, hbr. org/2019/02/the-3-elements-of-trust.

xxxi. Rousseau, Denise, et al. "Not So Different After All: A Cross-Discipline View Of Trust." Academy of Management Review, vol. 23, no. 3, 1998, pp. 393–404., doi:10.5465/amr.1998.926617.

xxxii. Rousseau, Denise, et al. "Not So Different After All: A Cross-Discipline View Of Trust." Academy of Management Review, vol. 23, no. 3, 1998, pp. 393–404., doi:10.5465/amr.1998.926617.

xxxiii. Edmondson, Amy. "Psychological Safety and Learning Behavior in Work Teams." Administrative Science Quarterly, vol. 44, no. 2, 1999, p. 350., doi:10.2307/2666999.

xxxiv. Lotto, Beau. Deviate: The Creative Power of Transforming Your Perception. Weidenfeld & Nicolson, 2018.

xxxv. Edmondson, Amy C. Teaming: How Organizations Learn, Innovate, and Compete in the Knowledge Economy. Jossey-Bass, 2012.

xxxvi. Soldan, Zhanna. "Group Cohesiveness and Performance: The Moderating Effect of Diversity." The International Journal of Interdisciplinary Social Sciences: Annual Review, vol. 5, no. 4, 2010, pp. 155–168., doi:10.18848/1833-1882/cgp/v05i04/51694.

xxxvii. Berns, Gregory. Iconoclast: A Neuroscientist Reveals How to Think Differently. Harvard Business Press, 2010.

xxxviii. Adolphs, Ralph. "The Biology of Fear." Current Biology, vol. 23, no. 2, 2013, doi:10.1016/j.cub.2012.11.055.

xxxix. Adolphs, Ralph. "The Biology of Fear." Current Biology, vol. 23, no. 2, 2013, doi:10.1016/j.cub.2012.11.055.

xl. REF: Meyer, Erin. The Culture Map: Decoding How People Think, Lead, and Get Things Done Across Cultures. PublicAffairs, 2014.

xli. Edmondson, Amy C. The Fearless Organization: Creating Psychological Safety in the Workplace for Learning, Innovation, and Growth. Wiley, 2018.

xlii. Baars, Bernard J. "Global Workspace Theory of Consciousness: Toward a Cognitive Neuroscience of Human Experience." Progress in Brain Research The Boundaries of Consciousness: Neurobiology and Neuropathology, 2005, pp. 45–53., doi:10.1016/s0079-6123(05)50004-9.

xliii. Adolphs, Ralph. "The Biology of Fear." Current Biology, vol. 23, no. 2, Jan. 2013, pp. 79–93., doi:10.1016/j.cub.2012.11.055.

xliv. Edmondson, Amy C. The Fearless Organization: Creating Psychological Safety in the Workplace for Learning, Innovation, and Growth. John Wiley & Sons, 2019.

xlv. Edmondson, Amy C. The Fearless Organization: Creating Psychological Safety in the Workplace for Learning, Innovation, and Growth. Wiley, 2018.

xlvi. Edmondson, Amy C. Teaming: How Organizations Learn, Innovate, and Compete in the Knowledge Economy. Jossey-Bass, 2012.

xlvii. Edmondson, Amy C. The Fearless Organization: Creating Psychological Safety in the Workplace for Learning, Innovation, and Growth. Wiley, 2018.

xlviii. Edmondson, Amy C. The Fearless Organization: Creating Psychological Safety in the Workplace for Learning, Innovation, and Growth. Wiley, 2018.

xlix. Edmondson, Amy C. The Fearless Organization: Creating Psychological Safety in the Workplace for Learning, Innovation, and Growth. Wiley, 2018.

l. Gazzaniga, Michael S., et al. Cognitive Neuroscience: The Biology of the Mind. Norton, 2014.

li. Edmondson, Amy C. The Fearless Organization: Creating Psychological Safety in the Workplace for Learning, Innovation, and Growth. Wiley, 2018.

lii. Edmondson, Amy. Teaming: How Organizations Learn, Innovate, and Compete in the Knowledge Economy. Wiley, J., 2012.

liii. Black, Stewart J., et al. Organizational Behavior. OpenStax Rice University, 2019.

liv. Edmondson, Amy. Teaming: How Organizations Learn, Innovate, and Compete in the Knowledge Economy. Wiley, J., 2012.

lv. Black, Stewart J., et al. Organizational Behavior. OpenStax Rice University, 2019.

lvi. Edmondson, Amy. Teaming: How Organizations Learn, Innovate, and Compete in the Knowledge Economy. Wiley, J., 2012.

lvii. Edmondson, Amy C. The Fearless Organization: Creating Psychological Safety in the Workplace for Learning, Innovation, and Growth. Wiley, 2018.

lviii. Owens, B.P., Johnson, M.D., & Mitchell, T.R. "Expressed Humility in Organizations: Implications for Performance, Teams, and Leadership." Organization Science 24.5 (2013): 1517–38.

lvix. Dweck, Carol S. Mindset: The New Psychology of Success. Ballantine Books, 2016.

lx. Edmondson, Amy C. The Fearless Organization: Creating Psychological Safety in the Workplace for Learning, Innovation, and Growth. John Wiley & Sons, 2019.

lxi. Edmondson, Amy. "Team Learning and Psychological Safety Survey." Administrative Science Quarterly, vol. 44, no. 2, June 1999, pp. 350–383., doi:10.13072/midss.111.

lxii. Edmondson, Amy. "Team Learning and Psychological Safety Survey." Administrative Science Quarterly, vol. 44, no. 2, June 1999, pp. 350–383., doi:10.13072/midss.111.

lxiii. Edmondson, Amy. "Team Learning and Psychological Safety Survey." Administrative Science Quarterly, vol. 44, no. 2, June 1999, pp. 350–383., doi:10.13072/midss.111.

lxiv. Edmondson, Amy. "Team Learning and Psychological Safety Survey." Administrative Science Quarterly, vol. 44, no. 2, June 1999, pp. 350–383., doi:10.13072/midss.111.

lxv. Bock, Laszlo. Work Rules!: Insights from Inside Google That Will Transform How You Live and Lead. John Murray Publishers, 2016.

lxvi. Bock, Laszlo. Work Rules!: Insights from Inside Google That Will Transform How You Live and Lead. John Murray Publishers, 2016.

lxvii. Bock, Laszlo. Work Rules!: Insights from Inside Google That Will Transform How You Live and Lead. John Murray Publishers, 2016.

lxviii. Bock, Laszlo. Work Rules!: Insights from Inside Google That Will Transform How You Live and Lead. John Murray Publishers, 2016.

lxix. Lotto, Beau. Deviate: The Creative Power of Transforming Your Perception. Weidenfeld & Nicolson, 2018.

lxx. Schmidt, Frank L., and John E. Hunter. "The Validity and Utility of Selection Methods in Personnel Psychology: Practical and Theoretical Implications of 85 Years of Research Findings." Psychological Bulletin, vol. 124, no. 2, Sept. 1998, pp. 262–274., doi:10.1037//0033-2909.124.2.262.

lxxi. Bock, Laszlo. Work Rules!: Insights from Inside Google That Will Transform How You Live and Lead. John Murray Publishers, 2016.

lxxii. Schmidt, Frank L., and John E. Hunter. "The Validity and Utility of Selection Methods in Personnel Psychology: Practical and Theoretical Implications of 85 Years of Research Findings." Psychological Bulletin, vol. 124, no. 2, Sept. 1998, pp. 262–274., doi:10.1037//0033-2909.124.2.262.

lxxiii. Bock, Laszlo. Work Rules!: Insights from Inside Google That Will Transform How You Live and Lead. John Murray Publishers, 2016.

lxxiv. Edmondson, Amy C. The Fearless Organization: Creating Psychological Safety in the Workplace for Learning, Innovation, and Growth. Wiley, 2018.

lxxv. Gazzaniga, Michael S., et al. Cognitive Neuroscience: The Biology of the Mind. Norton, 2014.

lxxvi. Raab, Markus, et al., editors. Performance Psychology: Perception, Action, Cognition, and Emotion. Academic Press Inc, 2015.

lxxvii. Raab, Markus, et al., editors. Performance Psychology: Perception, Action, Cognition, and Emotion. Academic Press Inc, 2015.

lxxviii. Ouellette, Judith A., and Wendy Wood. "Habit and Intention in Everyday Life: The Multiple Processes by Which Past Behavior Predicts Future Behavior." Psychological Bulletin, vol. 124, no. 1, 1998, pp. 54–74., doi:10.1037/0033-2909.124.1.54.

lxxix. Lange, Paul A. M. van, et al. Social Psychology. The Guilford Press, 2021.

lxxx. Zenger, Jack, et al. "Expand Learning: Going Beyond Sustainment to a Higher Level." Zenger | Folkman, 19 Aug. 2019, zengerfolkman.com/white-papers/expand-learning.

lxxxi. Zedeck, Sheldon. APA Handbook of Industrial and Organizational Psychology. American Psychological Association, 2011.

lxxxii. Frazier, M. Lance, et al. "Psychological Safety: A Meta-Analytic Review and Extension." Personnel Psychology, vol. 70, no. 1, 2016, pp. 113–165., doi:10.1111/peps.12183.

lxxxiii. Kahn, William A. "Psychological Conditions of Personal Engagement and Disengagement at Work." Academy of Management Journal, vol. 33, no. 4, 1990, pp. 692–724., doi:10.5465/256287.

lxxxiv. Fiske, Susan T., et al. "Universal Dimensions of Social Cognition: Warmth and Competence." Trends in Cognitive Sciences, vol. 11, no. 2, 2007, pp. 77–83., doi:10.1016/j.tics.2006.11.005.

lxxxv. Smith, Adam. "Cognitive Empathy and Emotional Empathy in Human Behavior and Evolution." The Psychological Record, vol. 56, no. 1, 2006, pp. 3–21., doi:10.1007/bf03395534.

Develop the Team

Acquiring and, by extension, passing knowledge represents the single most crucial ability of human beings; thousands of years ago, the duration of human beings' life span depended predominantly on that capability. Even in a modern context, where unexpected fatal events are limited to occasional situations, learning and teaching remain crucial to determine our hierarchical position in any social context.

Your ability to develop your team members increases the group's perception of your leadership effectiveness, and it has a profound impact on several crucial business aspects of your organization, such as but not limited to the discretionary effort and the intention to leave.[i] The team members are more likely to commit to challenging objectives and less likely to leave the company when they can continuously improve their professional competencies[ii] (Figure 7-1).

© Andrea Picchi 2022
A. Picchi, *Design Management*, https://doi.org/10.1007/978-1-4842-6954-1_7

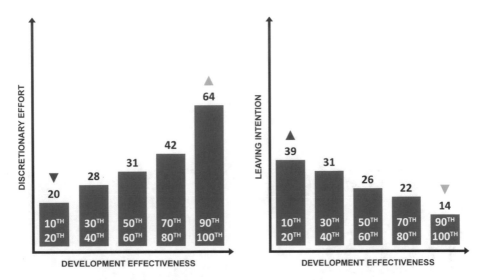

Figure 7-1. The correlation between development effectiveness, discretionary effort, and leaving intention

The process of helping an individual or a group in developing competencies is multifactorial in nature. However, the research identified a set of specific behaviors that you can adopt to promote experiential learning successfully.[iii] The following are the five behaviors adapted to the design context.

Elevate aspirations. Learning requires dedication to afford the significant cognitive energy demanded to overcome the *cognitive disequilibrium* and the level of ambiguity that this process generates. Your role is to share a vision of the future that can inspire a person to initiate the learning journey and persevere motivation during the inevitable challenging moment.

Build confidence. Learning requires self-efficacy to move from acquiring to implementing the new mental model of a specific behavior. Your role is to offer appreciative feedback to reinforce the new behavioral pattern and facilitate that transition.

Convey information. Learning requires comprehension to identify and fill behavioral gaps. Your role is to share the knowledge necessary to recognize areas of improvement, create a plan of action, and preserve intrinsic motivation.

Promote practice. Learning requires experiences to establish and consolidate the neural connections that produce a specific behavior. Your role is to provide opportunities to practice the new behaviors to preserve learning momentum.

Support practice. Learning requires examples to model a specific new behavior in a particular context and adapt it to different situations. Your role is to demonstrate and substantiate the desired behavior or to provide the opportunity to observe those actions from other individuals.

A crucial event of experiential learning is how an individual perceives, reflects on, and reacts to events. In this chapter, you will learn how to develop your team. You will learn how to establish the social and cognitive preconditions for learning and use the investigational tools of coaching. You will explore how to connect with the team to develop trustworthy, respectful, and psychologically safe relationships, discover their superpowers, and mitigate their shadow side. You will examine how to communicate about performance promoting social exchanges where feedback can be leveraged to initiate candid and productive conversations without generating psychological strain and being perceived as a social threat. You will also analyze the two dimensions of coaching and how to use this tool to support and improve both results and relationships.

The Three Fundamental Approaches to Development

From the perspective of a design manager and leader, learning intended as the act of passing and promoting the acquisition of knowledge directed to the development of new behaviors can be promoted assuming three distinct roles.[iv]

The Challenger: This approach provides high behavioral standards and developmental assignments that push individuals and groups outside their comfort zone.

The Expert: This approach provides behavioral references and collaborative opportunities that support the work of individuals and groups.

The Supporter: This approach provides behavioral reinforcement and encouragement for the development effort of individuals and groups.

These three approaches to development are directly connected with the three modalities of creative leadership that you will explore in Chapter 8: lead ahead with a strong viewpoint, lead alongside in direct contact with the team's activities, and lead behind through culture.

As a design manager and leader, you must concurrently develop and adopt all these three roles depending on the context in which your team is operating. Under this premise, the questionnaire in Table 7-1 can help you to understand your current behavioral tendency and competence in these situations.

Table 7-1. The survey questionnaire to assess the approach to development

Section 1: Behavioral Tendency[α]
For each section, indicate in which behavior you would most likely engage first and second.

- Commit to the achievement of goals.
- Offer your perspective to enhance outcomes.
- Achieve results and develop managing equally.

- Represent the organization to critical groups.
- Provide an active contribution to improve results.
- Remain in touch with individual and collective needs.

- Evangelize design to educate the organization.
- Demonstrate knowledge to develop respect.
- Encourage high levels of cooperation.

- Make changes to improve the organization.
- Anticipate and respond to problems.
- Manage creative tension and interaction.

- Promote new initiatives to receive support.
- Elevate confidence in your decisional capabilities.
- Achieve highly cooperative objectives.

- Recognize the need for change when present.
- Identify future trends, technologies, and opportunities.
- Nurture confidence in your design knowledge.

- Promote accountability for new challenges.
- Make effective decisions under pressure.
- Offer constructive and appreciative feedback.

- Define challenging objectives.
- Balance holistic and detailed perspectives.
- Coach and mentor team members.

(continued)

Table 7-1. (continued)

Section 2: Behavioral Competence[β]
Rate your level of effectiveness on the following behaviors.
- Create a climate of psychological safety that encourages continuous learning.
- Demonstrate concern about individual and group development.
- Act as a coach and a mentor to develop individuals and groups.
- Motivate individuals and groups to leave their comfort zone.
- Create a healthy, sustainable, and productive environment for diversity.
- Leverage diverse perspectives to accomplish objectives.
- Adapt to situations in response to individual and collective needs.
- Communicate insights to develop an understanding of issues.
- Remain in touch with individual and collective concerns.
- Offer candid and productive feedback to individuals and groups.
- Balance the achievement of results and the development of relationships.

[α] Closed question: Challenger (first behavior), Expert (second behavior), Supporter (third behavior).

[β] 5-Point scale: Outstanding strength, Strength, Competent, Needs some improvement, Needs significant improvement.

The first section of the questionnaire can help you to analyze your current conscious or unconscious behavioral tendency while developing others. If you can relate to all three behaviors presented, select the first and second that you would likely perform more naturally or frequently.

The second section of the questionnaire can help you to investigate your current behavioral competence in developing others. These behaviors are all directly or indirectly interrelated with the five clusters of leadership capabilities.

Developing individuals and groups is a joint effort; in the next section, you will explore how the counterpart needs to adopt a specific mindset to unlock experiential learning.

The Importance of a Growth Mindset

At the end of the 19th century, Santiago Ramón y Cajal, a pioneer of modern neuroscience, used the term *neuronal plasticity* for the first time to describe nonpathological changes in the structure of adult brains.[v] Two centuries later, there is clear evidence that cultural values and experiences shape neurocognitive processes and influence patterns of neural activation to the point where they affect neural structures.[vi] The neural pathways of our brain are far from being unchanging and immutable.[vii] Adopting a growth mindset in conjunction with experiential factors has the ability to shape the neural circuits underlying social and emotional behaviors from the prenatal period to the end of life.[viii] These factors include both *incidental influences* such as unexpected adversity and *intentional influences* that can be produced via specific interventions designed to promote healthily challenging conditions.[ix] When your team adopts a growth mindset, you notice a mental shift characterized by three attitudinal changes:[x]

- The group identifies success in expressing their full potential, learning, and improving.

- The group identifies setbacks as motivating because they are informative.

- The group begins to take charge of the practices that generate and sustain learning and, consequently, success.

The attitude adopted to approach experiential learning profoundly affects what is perceived as attainable and the overall performance of problem-solvers such as designers.[xi] While changing a person's mindset is possible by investing a significant amount of time and effort, this approach is often not feasible nor desirable within a business context. Hiring and releasing team members remains the most practical and cost-effective tool at your disposal to ensure the presence of that attitude within the team.

Assuming the presence of a learning-oriented mindset within the team, you need to model three critical behaviors to maximize the learning potential: connect with the team, communicate about performance, and coach the team. Figure 7-2 illustrates how these three interrelated behaviors must be implemented sequentially to fulfill the social preconditions necessary to unlock and sustain continuous development.

Adoption Plan

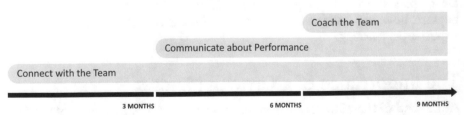

Figure 7-2. The three interconnected behaviors necessary to unlock and sustain continuous development

The first step is connecting personally with every team member to build a trustworthy, respectful, and psychologically safe environment. The second step is initiating healthy communications about performance to promote candid and productive exchanges of feedback. The third and last step is establishing a coaching for performance and development practice.

Adopting the nomenclature previously introduced for the reflective cycle in Chapter 4, you will explore the practice of coaching *on-action*, while in Chapter 9, you will learn how to adopt that mindset to nurture creative collaboration coaching *in-action*. Developing your team demonstrating genuine care increases the group's perceived engagement and happiness, which improves productivity.[xii] Spending quality time with your group individually and collectively requires dedication and perseverance, and it can be challenging under certain circumstances. Despite that, this practice represents one of the critical elements under your control to unlock and sustain high levels of performance.

Connect with the Team

Establishing trustworthy, respectful, and psychologically safe relationships with the team members allows you to satisfy the preconditions necessary to initiate candid and productive communications about performance and establish an effective coaching practice. As you will learn in Chapter 8, investing time in communicating individually with the team members is also essential to form a *personal* connection with team members that represents an instrumental factor in nurturing creative collaboration. When you try to establish a trustworthy and respectful relationship with an individual, that person consciously and unconsciously evaluates your communications using two primary criteria:[xiii]

Quality: The degree to which the subject relates to the interlocutor

Quantity: The frequency of your communications

In practice, the interlocutor considers how often you talk about what is vital to that person. You can leverage different types of social gatherings to fulfill these communicative needs, but the most efficient approach is to schedule regular one-on-one meetings with your team members and every key collaborator in your social system.

One-on-One Meetings

Face-to-face gatherings, whether in person or remote, are a type of business meeting that primarily concentrates on two objectives: build trust and respect and establish a mutually beneficial exchange of information with the interlocutor. The structure of an efficient one-on-one meeting is characterized by the following operational elements:

- It is scheduled.
- It is brief.
- It has a regular cadence.
- It has an official form of documentation.

One-on-one meetings are scheduled in advance with a high threshold for canceling or rescheduling; entertaining this form of communicative exchange improvising without a formal structure sends the socially undesired message that the interlocutor does not represent a priority for you.[xiv] They are brief due to their regular cadence, and in opposition to informal catch-ups, they have a form of official documentation to track the professional discourse.

These configurations do not allow you to take notes with your laptop without creating a physical barrier that divides you and your interlocutor, negatively affecting the levels of trust and psychological safety.[xv] Note-taking doesn't have to be elaborated or time-consuming as long as it allows you to recollect insights from past conversations concerning a specific objective. You can use any format that suits your style; the following example introduces a concise notation that can be used to supervise the development of a particular ability:

> [Name] [Feedback Valence], [Ability].
>
> Steve +, Ability to solve problems and analyze issues.

Similarly to other meetings, when the conversation requires it, share relevant information with the attendee before the meeting, and at the end of it, follow up with an email that summarizes the action points agreed during the discussion.

One-on-One Meetings with the Team

This type of meeting intends to support the achievement of the team's objectives. A one-on-one with a team member is typically implemented as follows:

- It is scheduled in advance.
- It has a duration of 30 minutes.
- It has a weekly cadence.
- It has a 15/15 agenda.
- It is documented using notes and action points.

The 30 minutes are divided into 15 initial minutes dedicated to the direct report to talk about professional or personal subjects and 15 following minutes assigned to you to discuss any topic related to the current week that requires a conversation. Encourage every team member to join the meeting with a list of challenges that you can help resolve; this approach sets the precondition for productive discussions, promotes the direct report's sense of ownership, and develops an attitude oriented to continuous development.

If, on some occasions, a direct report feels the need to discuss more complex issues beyond the 15-minute mark, consider it acceptable without interrupting the conversation. If the things on your list are not important and urgent, you can compress them in the remaining time or reschedule them for the following session. An agenda is a tool used to facilitate the dialogue, not the intrinsic objective of the conversation. Consider booking a coaching session if the direct report expresses the desire to discuss issues that require additional time and a more articulated dialogue to be elaborated and resolved. One-on-one meetings are also an effective way to capture the frame of mind of your team members. Opening the conversation always using the same question allows you to notice and capture changes in their mood or sentiment; the following is an example of a practical opening question:

What is on your mind?

Assuming the presence of trust, respect, and psychological safety, this type of question invites the interlocutors to candidly share what is paramount to them, granting them the autonomy to make a choice for themselves. If you want to craft your opening question, maintain a what-based structure and avoid using *why* or *how* to prevent triggering inquisitorial feelings in the interlocutors.[xvi] Listen to the response, note the answer, and use an emoji to quantify and track the frame of mind attached to it over time. Under candid and safe social conditions, this type of measurement can give you crucial insights into the mental health dimension of your team members.

One-on-Ones with Cross-Functional Collaborators

This type of meeting intends to support the achievement of cross-functional objectives. A one-on-one session with a collaborator is typically implemented as follows:

- It is scheduled in advance.
- It has a duration of 30 minutes.
- It has a biweekly, monthly, or quarterly cadence.
- It has a 15/15 agenda.
- It is documented using notes and action points.

This type of conversation with peers and senior figures in your organization is typically characterized by fluid communicative exchanges. The 15/15 agenda represents more an indicator of a time equally distributed between the parts than a guideline to follow necessarily. As long as the overarching conversation is healthily balanced, promoting a mutually beneficial collaboration, the communication can be considered satisfactory.

One-on-Ones with Your Line Manager

This type of meeting intends to support the achievement of your line manager's objectives and, via that person, the fulfillment of the organizational vision. Hopefully, your line manager is favorable to regular one-on-one conversations because this will significantly increase your probability of thriving in your role. In that case, you typically do not have control over the conversation structure, but you can confidently assume that you can use half of the time allocated to the meeting. Prepare a list of things that you want to discuss, labeling the items on it as follows:

- Report
- Decision
- Support

The items labeled as "report" are the ones that require a concise summary of the situation to maintain your manager informed concerning deadlines. The items labeled as "decision" are the ones that require your manager to commit to an action to move the workflow forward. The items labeled as "support" are the ones on which you need assistance from your manager. Set a reminder on your calendar to share the list with your manager one day before the meeting. Immediately after the end of the session, apply the same process introduced to "process and triage emails" in Chapter 4: identify the highest-priority items and commit to the most appropriate actions, complete immediately the activities that require less than two minutes, and schedule the others using the proper application.

Discover the Team Superpowers

A superpower represents the unique contribution that an individual offers to the team.[1] These human-centered behavioral characteristics can be delineated by one of more abilities and define the peculiarity that distinguishes each team member's contribution: what that given person does more effectively than anyone else within a given group. Analogously to any other capability, superpowers are contextual, and that peculiarity has two critical implications. First, a superpower is defined as such in relation to the abilities available in a given team, and what is perceived as a unique contribution in a group can be regarded as unspecial or even undesired in another one. Second, an individual can possess several superpowers that can be manifested under different circumstances.

---

[1] Superpowers are not innate or unique talents and are developed using dedication, deliberate practice, and commitment at the intersection of needs, passions, and requirements.

Activating a superpower, in conjunction with a clear professional purpose, allows a person to bring their whole self to work and unlock the perception of the meaning of work and the impact of work.[xvii] Discovering your team's superpowers is also results indispensable to pave a pathway to creative collaboration and unlock high levels of collective performance. The following are four behavioral benefits derived from the activation of your team's superpowers:

Increase creative confidence: Promoting self-awareness of individual characteristics and potential contribution to the team effort

Engender generosity: Fostering a consciousness of the opportunities to leverage individual characteristics in service of others

Promote diversity and inclusion: Providing a universal language to frame and effectively combine individual characteristics

Support personal development: Providing a shared framework to discuss and improve individual characteristics

In Chapter 6, you created behavioral standards to define the team's role requirements as a function of the abilities necessary to succeed in those specific positions. The superpower of a designer is the "unique way" to leverage one or more of these abilities in service of a particular objective. Figure 7-3 illustrates an example of a superpower delineated by one single ability.

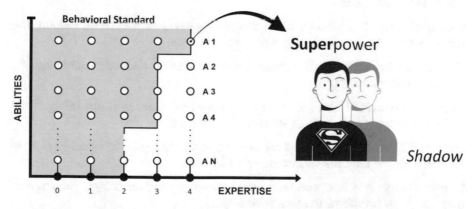

Figure 7-3. The inner connection between an ability, a superpower, and its shadow side

Figure 7-3 also illustrates that every superpower also possesses a dark entity called the *shadow side*. In the Jungian sense of the term, the shadow side of a superpower represents a latent aspect of the personality that the conscious part struggles to identify, accept, and integrate. The shadow constitutes an inseparable part of a superpower that typically manifests itself when an individual overemphasizes the application of a unique capability, originating unintentional negative consequences. The following is a summary of the 21 superpowers and their relative shadow side derived and adapted from the original work at SY/Partners.[xviii]

Connection: You sense latent needs and emotions, helping the team to relate with one another, stakeholders, and customers.

Shadow: Channeling external perspectives impedes the development of a personal viewpoint, limiting the contribution to the team.

Conciliation: You build trust and orchestrate mutually beneficial compromises, helping the team to preserve momentum when a formal agreement is needed.

Shadow: Promoting alignment inhibits divergent thinking hindering the ability to explore a situation creatively.

Detection: You identify missing and incorrect elements in a situation, helping the team to prevent mistakes and errors.

Shadow: Detecting issues consumes time and energy, reducing the level of focus and performance of the team.

Determination: You push yourself and the work until it is completed, helping the team to do the same and overcome difficulties.

Shadow: Committing to a result affects perception, reducing the ability to evaluate the quality of your work.

Examination: You discern the interrelating parts of a system, helping the team to navigate the decision-preparation process.

Shadow: Seeing various alternatives induces revisiting decisions, hindering the decision-making process.

Experimentation: You generate and prototype ideas without being afraid of failure, helping the team to undertake activities.

Shadow: Possessing a prototyping mindset increases the perceived stress level when the team spends time talking instead of taking action.

Exploration: You find connections between unrelated things, helping the team to solve problems in innovative ways.

Shadow: Concentrating on the generation of ideas reduces the ability to analyze the feasibility aspect of concepts.

Harmonization: You comprehend the qualities of other individuals, helping the team to identify effective ways to collaborate.

Shadow: Concentrating on maximizing the current effectiveness reduces the ability to discern growth opportunities.

Imagination: You see a future destination in your mind, helping the team to conserve ambition during the journey.

Shadow: Envisioning the benefits of completing a journey reduces the level of tolerance toward conversation oriented to tactical details.

Intervention: You analyze situations and opportunities, helping the team to navigate ambiguity and commit to a decision.

Shadow: Discerning the circumstances of a challenge reduces the level of tolerance for indecisiveness.

Investigation: You ask questions to uncover the essence of a problem, helping the team to reveal a previously unseen path forward.

Shadow: Being a decisive element of the problem-solving process reduces the level of creative autonomy of the team.

Invention: You leverage imagination and investigation, helping the team to generate original and unexpected possibilities.

Shadow: Being self-sufficient during the problem-solving process reduces the level of social interaction and collaboration.

Motivation: You connect the organizational purpose to objectives and goals, helping the team members to understand their contributions.

Shadow: Concentrating on individual contributions reduces the level of focus on what motivates the team as a collective entity.

Narration: You articulate compelling stories that balance information and emotion, helping the team to connect the design impact to the business value.

Shadow: Developing a connection to a specific narrative reduces the level of awareness of other forces surrounding the team.

Negotiation: You compromise and achieve balanced agreements, helping the team to conserve momentum during the journey.

Shadow: Discussing agreements increases the perceived stress level in individuals with a nonconfrontational attitude and an agreeable personality.

Orientation: You ask questions to uncover a new course of action, helping the team to realign with the organizational purpose.

Shadow: Maturing a cultural compass reduces the ability to perceive the need for a change.

Passion: You infuse energy with your optimistic outlook and wise approach to focus, supporting the team when there is a decline in morale.

Shadow: Concentrating on lifting the spirit of the team reduces the level of openness to difficult conversations.

Provocation: You lead the team outside its comfort zone, helping the group to achieve results beyond their expectations.

Shadow: Concentrating on achieving results affects the level of healthiness characterizing your relationships within the team.

Recalibration: You approach the work rationally and methodically, helping the team regain control during stressful moments.

Shadow: Being impassive originates tensions with individuals characterized by a predominantly extroverted personality.

Recognition: You ask questions to uncover similarities and patterns, helping the team to investigate challenges and attack problems.

Shadow: Focusing on isolating formal logic and its underlying connections reduces the capacity to identify promising but imperfect concepts.

Reduction: You process layers of information to find essential ideas, helping the team to leverage large amounts of data.

Shadow: Extracting meanings increases the possibility of introducing biases and assumptions.

Helping the team members to identify and activate their superpowers and integrate their shadow side significantly increases the group's collective intelligence[2] and the potential level of performance achievable.[xix] In the following two sections, you will explore these processes in detail.

Identify and Activate Superpowers

The process is composed of three steps, during which a team member reflects on their finest work and then uses that understanding to recognize and amplify the superpower.

Step 1: Reflect on the finest work. Ask the designer to consider three recent projects that benefited from that person's contributions and answer the following questions capturing the answers using notes or sketches:

What was I doing?

How did my superpower manifest?

What unique strengths did I leverage?

How did I feel?

Why did I feel that way?

---

[2] Collective intelligence is defined as a function of the mental abilities necessary for adaptation to, as well as shaping and selection of, any environmental context.

Step 2: Recognize the superpower. Ask the designer to reflect on the answers provided and use a pairwise comparison matrix to select the most appropriate superpower from the list of the 21 available. If the designer feels that the superpowers on the list are not representative of their unique contribution, they can propose the addition of a new peculiarity to the list. Once recognized, the superpower can then be formalized using the following structure:

My name is [name], and my superpower is [superpower].

Step 3: Amplify the superpower. At this point, you collaborate with the designer to create a strategy to develop the superpower. Open the session reviewing the definition of the superpower and the answers to the initial questions with the objective to promote further reflection. After this consideration, you concentrate on the following two things:

    a.   Develop a narrative.

    b.   Inform the developmental strategy.

Purposefully articulating a concise and compelling story around the superpower helps the designer to become more intentional about that specific unique contribution. Furthermore, providing insight to other team members regarding a given superpower promotes the acceptance and inclusion of that ability as part of the collective intelligence of the group. Ask the designer to answer the following questions:

What is the context of the challenge?

Why is my superpower needed?

How will my contribution help the team?

What will I say or do?

With a clear understanding of the superpower and the context of its application, invite the team member to reflect on how that person can leverage the superpower to enhance the collective intelligence of the team. Ask the designer to capture that thinking using the following structure:

When [contextual trigger]

I will use my superpower to [behavior]

helping the team to [objective]

// Superpower: Detection

During a design critique,

> *I will use my superpower to ensure that the problem is framed correctly,*
>
> *helping the team to explore relevant solutions.*

The more specific and actionable is the *contextual trigger*, the more effective this cue is in initiating the desired behavior.[xx] This reflection can also constitute the foundation of a possible developmental strategy; in that case, you can review it as part of your regular conversations during your coaching sessions.

Integrate the Shadow Side

What has the potential to make us great also possesses the capacity to represent our preeminent opponent. Integrating the shadow side of a superpower is crucial to developing this type of human-centered characteristic to its full potential. Comparably to identifying and activating the superpower, this process is also composed of three steps, during which a team member gathers feedback and then uses that information to reflect on and mitigate the shadow side.

Step 1: Gather feedback. Ask the designer to ask request input from the other team members to develop self-awareness around possible unintentional negative consequences derived from overemphasizing the superpower.

Step 2: Reflect on the shadow side. Ask the designer to consider the context within which the shadow manifested itself, ponder on the causes that allowed this side of the superpower to emerge, and answer the following questions capturing the answers using notes or sketches:

> *What is the context of the situation?*
>
> *Why did my shadow side emerge?*
>
> *What did I say or do?*
>
> *What effect did the shadow have on the team?*

Step 3: Mitigate the shadow side. At this point, you collaborate with the designer to create a strategy to mitigate the shadow side. Open the session reviewing the feedback received and the answers to the questions with the objective to promote further reflection. After this consideration, you concentrate on formalizing a way to integrate this ineffective side of the superpower. Ask the designer to capture that thinking using the following structure:

> *When [behavioral trigger]*
>
> *Instead of [default behavior]*
>
> *I will instead [new behavior]*

The behavioral trigger represents a critical element of this dynamic, and it is composed of four parts:[xxi]

- Context (time, location, event, etc.)
- Emotional state
- Actors
- Action

The more elements are used to define the trigger, the more specific and actionable that cue results in initiating the desired behavior;[xxii] the following is an example related to the shadow side of the superpower "experimentation":

> *When I feel stressed (emotional state) during our design review meeting (context) because the team (actors) discuss unimportant details rather than making decisions (action)*
>
> *Instead of [default behavior]*
>
> *I will instead [new behavior]*

Integrating the shadow side represents a more challenging task than identifying and activating its connected superpower because it forces the person to surface and resolve potential internal conflicts. In addition to reviewing the implementation of the integration strategy during your regular coaching sessions, invite the designer to connect with other individuals who experienced the same *superpower-shadow dichotomy* and can relate to that specific challenge by providing informed feedback. If you want to identify and activate your superpower and integrate its shadow side, you can follow the same process presented in these sections.

Communicate About Performance

An essential requirement to achieve and sustain high performance levels is the capacity to promote social exchanges where feedback can be leveraged to initiate candid and productive conversations about performance without generating psychological strain and being perceived as a social threat. Under the appropriate conditions, feedback constitutes a pivotal element of the reflective cycle at the foundation of experiential learning.[xxiii] This type of information promotes reflection-in-action and reflection-on-action, with the consequence of increasing self-awareness and the likeliness of unlocking behavioral change.[xxiv]

The conditions under which a communicative exchange about performance is also able to mitigate psychological strain and unlock learning are defined by two specific interconnected behavioral instantiations of trust and psychological safety:[xxv]

- The trustworthiness that underpins the social connection between the two interlocutors

- The psychological safeness that permeates the circumstances of the communicative exchange

Assuming that you preemptively established a trustworthy, and by extension respectful, connection with your team members, preserving a psychologically safe climate during a communication about performance must be considered one of your priorities.

Social Pain

When an individual encounters something unexpected, the limbic system is aroused, activating neurons and releasing hormones in the attempt to understand whether the new entity represents a chance for preservation or a potential danger.[xxvi] This physiological response initiates and governs the majority of our social behavior, trying to minimize threats and maximize comfort during our social exchanges.[xxvii] This dynamic encompasses physical and social dimensions, albeit as social cognitive neuroscience uncovered solely a couple of decades ago,[xxviii] social pain is "real" pain and can leave scars considerably more profound than its physical counterpart.[xxix] Indeed if someone asks you to think about the most painful moment in your life, you would probably recall the loss of a beloved person, not the day that you broke one of the bones in your body.

Social pain is defined as the mental and physical unpleasant sensation associated with concrete or potential threats to one's actual or possible social connections.[xxx] When the area of the brain dedicated to social cognition called *social network* frames a situation as a potential source of social threat and possibly pain, its activation is prioritized at the detriment of other systems, including the *analytical network*.[3] In this context, this dynamic impairs cognition and hinders experiential learning by inhibiting the ability to process the information contained in the feedback message.[xxxi] Under these conditions, it is imperative for you to minimize the perceived social threat and the experienced social pain every time you offer a feedback message.

---

[3] As introduced in Chapter 1, the social and analytical networks tend to operate at cross-purposes: resembling the two ends of a neural seesaw.

Minimize Social Pain

When you communicate about someone's performance, irrespective of the appreciative or constructive nature of the feedback characterizing the social exchange, there are five dimensions that can potentially trigger a threat or reward response.[xxxii] Your moral obligation as the person in charge of the team is to minimize the perceived social threat and experienced social pain associated with your communication across five specific dimensions.

Status: The dimension relative to the social hierarchy position within the team. The perception of an actual or potential reduction in hierarchical position within the group triggers a threat response.[xxxiii] In this case, minimize a threat response by offering the feedback using an empathetic tone and expressing an interest in assist, preventing the perception of an attempt to belittle the person within the group.

Affiliation: The dimension relative to the sense of belonging within the team. The perception of an actual or potential reduction in the sense of involvement within the group triggers a threat response.[xxxiv] In this case, minimize a threat response by offering the feedback, framing it as a natural learning event, preventing the perception of an attempt to exclude the person from the group.

Fairness: The dimension relative to the sense of equity of the feedback message received. The perception of an actual or potential reduction in impartiality triggers a threat response.[xxxv] In this case, minimize a threat response by offering the feedback articulating your message objectively using facts and observed behaviors, preventing the perception of relying on subjective opinions.

Certainty: The dimension relative to the sense of anticipation over the circumstances produced by the feedback message received. The perception of an actual or potential reduction in the ability to predict the environment triggers a threat response.[xxxvi] In this case, minimize a threat response by offering the feedback articulating your message with a clear behavioral direction, preventing the perception of a vague indication.

Autonomy: The dimension relative to the sense of control over the circumstances produced by the feedback message received. The perception of an actual or potential reduction in the ability to influence the environment triggers a threat response.[xxxvii] In this case, minimize a threat response by offering the feedback framing your message as a recommendation aimed to invite reflection, preventing the perception of a mandate.

During the experiential learning process, the mind requires an active discussion to process, interpret, and integrate the experience with the intent to extrapolate the insights necessary to inform future behavioral adaptations.[xxxviii] Minimizing the perceived social threat and experienced social pain contributes to the achievement of this objective by establishing the physiological preconditions for that event necessary to unlock the unique human's capability to theorize about the world, reflect on behaviors, rectify mistakes, and execute specific long-range intentions.[xxxix]

Promote Candid Conversations

Even in a trustworthy, respectful, and psychologically safe environment, promoting candid performance communications is, in practice, a continual delicate endeavor. Offering feedback represents a form of conversation that always involves a certain level of social fear and social pain for the same reason that makes this form of social exchange effective: it tends to undermine the *psychological homeostasis* of the person receiving the message.[xl] These unavoidable social dynamics always characterize every conversation entertained in your team to a certain degree. In Chapter 8, you will learn how to create stories, which represents an effective tool to reframe these sensations and promote behavioral change. Complementary to storytelling, there is also a strategy that you can implement to mitigate their effect to a level where candid and productive conversations about performance can be promoted and sustained successfully.[xli]

Adoption Plan

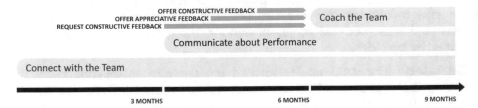

OFFER CONSTRUCTIVE FEEDBACK
OFFER APPRECIATIVE FEEDBACK
REQUEST CONSTRUCTIVE FEEDBACK

Coach the Team

Communicate about Performance

Connect with the Team

3 MONTHS   6 MONTHS   9 MONTHS

Figure 7-4. A strategy aimed to promote candid conversations

In Figure 7-4, you can see the three steps of that strategy intended to promote candid conversations: open the conversation requesting feedback on your work, then begin to offer appreciative feedback and subsequently constructive feedback.

Request Constructive Feedback

The hardest part of promoting candid conversations is initiating the process by inviting your team members to challenge you by offering you feedback on your work. At this stage, if social components such as trust, respect, and psychological safety are not present in your working environment, it will be improbable that an individual will manifest a socially risky behavior such as offering feedback and challenging the person in charge of the team. Opening the conversation using this approach allows you to *lead by example* by modeling the same level of social and psychological openness to constructive feedback that you strive to instill in every team member.

Begin by requesting feedback privately during one-on-one conversations to minimize the refraining effect of social considerations.[4] Subsequently, once an initial climate of trust, respect, and psychological safety has been established, begin to request feedback publicly in the presence of the entire team to reinforce that social atmosphere and enhance the sense of genuineness attached to your request.[xlii] In the following, you can see an example of three types of questions that you can ask in those circumstances:

Is there anything I can do to improve our collaboration?

Is there anything I can do to help you with your job?

Is there anything I can do to [task] more effectively?

Specific requests tend to be more effective in triggering a response, especially with individuals and groups that never challenged you with feedback in the past.[xliii] This approach will equally increase the opportunities to unlock the desired behavior in individuals characterized by predominantly introverted and predominantly extroverted personalities.[xliv] When you begin to receive feedback, it is crucial to promptly respond to the conversation and act on that information to validate your genuineness and reinforce the behavioral change.

Offer Appreciative Feedback

Once the team has developed enough confidence to share their feedback with you, you can begin to offer appreciative feedback, starting from top performers. Commencing from top performances allows you to redirect the group's attention toward the most desired behaviors[5] leveraging the social influence projected by positive role models.[xlv]

When you provide appreciative feedback, articulate the message focusing on connecting the achievement of a given team's objectives to the fulfillment of the organization's purpose. Offer appreciative feedback privately during one-on-one conversations but especially publicly in the presence of the entire team to enhance the sense of affiliation within the group and reinforce the trust and respect underlying your relationship with that given person.[xlvi] Furthermore, besides reinforcing desired behaviors, this approach also

---

[4] Revisit Chapter 4 to review how to create the intention to perform a behavior.
[5] A desired or undesired behavior must be identified using an objective observation that identifies a cause-effect relationship that promotes or hampers the performance and development of a given individual.

indirectly promotes the perception of two of the social needs of your team: the meaning of work and the impact of work. The following presents a quadripartite template that you can use to articulate an appreciative feedback message:

[context], [observed behavior or fact], [impact or consequence], [support].

// Example: User Research

[context] When you interview the users,

[observed behavior or fact] I noticed that you annotate observations and summarize insights.

[impact or consequence] When you do that, you help the team to learn quickly and efficiently, allowing everyone to make customer-centric decisions.

[support] Great job. Is there anything you need or that I can do to help you?

Offering appreciative feedback is not only crucial to reinforce desired behaviors and strengthen the underlying relationship between you and the team members, but it is also necessary to increase the credibility of your constructive messages and therefore reduce their potential capacity to trigger a threat response.[xlvii] Even during periods of tensions and conflicts, the overall ratio between appreciative and constructive conversations must maintain a net positive, with an indicative ratio of around 5:1, to preserve the trust, respect, and psychological safety that permeates your working environment.[xlviii]

Offer Constructive Feedback

Once the team has learned to process and interpret appreciative feedback, you can begin to offer constructive feedback, starting from top learners. Commencing from top learners allows you to redirect the group's attention toward the most mature growth mindsets leveraging the social influence projected by positive role models.[xlix]

When you provide constructive feedback, articulate the message focusing on triggering reflection. Offer constructive feedback privately and never publicly to minimize the likeliness of a threat response. Furthermore, besides correcting the undesired behavior, this approach also indirectly promotes the perception of high-quality standards within the team. The following example presents a pentapartite template that you can use to articulate a constructive feedback message:

[context], [observed behavior or fact], [impact or consequence], [recommendation], [validation], [support].

// Example: User Research

[context] When you interview the users,

[observed behavior or fact] I noticed that you do not summarize your observations during the session,

[impact or consequence] making it difficult for other persons to learn from the research.

[recommendation] Next time, consider extrapolating insights so we can create an accessible backlog of learning.

[validation] How does this sound to you?

[support] Is there anything that I can do to help you?

During the course "Managing at Apple" at Apple University, the professor plays a video from *The Lost Interview* of Steve Jobs articulating his view on offering effective feedback: "You need to do that in a way that does not call into question your confidence in their abilities but leaves not too much room for interpretation ... and that's a hard thing to do."[l] In that interview, Steve Jobs captured the essence of candidly and efficiently communicating about performance: promote behavioral change without undermining trust, respect, and psychological safety. A vital requirement of this form of conversation is to separate the behavior and its impact from the person. The content of the feedback must address the actions of that given person, not their professional value. When you offer feedback, you have to do it in a manner that makes clear to the other person that your confidence in their abilities remained intact.

Coach the Team

A century ago, Alfred Adler, psychotherapist and pupil of Sigmund Freud, said that "a man knows much more than he understands," asking us to believe in the power unleashed when a person realizes their full potential.[li] In departing from his teacher approach, Adler realized that we do not necessarily have to venture into the depth of our psychological history to develop ourselves because one way with which we determine our identity is via the meanings we attribute to situations, and therefore changing or expanding those meanings generates the possibility to reframe our actions and redefine ourselves.[lii] Today, neuroscientific research demonstrates that the experiences in our life shape the microscopic details of our brain that relentlessly rewrites the patterns in its neural networks, reshaping our identity without never reaching an endpoint.[liii]

Coaching, intended as a practice as well as a modality of leadership, constitutes an indispensable activity within the economy of a team because it contributes to the improvement of both results and relationships, increasing, among other factors, engagement, commitment, retention (Figure 7-5), and the perceived professionalism of the manager who establishes this practice.[liv]

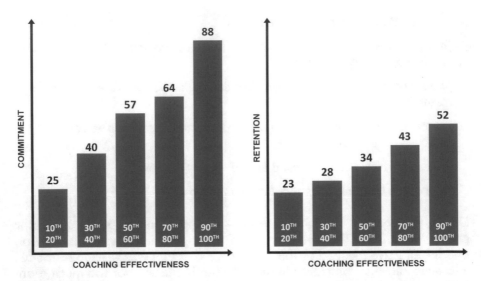

Figure 7-5. The positive correlation between coaching effectiveness and employee commitment and retention

Coaching sessions must have a reserved place in your calendar following a regular cadence, typically monthly, to provide continuous support to your team members and capitalize on the benefits associated with it. You can decide to allocate an additional slot on your calendar to this practice or use one of the already assigned one-on-one sessions: one coaching session every three one-on-one meetings.

The Cognitive Profile of a Coach

The nature of coaching demands you to develop a specific mindset and two interrelated capacities that allow you to connect and communicate productively with the coachee. As you will learn in Chapter 8, you will also need to adopt this mindset beyond the formal boundaries of a coaching session to *lead alongside* your team in situations where they can benefit from your expertise or perspectives.

Growth Mindset

A coach needs to believe that talents can be developed using deliberate practice[6] leveraging the ability of the brain to subtly adjust its neural networks in order to complete novel tasks.[iv] While a pure *growth* mindset is solely a

---

[6] Deliberate practice is a form of experimentation based on the intentional repetitions of small, related tasks with immediate feedback and revision.

theoretical concept, and, in practice, every individual has the tendency to devolve into a *fixed* mentality under specific circumstances,[lvi] as a design manager who manages like a leader, you have the moral responsibility to develop a *predominantly growth-oriented mentality* and instill in the coachees the capacity to perceive challenges and talents through the same prism. You can begin to foster a growth mindset in your coachees by making them aware of this attitudinal dualism and modeling the following behaviors:

- Reframe failure as a learning opportunity.

- Respond to failure with nonjudgmental curiosity.

- Emphasize the value of deliberate practice and grit in opposition to the idea of innate talents.

- Provide constructive feedback focusing on dedication, not on abilities.

- Provide appreciative feedback focusing on persistence, not achievements.

A growth mindset leverages *neuroplasticity*, the ability of the brain to generate neural growth under specific actions and activities,[lvii] which in turn drives intrinsic motivation and achievement.[lviii]

Developing and adopting this type of mentality allows you to shift the focus of the coaching relationship on the coachee's learning experience celebrating deliberate practice and *intelligent failure* independently from talents and achievements. It is crucial to remark that this practice is not expected to ignore achievement gaps, reward unproductive forms of failure, or pretend that trying equals success. Coaching is intended to encourage candid conversations about development and performance that promote experiential learning via reflection-in-action and reflection-on-action.

Nonreactive Empathy

A coach needs to sense and understand their emotional and behavioral substrate to regulate verbalization and expression productively. Fulfilling this prerequisite requires a balanced development of the four domains of emotional intelligence: self-awareness, self-management, social awareness, and social management. Reasoning and emotions are inherently intertwined, but an effective coach must develop *nonreactive empathy* to be emotionally invested in establishing a caring relationship with the coachee without being emotionally attached to preserve objectiveness and altruism.[lix]

Active Listening

Coaching is in the service of the coachee and cannot prescind from listening actively and comprehending what that particular person necessitates. A coach needs to cultivate an internal sense of *presence in the moment* that can identify the motives and emotions conveyed by the coachee's words while blocking external distractions. Failing to listen actively during a coaching session decreases the trust level attached to the coaching relationship and impedes the coach from identifying the most appropriate questions to ask.[ix] Active listening can be decomposed into four cardinal behaviors.

Behavior 1: Affirm your intention. Remark that you are listening without expressing agreement or disagreement. This behavior also provides an opportunity to develop the conversation further in a specific direction:

> *I hear what you are saying.*
>
> *I understand. Could you say more about ...*

Behavior 2: Reflect the emotional state. Mirror the sentiment shared by the coachee without expressing agreement or disagreement. This behavior encourages the interlocutor to deepen the verbal exploration:

> *You seem concerned about ...*
>
> *You seem to be ...*

Behavior 3: Draw distinctions. Observe the juxtaposition of two or more concepts shared by the coachee. This behavior identifies conflicts of beliefs, facilitates exploring options, and shifts the person's attention to consider which forces genuinely desire to resolve:

> *I hear two things, you [concept 1] and you [concept 2].*
>
> *Which is the most significant challenge here for you?*

Behavior 4: Validate your comprehension. Restate the essence of a concept by rephrasing what you hear. This behavior corroborates your perception of the facts and functions as a cognitive mirror to the coachee:

> *If I understand correctly, you are saying that ...*
>
> *Your main concern seems to be ... Is that correct?*

When you practice active listening, the intensity of focus characterizing a coaching conversation represents a far more crucial variable than the physical time allocated to it during the session. Deliberate practice will gradually develop your ability to maintain sustained periods of high intensity of focus.[7]

---

[7] Revisit Chapter 4 to review how to find your focus.

Stipulate the Social Contract

Every successful relationship is instituted on clear expectations.[lxi] Only when the coach and the coachee collaboratively craft and voluntarily agree on the terms of engagements instantiating a social contract, the coaching relationship has a concrete opportunity to be successful. The social contract that instantiates the coaching relationship originates from the conscious and deliberate design of roles, responsibilities, and objectives. Stipulating this form of social agreement decreases the likeliness of tensions and moments of irritation and, consequently, conflicts during the coaching conversations.[lxii] Specifically, in the context of a coaching relationship, the social contract is built on two main elements: the coachee's professional aspiration and the coach's behavioral expectations.

Professional aspirations. You cannot coach an individual with no intention to be coached or without a clear objective for the coaching relationship. An individual needs to demonstrate the intrinsic motivation necessary to embrace the responsibility to improve and purposefully commit to a coaching relationship. Under this premise, you can solely contribute to the definition of this part of the social contract by collaborating with that person using the NPR model to articulate the professional aspirations at the intersection of that person's needs, passions, and requirements. Using the following structure, you can collaborate with the coachee to articulate the envisioned success in behavioral terms specifying the *goal intention*:

> I intend to reach [aspiration].

Verbalizing and annotating the goal intention allows the coachee to be mindful of the journey and purposefully commit to the desired outcome at the beginning of each coaching session and, when necessary, redirect that aspiration in a different direction. Additionally, goal intentions, in and of itself, have a facilitating effect on behavior enactment.[lxiii] Based on the NPR model, in Table 7-2, you can see the designated decider in relation to the two dimensions of coaching that you will explore in the subsequent section.

Table 7-2. The decider for each element of the NPR model

	Needs	Passions	Requirements
Coach for Performance	Coach	Coachee	Organization (via coach)
Coach for Development	Coachee	Coachee	Organization (via coach)

If the coachee struggles to articulate their passions, you can take a step back and guide that person through the same process that you followed in Chapter 4 to find and activate your professional purpose. Assisting your team members in articulating their passions and aligning them with their performance or developmental needs and the organizational requirements is a necessary step to promote the expression of their boldest self at work. When completed successfully, this process also unlocks the perception of "meaning of work."[lxiv] In Chapter 9, you will learn how to identify which activities an individual perceives as "meaningful" and "impactful" and how to leverage that comprehension to establish objectives and goals for the team.

Behavioral expectations. The coachee's professional aspirations must be complemented by your behavioral expectations for the coaching relationship. Using the following structure, you can collaborate with the coachee to articulate the envisioned success in behavioral terms specifying the *implementation intention*:

If [situation], then I will [behavior].

The "if" component specifies when and where the person wants to act, while the "then" part identifies the designated action and closes the gap between setting a goal and realizing that aspiration.[lxv] Verbalizing and annotating the implementation intention allows the coachee to be mindful of the behavioral plan of action, forging a cognitive link between the anticipated situation specified in the "if" component and the intended behavior identified in the "then" part of the statement.[lxvi] Forming an implementation intention produces a highly activated mental representation of a critical future situation that enhances the cognitive accessibility of its interrelated behavioral patterns.[lxvii]

The Two Dimensions of Coaching

From the perspective of a person in charge of a team, also referred to as an *internal coach*, the coaching practice can adapt to the coachee's necessities and capabilities across two dimensions that define the type and modality of the session. The "type" can be oriented to performance or development, while the "modality" can be centered on teaching or facilitating (Figure 7-6).

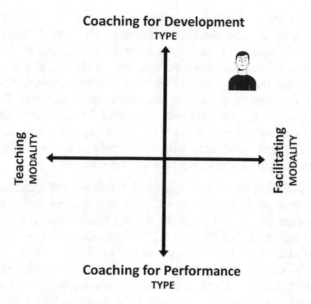

Figure 7-6. The two dimensions of coaching

Coaching for performance focuses on the short-term optimization of a specific behavior to improve its outcome. Typically, the coachee is a designer that needs to overcome a difficulty related to the day-to-day operations, and it is potentially connected with a weakness in the competency spectrum.

Coaching for development concentrates on the long-term evolution of a specific behavior or acquiring a new one to generate a novel outcome. Typically, the coachee is a designer that strives to advance to the next phase of the contribution model and desires to cultivate a strength or extend the competency spectrum. This type of coaching requires a developmental plan to direct and support the coachee's effort; you can apply what you have learned in Chapter 3 to create a developmental program for your team members.

Under these conditions, when the coachee has no knowledge and potential necessary to achieve the desired objective, the coach must assume a "teaching" role acting as a *thinking enabler*, directly assisting the individual in building the capabilities necessary to face the challenges and go beyond the blocks that hinder the progress. In this role, you impart wisdom and share knowledge with the coachee to help that person generate a course of action to achieve the desired objectives. Assuming a teaching role requires expertise in the subject at hand, and the upper limit of what is attainable is defined by the current level of knowledge of the coach.

In the opposite scenario, when instead the coachee has the knowledge and potential to achieve the desired objective, the coach must assume a "facilitating" role acting as a *thinking partner*, indirectly assisting the individual in using the capabilities necessary to face the challenges and see beyond the blocks that hinder the progress. In this role, you help the coachee to explore and question the mental processes and behavioral patterns that limit that person's perspective to unveil a novel course of action to achieve the desired objectives. Assuming a facilitating role requires expertise in coaching, and the upper limit of what is attainable with the coaching session is defined by the potential level of knowledge achievable by the coachee.

Setting the desired objective is essential to drive the coachee's cognitive effort, but as you will learn later in this chapter, it is not uncommon for that person to update the "goal intention" during the conversation as a result of an enhanced awareness of the situation. In these cases, you may need to modify your role to adapt to the new conditions. While coaching for performance is an essential tool that allows you to provide support to the design operations, your objective as the person in charge of the group is to gradually elevate every team member to a level where you can adopt a facilitating role and unlock the vaster benefits deriving from using this coaching modality.

The Structure of a Coaching Session

Irrespective of the model that you decide to adopt, a coaching session follows a specific sequence and fulfills three foundational needs:

1. Identify the desired outcome.

2. Uncover latent conflicts.

3. Articulate an achievable next step.

The "desired outcome" identifies what that coachee wants to obtain from the conversation. This objective, determined by the coachee at the beginning of the session, is derived from the "goal intention" and connected to the "implementation intention" and can be subject to evolution during the discussion as a result of a shift in perspective.

The "latent conflicts" represent a manifestation of the beliefs and emotional substrate that generate blocks and unresolved situations that deter the coachee from achieving the desired outcome.

The "achievable next step" represents a behavior that, congruent with the desired outcome, ensures progress between sessions advancing the coachee's journey toward the fulfillment of the professional aspirations. You can frame this outcome as the combination of a "goal intention" and one or more "implementation intentions" using the structures introduced in the previous section. When you collaborate with the coachee on the definition of this transitional step, you need to consider the three elements that need to converge simultaneously for a behavior to occur: motivation, ability, and a prompt.[lxviii]

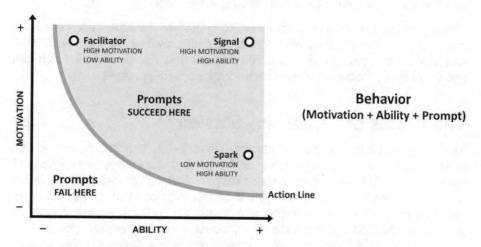

Figure 7-7. The Fogg Behavior Model

The model presented in Figure 7-7, developed by Brian Jeffrey Fogg at the Behavior Design Lab at Stanford University, illustrates the inverse relationship between motivation and ability.[lxix] An individual with low ability requires a high intrinsic motivation to perform a given behavior. Conversely, in a condition of high ability, the intrinsic motivation necessary to initiate that same given behavior is low.[lxx] This inverse correlation is visually identified in the model by the "action line."

Above the action line, an individual possesses the level of intrinsic motivation and ability required to perform a given behavior. Under these conditions, when that person encounters an efficacious prompt,[8] the behavior occurs.[lxxi] The type of *retrieval cues* provided is dictated by the level of motivation and ability of the person in question.[lxxii] There are three types of prompts: sparks, facilitators, and signals.[lxxiii]

Spark: This cue motivates a behavior when the person possesses high levels of motivation and ability.

Facilitator: This cue promotes a behavior when the person demonstrates high levels of motivation but low levels of ability.

---

[8] A prompt is a stimulus that serves as a retrieval cue used to guide memory recall.

Signal: This cue reminds a behavior when the person exhibits low levels of motivation but high levels of ability.

The "implementation intention" represents an example of a prompt that can assume different forms, from the design principles posted on the office walls to a feedback message delivered at the right moment.

When you and the coachee collaborate to articulate the achievable next step, you must carefully consider that person's current level of intrinsic motivation and abilities to influence the selection of a *minimal feasible behavior* that can ensure progress between sessions without hampering progression.

The Phases of a Coaching Session

The conversation that the coach establishes with the coachee is based predominantly on listening actively and inquiring accordingly. The intent of listening is to discover the coachee's viewpoint on a given situation surfacing the underpinning beliefs. Based on that information, the complementary purpose of inquiring is to explore, expand, and reframe those beliefs to promote reflection and generate or introduce a novel interpretation of that situation. A typical instance of a coaching session follows three phases punctuated by five steps (Figure 7-8).

Coaching Session

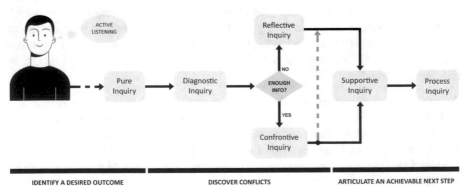

Figure 7-8. The three phases of a coaching session

Step 1: Pure inquiry. During this initial phase, you ask the coachee to introduce a subject for the conversation and specify a desired outcome for the session. If the coachee is reluctant or struggles to designate an objective for the session, ask that person to describe the sentiment characterizing the subject of the discourse and then disambiguate some of the words or phrases used to identify an intention. The following is an example with a topic represented by the expression "strong leadership presence":

> *When a person projects a strong leadership presence, what that person does differently?*
>
> *Imagine you projecting a strong leadership presence; what do you do?*
>
> *Imagine you projecting a strong leadership presence; how do you feel?*

Continue to investigate the subject and the coachee's sentiment until you can identify and agree on the desired outcome for the coaching session. Once you and the coachee have decided on an inceptive desired outcome for the session, ask what that person wants to discuss using a nonpresumptive question preparing the ground for active listening. After the initial answer, follow up asking for more information pushing the coachee to uncover more details:

> *Where do you want to begin?*
>
> *Which part of [outcome] do you want to discuss today?*

Step 2: Diagnostic inquiry. During this phase, you identify and frame the latent conflicts redirecting the conversation to focus on the underpinning beliefs and emotions characterizing the situation. Your objective is to progressively increase the focus of the discussion to identify challenges and uncover underlying conflicts. When the coachee introduces a condition, deconstruct it to investigate it effectively:

> *… we can look at this situation from three different perspectives:*
>
> *the content, the behavior, and the relationships.*
>
> *Where should we start?*

The "content" aspect refers to the challenge connected to the project or event that originated the condition. The "behavior" aspect refers to the challenge related to the compartmental patterns of the individuals involved in that situation. The "relationships" aspect refers to the challenge related to the social power and influential dynamics[9] between the individuals implicated in that circumstance. If there is not a predominant element, a common practice is to initiate the investigation from the "content" element and subsequently explore a possible involvement of "behaviors" and "relationships," exploring compartmental patterns and influential dynamics.

---

[9] Predominantly based on the interplay of roles and their underlying sources of power: role, expertise, and relationship, as introduced in Chapter 5.

Coaching presents the same wicked problem-solving nature that characterizes numerous design challenges, and both coaching and designing leverage questions as the atomic unit of exploration. During the conversation, predominantly when using diagnostic inquiry, you have to use an open-to-close pattern (Figure 7-9) to purposefully craft questions to explore the conflict domain, decipher its nature, and then validate your comprehension.

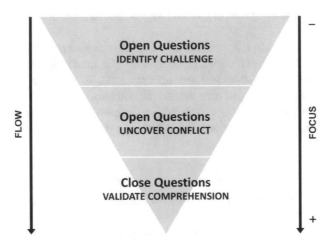

Figure 7-9. The open-to-close question pattern

During this phase, "what"- and "how"-based questions are preferred to investigate objective facts over "why"-based ones that instead trigger answers based on a subjective interpretation of experiences.[lxxiv] When you feel that a pivotal aspect of the situation surfaced, it is crucial to assist the coachee in framing the associated challenge:

> *Let's assume that ...*
>
> *What would be the challenge for you in this situation?*

When you identified the challenge and uncovered the conflict, validate your comprehension using a closed question and help the coachee identify how that conflict is blocking their progression:

> *Based on what you are saying, ... Is that correct?*
>
> *How is this conflict affecting your ability to achieve your outcome?*

Assisting the coachee in articulating the challenge and the underlying conflict allows you to set the preconditions for moving to the next step leveraging *confrontive inquiry* and *reflective inquiry*.

Step 3a: Confrontive inquiry. During this phase, if you gather enough information to develop a novel interpretation of the situation that can produce a moment of reflection and the coachee cannot overcome the challenge autonomously, you validate and introduce your perspective in a nonbelligerent way. Confrontive inquiry is more common when you assume a teaching role, and the coachee is in a position to benefit from a novel viewpoint that underlies a hypothesis that can challenge that person's narrative. This scenario is the only occasion during a coaching session where the coach utilizes advocacy to proactively introduce an external interpretation into the conversation:

> *What if ... [novel interpretation]*
>
> *How might you ... [outcome] if ... [novel interpretation]*

A message conveyed using confrontive inquiry can potentially challenge the coachee's beliefs generating a certain level of psychological discomfort.[lxxv] This type of communication can be perceived across cultures with different degrees of antagonism.[lxxvi] The degree of cognitive distress perceived is a function of the person's culture, intended as the characteristic attitudes and behaviors that identify a distinct social group.[lxxvii] In Figure 7-10, you can see the distribution of behavioral and emotional attitudes across cultures.

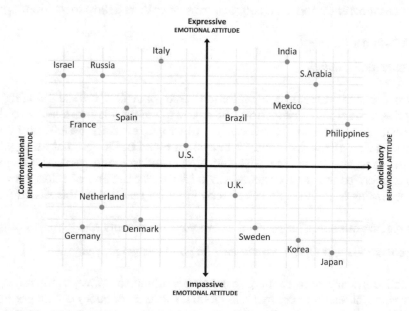

Figure 7-10. The distribution of behavioral and emotional attitudes across cultures

Every time you challenge the coachee's beliefs, perplexing that person's brain by undermining the perception of a situation in a way that feels uncomfortable, confusing, and sometimes even embarrassing, you have to demonstrate nonreactive empathy. Nonreactive empathy is characterized by a curious and caring attitude that stays with the inquiry patiently.[lxxviii]

In those situations, if you remain in a calm and neutral state, preserving a trustworthy environment where the coachee can safely process emotions, that initial reaction tends to fade as a new perspective on the situation develops on their mind. You cannot avoid generating psychological discomfort with your statements and inquiry if you want to assist individuals to perceive the world around them in a more expansive way, but you can mitigate those moments by doing it in a nonreactive empathetic manner.[lxxix]

Step 3b: Reflective inquiry. During this phase, you introduce a statement to provoke critical thinking and a moment of reflection followed by an inquiry to internally shift the attention to critically analyze thoughts and actions and envision new possibilities:

Reflective Inquiry = Reflective Statement + Contextual Inquiry

Reflective inquiry is more common when you assume a facilitating role, and the coachee is in a position where they can benefit from a cognitive mirror that reflects the consequences of thinking and behavioral patterns. This scenario allows the coachee to obtain a broader perspective on the situation, shift the perspective, reframe those behavioral patterns, and redefine that given circumstance and, as a consequence of that, redefine their identity:[lxxx]

> [reflective statement].
>
> [contextual inquiry].

When a coach introduces a statement that provokes critical thinking and provides a moment of reflection, the following inquiry is more likely to drive the coachee's attention to discern gaps in logic, evaluate beliefs, and clarify emotions that affect decisions.[lxxxi] Imagine a Head of Design that struggles to comprehend the difference between high standards and perfectionism that always offers to the team unnecessary criticisms at the end of a design review. The following is an example of reflective inquiry tailored to this situation:

> [reflective statement] *It seems that everyone disappoints you.*
>
> [contextual inquiry] *Will anyone ever be good enough for you?*

Reflective inquiry represents an assertive manner to convey a message. For this reason, similarly to confrontive inquiry, this type of query can be perceived as antagonistic by individuals with a predominantly nonconfrontational culture, triggering a defensive state that hinders critical thinking and reflection.[lxxxii] In this case, if you use reflective inquiry in a nonreactive empathetic manner accompanied by a comforting tone of voice, you can mitigate that reaction and positively affect how the interlocutor processes your message.[lxxxiii]

When you introduce a reflective statement, you frame a situation and project it back to the coachee to provoke critical thinking. There are three types of reflective statements:[lxxxiv]

- Summarize
- Paraphrase
- Encapsulate

When you "summarize," you frame the situation using keywords from the coachee's narrative to trigger the perception of the beliefs and emotions intrinsic to that person's story. Summarizing facilitates the coachee's effort of developing awareness of a conflict, fulfilling the preconditions to investigate its nature and implications:[lxxxv]

So you are saying … [summarize]

When you "paraphrase," you frame the situation using incisive words to enhance the coachee's perception of the beliefs and emotions intrinsic to that person's story. Paraphrasing facilitates the coachee's effort of investigating the nature of a conflict, deciphering the assumptions behind their words and emotions:[lxxxvi]

I hear that you … [paraphrase]

When you "encapsulate," you frame the situation capturing the crucial elements of the coachee's narrative succinctly to promote a reframing of the beliefs and emotions intrinsic to that person's story. Encapsulating facilitates the coachee's effort of processing the conflict, exploring a possible course of action to its resolution:[lxxxvii]

It sounds like you … [encapsulate]

When you introduce a contextual inquiry, you drive critical thinking to elaborate a new course of action. While this type of query is dependent on the context of the social exchange, concise sentences tend to have a more profound impact.

Step 4: Supportive inquiry. During this phase, you must consolidate the new reframed perspective by asking the coachee to articulate what that person now "sees" differently. An individual reframes the comprehension of a situation when an insight expands the perception of self in relation to the conflict, and therefore the identity, or when that person discovers a novel interpretation of that situation that they now believe is true.[lxxxviii] At that point, you must invite the coachee to verbally articulate these uncovered insights and newly formed realizations to crystallize the change that they provoked in the cognitive frame:[10]

> *What do you see now that changed your perspective?*
>
> *What does [reframed perspective] mean to you?*

After the coachee has formalized how the expanded awareness of the situation changed their cognitive frame, you must invite that person to commit to a *minimum feasible behavior* to ensure progress between sessions:

> *What will you do now? By when?*
>
> *What can hamper your commitment? How can I help you?*

Step 5: Process inquiry. During this closing phase, you focus the conversation on the coaching relationship by asking for feedback on the helpfulness of the session:

> *Was the conversation useful?*
>
> *… what was most valuable for you?*

Having a clear flow in mind allows you to "be mindful of the process," knowing where you are during the coaching conversation, what methods to use, and the objective of that given stage. Being mindful of the process also constitutes one of the seven foundation principles of human-centered design[11] and represents another point of contact between the practice of coaching and designing. Implementing this principle also enables you to develop the awareness necessary to respond to an eventual evolution of the desired outcome during the session.

---

[10] A cognitive frame is a set of parameters defining either a particular mental schema or the more comprehensive cognitive structure by which an individual perceives and evaluates the world.

[11] As defined at the Hasso Plattner Institute of Design at Stanford University: focus on human values, embrace experimentation, bias toward action, show don't tell, craft clarity, be mindful of the process, radical collaboration.

Respond to an Evolution of the Desired Outcome

As we previously introduced, the desired outcome can, on some occasions, evolve or pivot during the conversation as a result of a shift in the coachee's perspective. When you reach a point where the coachee shifts perception and uncovers a new need, investigate the willingness to explore this new direction with a question similar to the following examples:

Can we explore …

Can you tell me more about …

// Diagnostic Inquiry

Let's assume that …

What would be the challenge for you in this situation?

If the coachee is willing to explore this often more profound and personal need, agree on a new desired outcome and mentally reset the coaching flow to the diagnostic inquiry phase proceeding forward from that step.

Understand the Conative Mentality

The practice of coaching is intrinsically connected with learning. Whether the session is focused on performance or development, uncovering and resolving conflict always involves acquiring new capabilities that allow the coachee to move beyond a given challenge. Often at the end of a coaching session, you may need to assign a task[12] in preparation for the subsequent meeting; in order to do it effectively, it is crucial for you to understand the possible conative mentalities that a person can demonstrate in that specific context.

Conation derives from the Latin word *conari*, which means "to try," and denotes the *modus operandi* or learning attitude used to process challenges and solve problems by an individual. Designers are trained to solve business problems using "design thinking," but the mindset with which they process challenges outside that scope can vary, voluntarily or involuntarily, based on the context and prior experiences.[lxxxix] Your objective as a coach is to assist your team members in approaching learning using "design thinking" and a "builder" attitude, but it is likely that some of them unconsciously process learning opportunities following a different way to construct their intentions and actions.

One of the first things that a student learns at the d.school at Stanford is that human beings construct their intentions and actions using four main types of mutually nonexclusive thinking.[xc]

---

[12] Revisit Chapter 3 to review the different types of activities and their diverse information retention rates.

Business thinking: This mindset optimizes its way forward by increasing or decreasing the solution to a given problem in accordance with a certain logic.

Research thinking: This mindset analyzes its way forward by processing scientific data and isolating patterns to uncover and solve problems.

Engineering thinking: This mindset solves its way forward by acquiring the information necessary to model the problem and solve it.

Design thinking: This mindset builds its way forward by acquiring empirical data and creating prototypes to learn more about the problem and solve it.

The first three mentalities are predominantly effective on tame problems characterized by static criteria, while design thinking is better suited for wicked problems where the criteria are dynamic; you cannot recognize success until you encounter it, and once you have achieved a satisfactory solution, it's not reusable.[xci] Wicked problems are overwhelmingly human problems, which include acquiring and developing abilities. Table 7-3 summarizes the differences between tame and wicked problems.[xcii]

Table 7-3. The differences between tame and wicked problems

	Tame Problems	Wicked Problems
Problem	The problem is not unique and can be standardized.	The problem is unique and cannot be standardized.
	The precise definition of the problem also unveils the solution.	The ambiguous frame of the problem does not unveil the solution.
Solution	The solution is true or false, and it's assessed according to criteria revealing the degree of effect.	The solution is not true or false, and it's assessed as "better" or "worse" or "satisfactory."
	The solution can be created entirely, and it's reusable.	The attempt to create a solution changes the problem, and it's not reusable.
Stopping Rule	The problem is static and does have an end state.	The problem is dynamic, evolves continuously, and does not have an end state.
	The problem-solving process concludes when the problem is solved.	The problem-solving process is never concluded.
Stakeholders	There is agreement around the definition of the problem.	There is no agreement around the definition of the problem.
	Monodisciplinary stakeholders define the problem in agreement with the support of scientific data.	Multidisciplinary stakeholders frame the problem by integrating diverse perspectives with the support of empirical data.
Example	Given a specific software, animate a prototype.	Democratize access to healthcare.

As you will learn in Chapter 9, promoting diversity and inclusion within a team, among other things, means nurturing and leveraging different mentalities and understanding that an individual needs to adopt the most effective mindset under the requirement of the current situation. When the objective is to acquire abilities, the design mindset represents the only way to unlock experiential learning and produce successful results. Despite that, developing a new state of mind requires time; therefore, you need to be aware of the mentality that the coachee initially brings into the coaching conversation to respond to that way of thinking accordingly. The previously introduced four types of thinking delineate four distinct learning profiles characterized by different cognitive requirements that tend to approach the learning experience differently.[xciii]

Optimizer: This profile uses business thinking to form a holistic view of the challenge as a way to understand the designated task. The most powerful learning moments for the optimizer occur when that holistic view is created. With this top-down approach, you have to provide a high-level representation of the challenge to allow that person to understand the course of action required. Providing the opportunity to watch another person performing the entire task is a productive way to engage this mentality.

Analyzer: This profile uses research thinking to form an atomistic view of the challenge as a way to understand the designated task. The most powerful learning moments for the optimizer occur when that atomistic view is created. With this bottom-up approach, you have to provide a low-level representation of the challenge to allow that person to understand the course of action required. Providing enough time to prepare in detail prior to the task is a productive way to engage this mentality.

Solver: This profile uses engineering thinking to resolve the challenge preliminarily as a way to understand the designated task. The most powerful learning moments for the solver occur at the end of the resolution prior to the performance. With this outside-in approach, you have to provide the components of the solution to allow that person to understand the course of action required. Providing access to detailed relevant information is a productive way to engage this mentality.

Builder: This profile uses design thinking to explore the challenge nonpresumptively as a way to understand the designated task. The most powerful learning moments for the solver occur at the end of the reflection after the performance. With this inside-out approach, you have to provide the condition for exploration to allow that person to understand the course of action required. Providing a simple initiating step to complete is a productive way to engage this mentality.

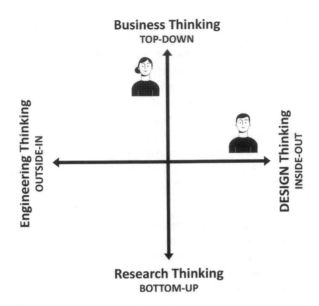

Figure 7-11. The four learning profiles

Figure 7-11 illustrates these four learning profiles. As a design manager and leader, you must investigate the coachee's learning mentality inside and outside the coaching sessions to adapt your style of communication, the information provided during the conversations, and the strategy underlying the performance or development-oriented plan of action. Use *active listening* during the coaching sessions and *active observation* outside them in contexts that include an element of learning to investigate the way of thinking utilized by your team members. Understanding the diverse conative mentalities inhabiting your team enables you to maximize the potential results of the coaching sessions, allowing you to help the coachee to identify appropriate "goal intentions" and "implementation intentions" and, as you will learn in Chapter 8, leverage the diversity of thinking in your group to cultivate creative collaboration.

Delegate Work to Promote Learning

As a design manager and leader, delegation represents a crucial tool that you can use to improve your and your team's performance. Delegation can be framed as the transfer of the responsibility of a given task, for which you retain accountability, to an individual, defining an objective and providing to that person the resources and role power necessary to complete the work. Delegationx implements the following two main intentions:

- Reduce your workload.
- Develop a team member.

When delegation is implemented purposefully, these two scenarios present a degree of overlap; you reduce your workload, and the designated team member works on something that promotes learning. Typically, activities that can be transferred are meetings, presentations, or entire projects that do not represent a core responsibility of your role. Delegating your core activities negatively impacts the perception of your level of expertise and relationship power within the organization and must be avoided.

Delegating a task also requires a deliberate approach because it represents an event that can trigger a threat response producing social pain.[xciv] The following six steps implement this intention, providing you a blueprint that can help you to reduce the likeliness of triggering a threat response.

Step 1: Identify the appropriate person-task combination. In this initial phase, you can adopt two different perspectives. You can select a job that you need to transfer to reduce your workload or identify a task that you decided to delegate to develop a team member. In both scenarios, the NRP model introduced in Chapter 3 can help you to select an activity at the intersection of the designer's needs (N), passions (P), and requirements (R).

Step 2: Introduce the delegation proposal. In this phase, you provide context to the request emphasizing the potential opportunity to grow and the peculiarities of that specific person-task combination. If you have already introduced the concept of the NRP matrix to the team, you can use that model to illustrate how that opportunity can advance the ability required by that specific task.

Step 3: Ask for specific acceptance. In this phase, you directly ask the interlocutor to accept the responsibility to complete the task, whether occasional or recurrent; the following is an example:

> *Would you take over [task]?*

If the person declines the request, honor the response, and if possible, work with that person to identify an alternative person-task combination.

Step 4: Describe the task. In this phase, if the person accepts the request, you provide a detailed description of the responsibility; if you are aware of the conative mentality of the delegatee, you can use that information to provide a more engaging explanation. This description also needs to include a structured example of how you have completed that task in the past:

> *Here is what I usually do ...*

Step 5: Clarify expected behaviors and outcomes. In this phase, you introduce the expected result, elucidate the specific actions required by the task, and define the quality standards to respect. If present, you also mention eventual deadlines to consider. At this point, when necessary, you also provide some form of preparatory training.

Step 6: Provide continuous support. In this phase, you establish a regular check-in to assist the delegatee and ensure the execution of the task in compliance with the agreed expected behaviors, outcome, and standards. Depending on the complexity of the task delegated, you may need additional time to discuss the delegated task in detail; in that case, you can use the weekly one-on-one meeting with that person. A form of control is also made necessary by the fact that when you delegate the responsibility of a task, you always retain the accountability for that specific job.

It is essential to observe that the amount of responsibility delegated with the task must be accompanied by the proportionated degree of role power necessary to complete the job. This requirement means that the team and the stakeholders must be aware of the delegated task to respond to this form of social change accordingly.

Prevent Reverse Delegation

A delegation that promotes development is a delegation that pushes the delegatee outside of the comfort zone into the learning dimension permeated by eustress. In that uncomfortable mental state, some individuals can react by deciding to reconsider their commitment or asking for substantial assistance. Assuming that the task delegated is appropriately calibrated to the delegatee's capacities, you must prevent any form of reverse delegation using a nonreactive empathetic response. In this situation, block the eventual attempt to shift the responsibility back to you and use active listening and the appropriate form of inquiry to promote reflection and help the delegatee to identify a possible behavioral direction to follow.

Control Complexity

Formally defining the concept of complexity and its opposite, simplicity, would require the introduction and combination of different notions to capture its meaning comprehensively. From a design manager and leader perspectivec, complexity is a function of both systems and human beings that manifest context-dependent and subject-related peculiarities.[xcv] Complexity is related to systems via nonholonomic constraints, while it is related to human beings via their perceptual process, which involves their prior experiential knowledge via their reflective assumptions and current objectives.[xcvi] With this conceptual distinction in mind, complexity, intended as the cognitive effort aimed to create relationships that connect actions and responses to a given context and subsequently assign behavioral meanings to these associations,[xcvii] can be separated into two interrelated entities:

Objective complexity: The degree of relational entanglement physically present in a stimulus[xcviii]

Subjective complexity: The perception of the degree of relational entanglement physically present in a stimulus[xcix]

In combination with other cognitive factors such as expertise and motivation, objective and subjective complexity directly influence task performance.[c] Given a degree of objective complexity, when you decide to delegate a task, you must consider that when a job organically created at level N cascades at level N-1 on the hierarchy,[13] it increases its subjective complexity considerably. Figure 7-12 illustrates this dynamic depicting the relationship between delegation and subjective complexity.

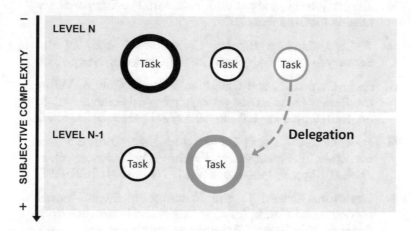

Figure 7-12. The relationship between delegation and subjective task complexity

As a corollary, a subjectively high complexity task for you at level N commonly represents an impossible task for one of your direct reports at level N-1. As you will learn in Chapter 9, this type of task pushes an individual outside the comfort zone into an unhealthy area called distress. Conversely, a subjectively low complexity task for you at level N commonly represents a potentially suitable candidate for a delegation. In Chapter 9, you will learn how this type of task pushes an individual outside the comfort zone into a healthy area called eustress, also identified as the "learning zone."

Irrespective of the perceived level of complexity of a task, pushing an individual outside the comfort zone can always potentially trigger a threat response that generates social pain and hinders critical thinking and reflection.[ci] As a consequence of this dynamic, during the delegating process, you must leverage mindful communication to minimize social pain and nonreactive empathy to recognize and manage any potential sign of a threat response.

---

[13] Hierarchy in this context is intended as a proxy for a person's expertise and level of development of the five clusters of capabilities composing the Leadership Tent model.

Endnotes

i. Zenger, John H., and Joseph R. Folkman. The Extraordinary Leader. McGraw-Hill, 2009.

ii. Zenger, John H., and Joseph R. Folkman. The Extraordinary Leader. McGraw-Hill, 2009.

iii. Zenger, John H., and Joseph R. Folkman. The Extraordinary Leader. McGraw-Hill, 2009.

iv. Zenger, John H., and Joseph R. Folkman. The Extraordinary Leader. McGraw-Hill, 2009.

v. Y Cajal, Santiago Ramón. Comparative Study of the Sensory Areas of the Human Cortex. Clark University, 1899.

vi. Park, Denise C., and Chih-Mao Huang. "Culture Wires the Brain." Perspectives on Psychological Science, vol. 5, no. 4, 2010, pp. 391–400., doi:10.1177/1745691610374591.

vii. Park, Denise C., and Chih-Mao Huang. "Culture Wires the Brain." Perspectives on Psychological Science, vol. 5, no. 4, 2010, pp. 391–400., doi:10.1177/1745691610374591.

viii. Davidson, Richard J., and Bruce S. McEwen. "Social Influences on Neuroplasticity: Stress and Interventions to Promote Well-Being." Nature Neuroscience, vol. 15, no. 5, 2012, pp. 689–695., doi:10.1038/nn.3093.

ix. Davidson, Richard J., and Bruce S. McEwen. "Social Influences on Neuroplasticity: Stress and Interventions to Promote Well-Being." Nature Neuroscience, vol. 15, no. 5, 2012, pp. 689–695., doi:10.1038/nn.3093.

x. Dweck, Carol S. Mindset: The New Psychology of Success. Ballantine Books, 2016.

xi. Dweck, Carol S. Mindset: The New Psychology of Success. Ballantine Books, 2016.

xii. Zak, Paul J. Trust Factor: The Science of Creating High-Performance Companies. Amacom, 2017.

xiii. Hook, Derek, et al. The Social Psychology of Communication. Palgrave Macmillan, 2011.

xiv. Hook, Derek, et al. The Social Psychology of Communication. Palgrave Macmillan, 2011.

xv.   Edmondson, Amy C. The Fearless Organization: Creating Psychological Safety in the Workplace for Learning, Innovation, and Growth. Wiley, 2018.

xvi.   Fink, George. Stress: Concepts, Cognition, Emotion and Behavior. Academic Press, 2016.

xvii.   Cuddy, Amy. Presence: Bringing Your Boldest Self to Your Biggest Challenges. Orion, 2016.

xviii.   Superpowers. SY/Partners, superpowers.sypartners.com.

xix.   Robbins, Mike. Bring Your Whole Self to Work: How Vulnerability Unlocks Creativity, Connection, and Performance. Hay House Business, 2021.

xx.   Nelson, Richard R., and Sidney G. Winter. An Evolutionary Theory of Economic Change. The Belknap Press of Harvard University Press, 2004.

xxi.   Nelson, Richard R., and Sidney G. Winter. An Evolutionary Theory of Economic Change. The Belknap Press of Harvard University Press, 2004.

xxii.   Nelson, Richard R., and Sidney G. Winter. An Evolutionary Theory of Economic Change. The Belknap Press of Harvard University Press, 2004.

xxiii.   Nelson, Richard R., and Sidney G. Winter. An Evolutionary Theory of Economic Change. The Belknap Press of Harvard University Press, 2004.

xxiv.   HBR Guide to Delivering Effective Feedback. Harvard Business Review Press, 2016.

xxv.   Gottman, John Mordechai, and Joan DeClaire. The Relationship Cure: A Five-Step Guide to Strengthening Your Marriage, Family, and Friendships. Harmony Books, 2002.

xxvi.   Gordon, Evian. Integrative Neuroscience: Bringing Together Biological, Psychological and Clinical Models of the Human Brain. Harwood Academic Publishers, 2000.

xxvii.   Gordon, Evian. Integrative Neuroscience: Bringing Together Biological, Psychological and Clinical Models of the Human Brain. Harwood Academic Publishers, 2000.

xxviii.   Lieberman, Matthew D. "Social Cognitive Neuroscience: A Review of Core Processes." Annual Review of Psychology, vol. 58, no. 1, 2007, pp. 259–289., doi:10.1146/annurev.psych.58.110405.085654.

xxix. Lieberman, Matthew D. Social: Why Our Brains Are Wired to Connect. Oxford University Press, 2015.

xxx. Lieberman, Matthew D. Social: Why Our Brains Are Wired to Connect. Oxford University Press, 2015.

xxxi. Sandi, Carmen. "Stress and Cognition." Wiley Interdisciplinary Reviews: Cognitive Science, vol. 4, no. 3, 2013, pp. 245–261., doi:10.1002/wcs.1222.

xxxii. Rock, David, and Linda J. Page. Coaching with the Brain in Mind: Foundations for Practice. Wiley, 2009.

xxxiii. Zink, Caroline F., et al. "Know Your Place: Neural Processing of Social Hierarchy in Humans." Neuron, vol. 58, no. 2, 2008, pp. 273–283., doi:10.1016/j. neuron.2008.01.025.

xxxiv. Eisenberger, N. I., et al. "Does Rejection Hurt? An FMRI Study of Social Exclusion." Science, vol. 302, no. 5643, 2003, pp. 290–292., doi:10.1126/science.1089134.

xxxv. Tabibnia, Golnaz, and Matthew Lieberman. "Fairness and Cooperation Are Rewarding: Evidence from Social Cognitive Neuroscience." Annals of the New York Academy of Sciences, vol. 1118, no. 1, 2007, pp. 90–101., doi:10.1196/annals.1412.001.

xxxvi. Peters, Achim, et al. "Uncertainty and Stress: Why It Causes Diseases and How It Is Mastered by the Brain." Progress in Neurobiology, vol. 156, 2017, pp. 164–188., doi:10.1016/j.pneurobio.2017.05.004.

xxxvii. Chorpita, Bruce F., and David H. Barlow. "The Development of Anxiety: The Role of Control in the Early Environment." Psychological Bulletin, vol. 124, no. 1, 1998, pp. 3–21., doi:10.1037/0033-2909.124.1.3.

xxxviii. Baker, Ann C., et al. "In Conversation: Transforming Experience into Learning." Simulation & Gaming, vol. 28, no. 1, 1997, pp. 6–12., doi:10.1177/1046878197281002.

xxxix. Dennett, Daniel C. Kinds of Minds: Toward an Understanding of Consciousness. Basic Books, 1996.

xl. Alicke, Mark D., et al. "The Motivation to Maintain Favorable Identities." Self and Identity, vol. 19, no. 5, 2019, pp. 572–589., doi:10.1080/15298868.2019.1640786.

xli. Scott, Kim. Radical Candor. Pan Books, 2019.

xlii. Zak, Paul J. Trust Factor: The Science of Creating High-Performance Companies. Amacom, 2017.

xliii. Scott, Kim. Radical Candor. Pan Books, 2019.

xliv. Nussbaum, Michael. "How Introverts versus Extroverts Approach Small-Group Argumentative Discussions." The Elementary School Journal, vol. 102, no. 3, Jan. 2002, pp. 183–197., doi:10.1086/499699.

xlv. Morgenroth, Thekla, et al. "The Motivational Theory of Role Modeling: How Role Models Influence Role Aspirants' Goals." Review of General Psychology, vol. 19, no. 4, 2015, pp. 465–483., doi:10.1037/gpr0000059.

xlvi. Zak, Paul J. Trust Factor: The Science of Creating High-Performance Companies. Amacom, 2017.

xlvii. Gottman, John Mordechai, and Joan DeClaire. The Relationship Cure: A Five-Step Guide to Strengthening Your Marriage, Family, and Friendships. Harmony Books, 2002.

xlviii. Gottman, John Mordechai, and Joan DeClaire. The Relationship Cure: A Five-Step Guide to Strengthening Your Marriage, Family, and Friendships. Harmony Books, 2002.

xlix. Morgenroth, Thekla, et al. "The Motivational Theory of Role Modeling: How Role Models Influence Role Aspirants' Goals." Review of General Psychology, vol. 19, no. 4, 2015, pp. 465–483., doi:10.1037/gpr0000059.

l. Cringely, Robert X. Steve Jobs: The Lost Interview, Magnolia Home Entertainment, 23 Oct. 2012, www.magpictures.com/stevejobsthelostinterview.

li. Adler, Alfred. Social Interest: A Challenge to Mankind. Martino Fine Books, 1938.

lii. Adler, Alfred. Social Interest: A Challenge to Mankind. Martino Fine Books, 1938.

liii. Eagleman, David. The Brain: The Story of You. Canongate Books, 2016.

liv. Zenger, Jack, and Joe Folkman. "How Developing a Coaching Culture Pays Off." Zenger | Folkman, 19 Aug. 2019, zengerfolkman.com/white-papers/how-developing-a-coaching-culture-pays-off/.

lv.   Gratton, Caterina, et al. "Evidence for Two Independent Factors That Modify Brain Networks to Meet Task Goals." Cell Reports, vol. 17, no. 5, 2016, pp. 1276–1288., doi:10.1016/j.celrep.2016.10.002.

lvi. Dweck, Carol. "What Having a 'Growth Mindset' Actually Means." Harvard Business Review, 13 Jan. 2016, hbr. org/2016/01/what-having-a-growth-mindset-actually-means.

lvii. Hallowell, Edward M. Shine: Using Brain Science to Get the Best from Your People. Harvard Business Review Press, 2011.

lviii. Dweck, Carol S. Mindset: The New Psychology of Success. Ballantine Books, 2016.

lix. Batista, Ed. "Coaching Your Employees." Harvard Business Review, 10 Nov. 2015, hbr.org/webinar/2014/09/coaching-your-employees.

lx.   Stine, Mary, et al. "The Impact of Organizational Structure and Supervisory Listening Indicators on Subordinate Support, Trust, Intrinsic Motivation, and Performance." International Journal of Listening, vol. 9, no. 1, 1995, pp. 84–105., doi:10.1080/10904018.1995.10499143.

lxi. Gottman, John Mordechai, and Joan DeClaire. The Relationship Cure: A Five-Step Guide to Strengthening Your Marriage, Family, and Friendships. Harmony Books, 2002.

lxii. Dutton, Jane E., and Belle Rose Ragins. Exploring Positive Relationships at Work: Building a Theoretical and Research Foundation. Psychology Press, 2017.

lxiii. Gollwitzer, Peter M., and Paschal Sheeran. "Implementation Intentions and Goal Achievement: A Meta-Analysis of Effects and Processes." Advances in Experimental Social Psychology, 2006, pp. 69–119., doi:10.1016/s0065-2601(06)38002-1.

lxiv. Scott, Kim. Radical Candor. Pan Books, 2019.

lxv. Achtziger, Anja, et al. "Implementation Intentions and Shielding Goal Striving From Unwanted Thoughts and Feelings." Personality and Social Psychology Bulletin, vol. 34, no. 3, 2008, pp. 381–393., doi:10.1177/0146167207311201.

lxvi. Gollwitzer, Peter M., and Paschal Sheeran. "Implementation Intentions and Goal Achievement: A Meta-Analysis of Effects and Processes." Advances in Experimental Social Psychology, 2006, pp. 69–119., doi:10.1016/s0065-2601(06)38002-1.

lxvii. Gollwitzer, Peter M., and Veronika Brandstätter. "Implementation Intentions and Effective Goal Pursuit." Journal of Personality and Social Psychology, vol. 73, no. 1, 1997, pp. 186–199., doi:10.1037/0022-3514.73.1.186.

lxviii. Fogg, Brian Jeffrey. "A Behavior Model for Persuasive Design." Persuasive 09: Proceedings of the 4th International Conference on Persuasive Technology, Apr. 2009, pp. 1–7., doi:10.1145/1541948.1541999.

lxix. Fogg, Brian Jeffrey. "A Behavior Model for Persuasive Design." Persuasive 09: Proceedings of the 4th International Conference on Persuasive Technology, Apr. 2009, pp. 1–7., doi:10.1145/1541948.1541999.

lxx. Fogg, Brian Jeffrey. Tiny Habits: The Small Changes That Change Everything. Virgin Books, 2020.

lxxi. Fogg, Brian Jeffrey. "A Behavior Model for Persuasive Design." Persuasive 09: Proceedings of the 4th International Conference on Persuasive Technology, Apr. 2009, pp. 1–7., doi:10.1145/1541948.1541999.

lxxii. Fogg, Brian Jeffrey. "A Behavior Model for Persuasive Design." Persuasive 09: Proceedings of the 4th International Conference on Persuasive Technology, Apr. 2009, pp. 1–7., doi:10.1145/1541948.1541999.

lxxiii. Fogg, Brian Jeffrey. "A Behavior Model for Persuasive Design." Persuasive 09: Proceedings of the 4th International Conference on Persuasive Technology, Apr. 2009, pp. 1–7., doi:10.1145/1541948.1541999.

lxxiv. Reisberg, Daniel. Cognition: Exploring the Science of the Mind. W. W. Norton & Company, 2019.

lxxv. Lyons, Nona, editor. Handbook of Reflection and Reflective Inquiry. Springer, 2016.

lxxvi. Meyer, Erin. The Culture Map: Decoding How People Think, Lead, and Get Things Done Across Cultures. PublicAffairs, 2014.

lxxvii. VandenBos, Gary R. APA Dictionary of Psychology. American Psychological Association, 2015.

lxxviii. Reynolds, Marcia. Coach the Person, Not the Problem: A Guide to Using Reflective Inquiry. Berrett-Koehler, 2020.

lxxix. Reynolds, Marcia. Coach the Person, Not the Problem: A Guide to Using Reflective Inquiry. Berrett-Koehler, 2020.

lxxx. Reynolds, Marcia. Coach the Person, Not the Problem: A Guide to Using Reflective Inquiry. Berrett-Koehler, 2020.

lxxxi. Reynolds, Marcia. Coach the Person, Not the Problem: A Guide to Using Reflective Inquiry. Berrett-Koehler, 2020.

lxxxii. Reynolds, Marcia. Coach the Person, Not the Problem: A Guide to Using Reflective Inquiry. Berrett-Koehler, 2020.

lxxxiii. Reynolds, Marcia. Coach the Person, Not the Problem: A Guide to Using Reflective Inquiry. Berrett-Koehler, 2020.

lxxxiv. Reynolds, Marcia. Coach the Person, Not the Problem: A Guide to Using Reflective Inquiry. Berrett-Koehler, 2020.

lxxxv. Reynolds, Marcia. Coach the Person, Not the Problem: A Guide to Using Reflective Inquiry. Berrett-Koehler, 2020.

lxxxvi. Reynolds, Marcia. Coach the Person, Not the Problem: A Guide to Using Reflective Inquiry. Berrett-Koehler, 2020.

lxxxvii. Reynolds, Marcia. Coach the Person, Not the Problem: A Guide to Using Reflective Inquiry. Berrett-Koehler, 2020.

lxxxviii. Reynolds, Marcia. Coach the Person, Not the Problem: A Guide to Using Reflective Inquiry. Berrett-Koehler, 2020.

lxxxix. Burnett, William, and David J. Evans. Designing Your Life: How to Build a Well-Lived, Joyful Life. Alfred A. Knopf, 2016.

xc. Burnett, William, and David J. Evans. Designing Your Life: How to Build a Well-Lived, Joyful Life. Alfred A. Knopf, 2016.

xci. Buchanan, Richard. "Wicked Problems in Design Thinking." Design Issues, vol. 8, no. 2, 1992, pp. 5–21., doi:10.2307/1511637.

xcii. Buchanan, Richard. "Wicked Problems in Design Thinking." Design Issues, vol. 8, no. 2, 1992, pp. 5–21., doi:10.2307/1511637.

xciii. Burnett, William, and David J. Evans. Designing Your Life: How to Build a Well-Lived, Joyful Life. Alfred A. Knopf, 2016.

xciv. Fink, George. Stress: Concepts, Cognition, Emotion and Behavior. Academic Press, 2016.

xcv. Flood, R. L. "Complexity: A Definition by Construction of a Conceptual Framework." Systems Research, vol. 4, no. 3, 1987, pp. 177–185., doi:10.1002/sres.3850040304.

xcvi. Flood, R. L. "Complexity: A Definition by Construction of a Conceptual Framework." Systems Research, vol. 4, no. 3, 1987, pp. 177–185., doi:10.1002/sres.3850040304.

xcvii. Flood, R. L. "Complexity: A Definition by Construction of a Conceptual Framework." Systems Research, vol. 4, no. 3, 1987, pp. 177–185., doi:10.1002/sres.3850040304.

xcviii. Flood, R. L. "Complexity: A Definition by Construction of a Conceptual Framework." Systems Research, vol. 4, no. 3, 1987, pp. 177–185., doi:10.1002/sres.3850040304.

xcix. Flood, R. L. "Complexity: A Definition by Construction of a Conceptual Framework." Systems Research, vol. 4, no. 3, 1987, pp. 177–185., doi:10.1002/sres.3850040304.

c. Maynard, Douglas C., and Milton D. Hakel. "Effects of Objective and Subjective Task Complexity on Performance." Human Performance, vol. 10, no. 4, 1997, pp. 303–330., doi:10.1207/s15327043hup1004_1.

ci. Sandi, Carmen. "Stress and Cognition." Wiley Interdisciplinary Reviews: Cognitive Science, vol. 4, no. 3, 2013, pp. 245–261., doi:10.1002/wcs.1222.

Managing Design Teams and Workgroups

Cultivate Creative Collaboration

In the middle of the 20th century, the lifetime of a company was circa 65 years; today, it is approximately a decade.[i] The business landscape changes rapidly in the modern world, leading to disruption from multiple angles; organizations worldwide are under pressure to remain relevant and competitive. Thriving in this new ecosystem requires embracing creativity and creative collaboration to keep individuals and organizations inspired, motivated, and visionary. Under these conditions, companies began to hire creative problem-solvers and support innovative ways of working with the intent to push the boundaries of their sectors.

Within psychological research, a typical division separates creativity into scientific and artistic. If scientists help us understand the world, artists create a world that we can understand. Creative problem-solving, which, among other domains, includes architecture, experience, product, and service design, represents a combination of scientific and artistic creativity. Highly creative problem-solvers show greater openness to novel experiences, are attracted

© Andrea Picchi 2022
A. Picchi, *Design Management*, https://doi.org/10.1007/978-1-4842-6954-1_8

to uncertainty and complexity, and display heightened intellectual functions[i] and aesthetic sensibilities.[ii] Creative problem-solvers bear the same relationship to society that dreams have with real life, transforming what we cannot understand into something that, at least, we can begin to see.

From a cognitive perspective, a world is an explored area inside an unexplored territory, a dimension defined by the juxtaposition of known and unknown elements. Creative problem-solvers tend to gravitate on the edge that delineates these two regions, in Jungian terms, on the fine line between the "chaos" and the "order," obeying a compulsive need to expand the domain of the order into the realm of the chaos, redefining other individuals' perception and pushing the frontier of human understanding. For human beings with this kind of personality, acting creatively represents a psychological imperative, not merely one of the possible behavioral options, and frustrating that cognitive need makes their life despondent.

Advancing the proposition that everyone is or can be creative is deceptive. In this chapter, you will learn how to cultivate creative collaboration. You will explore the nature of creativity with its cognitive and social components, analyze its behavioral dynamics, and learn how to promote and support the creative problem-solving process. You will also learn how to lead for creativity and examine the three leadership modalities required to manage a design team like a leader cultivating the cognitive and social preconditions that support the behaviors necessary to face uncertainty and curiously venture into the unknown, unlock experiential learning, and sustain creative collaboration.

The Nature of Creativity

Design is the practice of generating value through problem-solving.[iii] Given a problem domain, defined as a situation for which an immediate solution is unclear, the design practice provides the tools to leverage creativity to create and make choices to fulfill the needs distinguishing that specific domain. Creativity, at its core, is problem-solving with two distinctive characteristics:[iv]

Originality: The degree of uniqueness that a solution exhibits against a problem domain

Appropriateness: The degree of suitability that a solution exhibits against the needs of a problem domain

---

[i] Intellectual functions define the mental operations and resources involved in the acquisition of information with the intent to bend, break, and blend its constituents, developing concepts and hypotheses. Memory, imagination, and empathy can also be considered intellectual functions.

Originality is necessary but not sufficient to ensure effectiveness; a merely original solution might also be completely useless. An original idea must also be appropriate to be considered creative[v] (Figure 8-1). The need for appropriateness represents the foundational discriminator between design, intended as creative problem-solving, and art. Whether original or not, art does not need to be necessarily appropriate. The creative problem-solving process represents the most profound form of behavioral adaptation of the human species, and it directly connects with our evolutionary biological need to survive, forging our existential path into the future.

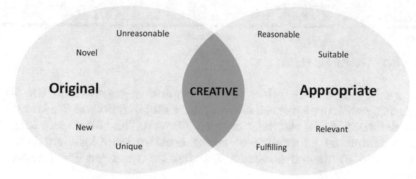

Figure 8-1. The two characteristics of creativity: originality and appropriateness

The Four Dimensions of Creativity

If creativity represents the most profound form of evolutionary adaptation, does this mean that everyone is a designer? Answering this question requires the introduction of a model to conceptualize how creativity can evolve during the life of a human being. This model is adapted from the work of two professors of psychology, distinguished experts in creativity, James Kaufman at the California State University and Ronald Beghetto at Arizona State University.[vi] The Four C Model (Figure 8-2) captures the potential developmental trajectory of creativity as a function of the creative output across a person's life span. During this temporal interval, not every individual necessarily reaches the last developmental stages of the model and becomes an eminent creative creator, and not every person traverses its steps with the same velocity.

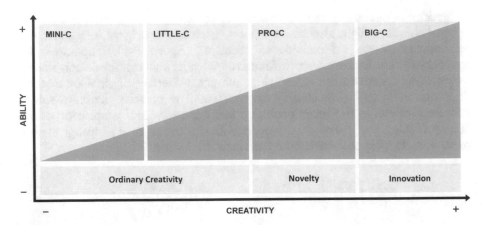

Figure 8-2. The Four C Model

The "mini-c" dimension refers to the empiric engagement in the world directed to construct a *perceptual history* of reality that defines the knowledge and understanding of a particular problem domain. This social space identifies the developmental experience of young creative individuals and students characterized by mental constructions that have not yet been expressed tangibly.

The "little-c" dimension refers to the production of creative solutions to ordinary problems. This social space identifies the nonprofessional experience of individuals predominantly focused on everyday activities.

The "pro-c" dimension refers to the production of creative solutions that generate novelty within a particular problem domain. This space defines the professional experience of designers that produce incremental improvements in their fields and have not reached eminent status.

The "big-c" dimension refers to the production of creative solutions that generate innovation within a particular problem domain. This space identifies the professional experience of designers that produce radical improvements in their fields and have reached eminent status.

With these four dimensions in mind, is everyone able to solve everyday problems creatively? Yes. Is everyone able to solve complex challenges at the intersection of human, technological, and business needs? No. Professional designers invest decades in developing domain-relevant competencies and creativity-relevant abilities such as, among many others, reframing problems, exploring uncertainty, managing ambiguity, and preserving self-discipline during the inevitable moments of psychological discomfort that characterize a complex problem-solving process. The inappropriate statement "everyone is a designer" that commonly attempts to relegate the designer to a facilitator role is fundamentally incorrect; when it's not deliberately used to deceive with the intent to sell services, it demonstrates a superficial and unsophisticated comprehension of the creative process.

The Cognitive Phases of Creativity

Despite being presented using the singular noun, the creative problem-solving "process" is not unitary in any sense.[vii] From a process perspective, it includes preparatory activities such as problem finding and problem framing, but more significantly, from a cognitive viewpoint, it involves operations that underlie different types of brain networks affecting rational and emotional contexts, including, but not limited to, perception, motivation, learning, planning, and communication.[viii]

The creative problem-solving process is based on processing and connecting information in novel ways within the context of a given vision; it never invents something new out of nothing without a preexisting input.[ix] The outcome of the creative process can be innovative, but the single elements composing that given solution, by definition, are not novel.[x]

Even while particular skills can be restricted to local brain regions, the neural activity underpinning the creative problem-solving process is not confined to a specific area of the brain: distant neural networks continuously communicate, negotiate, and cooperate to understand the world and generate efficacious approaches to adapt to the environment.[xi] The creative problem-solving process can be abstracted into three overarching cognitive phases: acquisition, elaboration, and implementation of information.[xii]

As illustrated in Figure 8-3, an instance of the creative problem-solving process initiates when the brain in a state of cognitive equilibrium uses attention to isolate and "acquire" a given set of information, whether from the external environment or accessing the long-term memory.[xiii]

Creative Process

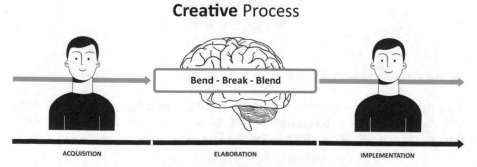

Bend - Break - Blend

ACQUISITION　　　ELABORATION　　　IMPLEMENTATION

Figure 8-3. The cognitive phases of the creative problem-solving process

The acquisition of novel information perturbates the mental models underpinning the current knowledge and beliefs, generating a series of psychological and physiological responses that, among others, include the perception of ambiguity. In that uncertain state of cognitive disequilibrium, the brain is impelled to resolve that condition of intellectual incongruity.[xiv]

The brain uses its resources to "elaborate" the novel information, integrating it with the relevant experience retrieved and the emotions triggered during the acquisition phase. At this point of the process, the brain leverages divergent and convergent thinking to reflect in-action[2] and perform three primary macro operations:[xv]

Bending: The act of restructuring information

Breaking: The act of deconstructing information

Blending: The act of constructing information in novel ways

During this phase, in addition to the social and analytical network, another highly evolved neuronal network system of the brain is engaged to generate mental representation without sensory stimulation from the environment and simulate "what if" and "how might we" scenarios using divergent and convergent thinking:[xvi] the semantic network. Most neurons exhibit a multipurpose nature, and these three interrelated systems share an area of overlaps.[xvii] The semantic network (Figure 8-4) is identified by noncontiguous regions of the brain[xviii] that leverage active attention and working memory to coordinate thinking in accordance with internal goals, forming new neural connections that generate and represent new possibilities.[xix]

Semantic Network

Figure 8-4. The semantic network

This neural network represents and stores conceptual knowledge in the form of neural nodes linked to each other via connections such that closely related concepts are directly connected.[xx] The connections between nodes

---

[2] As introduced in Chapter 4, reflection-in-action is also known as reframing.

(Figure 8-5) can be either *excitatory*, where the activation of one node increases the activation of a proximal node, or *inhibitory*, where the activation of one node decreases the activation of a proximal node.[xxi]

Semantic Network

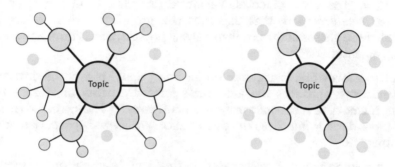

Figure 8-5. The excitatory (left) and inhibitory (right) connections of the semantic network

This neural trait determines the vastity of the activable nodes and, consequently, the amount of knowledge that can be retrieved from the semantic network at any point in time; it tends to be highly developed among designers with a generalist background and distinguishes highly creative problem-solvers.[xxii] Depending on the necessities of your team, when you define its composition, you need to balance accordingly the ability of *generalists* to surface a vast set of diverse possibilities with the capacity of *specialists* to explore them in depth.

The instance of the creative problem-solving process terminates when the brain "implement" the new possibility generated, restoring a state of cognitive equilibrium and creating an output that triggers a response in the external world that sets the preconditions to reflect on-action and unlock experiential learning.[xxiii] While the brain can passively process information, the level of comprehension demanded by the creative problem-solving process requires an active engagement with the world, an act that makes meaningful connections between prior knowledge and novel information within the context of a specific objective.[xxiv] This overarching creative problem-solving experience has the effect of reconfiguring the brain continuously, constituting one of the foundational elements of neuroplasticity.[xxv]

The Social Component of Creativity

In opposition to the deceiving myth of the lone creator, navigating the uncertainty that permeates the creative process is an inherently socially enhanced act.[xxvi] Uncertainty represents a state that our brain is evolved to untangle driven by an innate tendency to continuously increase the probability of surviving. Resolving uncertainty represents a cardinal aspect of short-, medium-, and long-term behavioral adaptations: learning, development, and evolution.

When a group collaborates creatively, establish a brain-to-brain connection that allows the team to navigate the problem-solving process as a joint entity.[xxvii] In that state of collective interaction, the team can leverage a *distributed neural network* that enhances the group's creative performance across all three phases of the problem-solving process: acquisition, elaboration, and implementation.[xxviii]

When the information is acquired during the "input" phase, the distributed neural network enhances the individual's limited capacity of attention and working memory to retain information concurrently.[xxix] The increased richness of the input directly improves the potential originality and appropriateness of the output.[xxx]

When the information is elaborated during the "elaboration" phase, the distributed neural network impacts the creative process on two different levels. Initially, when the data is integrated with the previous experiences, the distributed neural network enhances the ability of the semantic network to form new neural nodes by increasing its capacity to represent and store conceptual knowledge.[xxxi] Subsequently, when the data is elaborated, bending, breaking, and blending information, the distributed neural network enhances the degree of diversity of the conceptual knowledge accessible, increasing the number of potential operations executable.[xxxii]

When the information is implemented during the "output" phase, producing an outcome that triggers a response in the external world, the distributed neural network enhances the collective capacity to implement the envisioned solution.[xxxiii]

When a group explores a problem domain, the distributed neural network with its shared and expanded working memory capacity that characterizes the collective interaction enhances the ability of the group to confront two neurologically distinct peculiarities of uncertainty: risk and ambiguity.[xxxiv] Figure 8-6 illustrates the two areas of the brain activated when a person perceives risk and ambiguity.

Prefrontal Cortex
LATERAL

Parietal Cortex
POSTERIOR

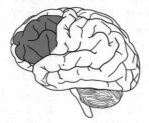 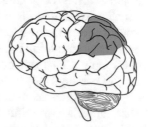

Figure 8-6. The lateral prefrontal cortex and the posterior parietal cortex

Making decisions under a condition of risk activates the posterior parietal cortex of the brain; in contrast, making decisions under a state of ambiguity activates the lateral prefrontal cortex of the brain.[xxxv] Navigating ambiguity not merely represents a unique and more complex form of risk but constitutes a different cognitive effort that leverages distinct neural mechanisms.[xxxvi] Facing risk requires navigating uncertainty with known probabilities, while facing ambiguity requires navigating uncertainty with unknown probabilities.[xxxvii] When a group collaborates creatively in a cognitively joint state, it is more capable of navigating uncertainty in all its forms, increasing the probability of framing a problem effectively and exploring it by producing an abundance of ideas with a high degree of originality and appropriateness.[xxxviii]

The Behavioral Dynamics of Creativity

The ability to be creative is potential in nature because it cannot prescind from the willingness to overcome the psychological struggle necessary to "acquire," "elaborate," and "implement" information. For this reason, this cognitive process is influenced by environmental elements such as trust and psychological safety and individual factors like personality and intrinsic motivation. However, this mental attitude can be promoted and sustained by fostering specific behaviors oriented toward a curious, contributory, and learning-oriented investigation of the unknown.

The Cognitive Path of Least Resistance

Assuming the presence of an environment permeated by trust and psychological safety, the behavioral propensity to demonstrate curiosity, seek contribution, and prioritize learning serve as the foundation of all forms of creative collaboration.[xxxix] Under these conditions, given a problem to solve, the higher the degree of complexity, the higher is the level of trust, respect, and psychological safety necessary to minimize the *cognitive cost of taking action* and promote these three behaviors.[xl]

This dynamic naturally originates from the fact that navigating complex and uncertain challenges demands a significant metabolic commitment that our brain generally tends to evade for evolutionary reasons.[xli] The central nervous system tissue responsible for the higher cognitive functions of our brain, while accounting for only 2% of our body mass, consumes circa 20% of the energy depleted by the resting body.[xlii] Approximately 75% of this total brain energy consumption is directed to neurons, while 25% is utilized by glial cells, their partners that, among other functions, provide neurons with a scaffold for neural development and recovery from neural injury.[xliii] A single cortical neuron utilizes approximately 4.7 billion adenosine triphosphate (ATP)[3] molecules per second in a resting human brain.[xliv]

This necessary and disproportionate energy consumption required by neural mechanisms includes the ones responsible for acquiring, elaborating, and processing information during the creative problem-solving process. The only way for the brain to sustain this strenuous metabolic requirement is to leverage various strategies to modulate the homeostatic regulation of energy.[xlv] This energy management strategy initiates a complex chain of biological adaptations that trigger responses into the psychological and social sphere that can interfere with the creative act.[xlvi] One of these adaptations includes influencing how sensory information is processed to reduce cognitive dissonance by increasing the *psychological threshold to act* and requiring a higher reward system activation.[xlvii] As a consequence of these adaptations, human beings are biased toward perceiving uncertain, risky, and ambiguous situations as less appealing; this dynamic is subconscious and inevitable.[xlviii]

Cultivating creative collaboration means fighting the fundamental and evolutionary trait of the brain to operate efficiently, taking a less cognitively demanding pathway that promotes behaviors oriented toward certainty and clarity and away from uncertainty and ambiguity. Unfavorably, creativity emerges when we are willing to afford the cognitive effort required to diverge from the path of least resistance, exploring the conceptual knowledge represented in our semantic network to manipulate its nodes and form new neural connections that create new ideas and possibilities. In the following section, you will explore the behaviors that you, as a design manager and leader, proactively need to employ to provide continuous support to the creative problem-solving process.

---

[3] Adenosine triphosphate (ATP) is the principal molecule for storing and transferring energy in living cells.

Promote Creative Problem-Solving

Assembling a team does not correspond to creating the precondition necessary to unlock creative collaboration. Assuming the presence of an environment permeated by trust and psychological safety, this type of social interaction requires a specific mental attitude substantiated by three core behaviors: demonstrate curiosity, prioritize learning, and seek contribution[xlix] (Figure 8-7).

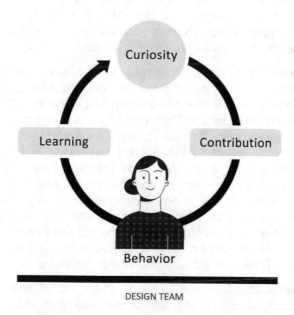

Curiosity

Learning   Contribution

Behavior

DESIGN TEAM

Figure 8-7. The three essential behaviors that promote creative problem-solving

Demonstrate curiosity: Explore challenges driven by a desire to uncover new knowledge using a nonpresumptive inquisitive attitude. Increasing curiosity is directly correlated to an increase in creativity.[l] As Richard Feynman explained during the interview for *The Pleasure of Finding Things Out* filmed in 1981 by BBC, "when we go out and investigate something, we shouldn't pre-decide what we are trying to do except to find out more about it." You can promote curiosity using a two-step strategy: ask questions to open an information gap and then provide intriguing but incomplete information to trigger a relevant reaction. Asking specific questions to an individual that redirects that person's attention to an information gap generates the feeling of deprivation that defines curiosity.[li] Once you have shifted the focus to an information gap, exposing that person to events with an anticipated but unknown resolution promotes behaviors oriented toward the closure of that lack of knowledge.[lii] Curiosity represents a situational-based response that cannot develop without a certain level of understanding of the context surrounding the information gap.[liii] Later in this chapter, you will learn more ways to promote curiosity, exploring how to lead ahead with a strong viewpoint.

Seek contribution: Solve challenges seeking to involve other persons to enhance individual cognitive capacities and mitigate subjective biases. Increasing contribution is directly correlated to an increase in creativity.[liv] You can promote contribution by developing the ability to interact as a collective entity defining social norms intended to optimize the sharing of information, the coordination of behaviors, and the reflection on the effectiveness of comportments.[lv] Later in this chapter, you will learn more ways to promote contribution, exploring how to lead behind through culture.

Prioritize learning: Frame challenges as an opportunity to acquire knowledge instead of a series of tasks to complete. Increasing learning is directly correlated to an increase in creativity.[lvi] As Heraclitus, the pre-Socratic philosopher, reminded us almost 2500 years ago, our only objective should be to expect the unexpected. You can promote a learning-oriented behavior using a strategy based on a distributed approach. Initially, by creating psychological safety, you can establish the preconditions necessary for the manifestation of learning-oriented behaviors. Subsequently, by defining individual and collective objectives based on their capacity to increase knowledge, you can promote learning-oriented behaviors, and by establishing a practice that celebrates *intelligent failure*, you can reinforce them.[lvii] Later in this chapter, you will learn more ways to promote learning, exploring how to lead alongside in direct contact with the team's activities.

When manifested simultaneously, these three interconnected behaviors generate a compound effect that amplifies their individual attitudinal impact.[lviii] Curiosity originates the desire to learn, which consequently creates the propensity to seek contribution, which in turn stimulates curiosity.[lix]

Support Creative Problem-Solving

Investing time and effort in promoting creative problem-solving is a necessary but not sufficient condition to ensure the long-term success of your team. A curious, participatory, and learning-oriented attitude needs proactive, continuous support to be sustained and established as a set of consolidated behaviors. As a design manager and leader, your responsibility is to manifest the following six behaviors to create a supportive environment for creative problem-solving (Figure 8-8).

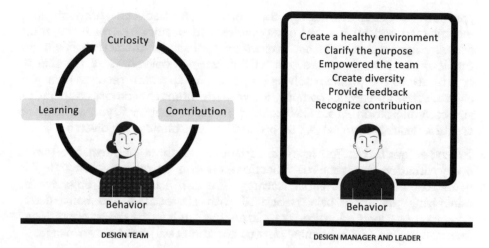

Curiosity

Learning

Contribution

Behavior

DESIGN TEAM

Create a healthy environment
Clarify the purpose
Empowered the team
Create diversity
Provide feedback
Recognize contribution

Behavior

DESIGN MANAGER AND LEADER

Figure 8-8. The essential behaviors (right) that support creative problem-solving

Create a healthy environment: You have to accomplish high performances sustainably. You can achieve this objective from two different angles. On your side, you have to develop your abilities by balancing results-oriented and social-oriented skills. On the group's side, you have to support the team members in developing their best selves by providing them the opportunity to deliberately design their professional lives in harmony with their personal needs. In Chapter 9, you will learn to manage stress to maximize achievements and avoid exhaustion.

Clarify the purpose: You have to embody why the design team exists beyond making profits and articulate how that resolution translates in operational terms. You can achieve that objective by connecting the organizational purpose and vision to the design challenges that the team needs to face. Later in this chapter, you will learn to frame problems making that connection visible.

Empower the team: You have to provide what is necessary for the team to be autonomous and take ownership of their contribution. You can achieve this objective by supporting a distributed decision-preparation and decision-making system. On that premise, your active participation in the design operations needs to be situational according to the group's needs switching from leading ahead with a strong viewpoint, leading alongside in direct contact with the team's activities, and leading behind through culture. You will explore these three leadership modalities later in this chapter.

Create diversity: You have to include and protect different perspectives as an integral part of the team's collective intelligence. You can achieve this objective from two different angles. From the inside, you have to ensure that all designers are able to contribute to the problem-solving process. This approach requires managing the interaction between predominantly extroverted and introverted personalities and promoting creative tension, allowing frictions of ideas but preventing conflicts of characters. From the outside, you have to ensure that all nondesigner collaborators are able to contribute to the problem-solving process. This approach requires managing the collective intelligence composition with the intent to promote cooperation and communication across different disciplines. In Chapter 9, you will learn to create a healthy, sustainable, and productive environment for diversity.

Provide feedback: You have to communicate about performance during every situation that requires reflection-in-action or reflection-on-action in order to unlock experiential learning. You can achieve this objective by minimizing the social pain associated with appreciative and constructive conversations, institutionalizing candid discussions into the design operations.[4] Revisit Chapter 7 to reexamine how to communicate about performance.

Recognize contribution: You have to demonstrate gratitude for the individual and collective participation of the group. You can achieve this objective by acting on two different levels. Privately, you can emphasize the impact of the individual *superpowers* present within the team. Publicly, you can acknowledge the effect that the amalgamation of these abilities had on the team and organizational success. In Chapter 9, you will learn to celebrate individual and collective contributions.

Previously in the book, we already introduced many of the abilities, processes, and events necessary to instantiate these six behaviors. Later in this chapter, instead, you will learn how to create and use rituals to initiate or support these six behaviors and establish or reinforce the beliefs underpinning them. In the next section, you will explore how to leverage different leadership modalities to sustain curiosity, contribution, and learning, minimizing the cognitive cost of diverging from the path of least resistance and unlocking your team's creative potential.

Lead for Creativity

Unlocking an organization's creative potential represents a competitive imperative in today's business landscape. This effort requires the capacity to diverge from the cognitive path of least resistance bending, breaking, and blending information into novel audacious concepts during the resolution of problems. As a person in charge of a team, you have three interconnected behavioral strategies at your disposal that you can concurrently leverage to sustain creative collaboration.

Lead ahead: Foster curiosity framing problems.

---

[4] Revisit Chapter 7 to review how to communicate about performance.

Lead alongside: Support learning guiding experiments.

Lead behind: Encourage contribution creating rituals.

Each strategy is connected with a precise modality of creative leadership. Each modality instantiates a specific mindset and a set of behaviors that allows you to guide your team through uncertainty collaboratively, leveraging curiosity to conduct your team through the creative problem-solving process in accordance with the particular needs of the situation.

Lead Ahead

From this perspective, you lead with a strong viewpoint articulating a destination that motivates the team to follow you intellectually and emotionally. In this role, you adopt the mindset of an "explorer" to frame challenges, take informed risks, and support the attitude necessary to face uncertainty and curiously venture into the unknown.

When: This modality of leadership is typically required when you need to indicate a direction of travel at the beginning of a journey or when an uncertain situation requires bold decisions during the process.

Frame Problems

A crucial aspect of leading with a strong viewpoint is framing problems purposefully; this act represents a core activity of the problem-solving process (Figure 8-9) and a pillar of a culture based on continuous learning.

Problem-Solving Process

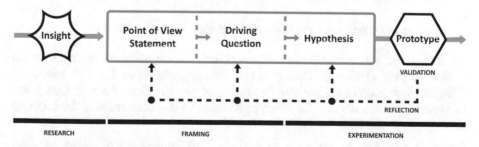

Figure 8-9. Framing within the broader context of the problem-solving process

A problem frame is the atomic unit of comprehension: an actualization of the design challenge. Framing a problem is a process composed of three steps:

1. Define a point of view statement.

2. Identify a driving question.

3. Invite reflection.

Step 1: Define a point of view statement. You consider the mission of the company, and you describe the challenge by crafting a statement that specifies a type of user, a relevant user's need, and a transformative direction without dictating a particular solution. The following structure can help you to describe a challenge effectively:

[user] needs to [need] because [insight]

This affirmation expresses distilled findings in an intriguing format and defines the perspective through which a team attacks a problem. After the first iteration, you can critically analyze the point of view statement (POV) by asking yourself the following four questions adapted from the original work of David Labaree of the Stanford School of Education:

// Perspective: What is the point?

Is the POV user-centered, need-based, and insight-driven?

// Relevance: Who says?

Is the POV supported by distilled findings from users?

// Value: What is new?

Does the POV describe something valuable for the user?

// Significance: Who cares?

Does the POV describe something worth doing?

At this point, you can revisit and eventually rephrase the statement until it describes a challenge that can unlock tangible value for the users. In Chapter 9, you will explore the five types of design value. An effective point of view statement allows the team to ideate in a directed manner by creating driving questions based on that perspective.

Step 2: Identify a driving question. You consider the point of view statement that describes the challenge, and you decompose it into one or more questions that can be leveraged to unfold the solution space. This type of interrogation is adequately broad to generate a wide range of solutions but sufficiently narrow to impose the necessary boundaries to stimulate creative thinking. Driving questions are ambitious, actionable, invite participation, and promote curiosity, encouraging a nonpresumptive inquisitive attitude. Crafting driving questions is a deliberate practice that requires time to develop, but it represents a central competence for a design manager who manages like a leader. The following are five rules that can help you to craft this form of creative prompt.

Rule 1: The question must not include the solution. A question needs to generate a significant amount of radical concepts, typically within 10–20 ideas:

// No

How might we launch a campaign and app to inspire teens to adopt healthy eating habits?

// Yes

How might we leverage technology to inspire teens to adopt healthy eating habits?

Rule 2: The question must be ambitious and inspiring. A question needs to achieve a balance between being broad and inspiring but not too abstract and being narrow and actionable but not too directive:

// No

How might we grow our business?

// Yes

How might we extend our position in [country] to products beyond [market]?

The following two rules are often applied after the first iteration of the problem-solving process when you have expanded your knowledge of the problem domain, and you are in a position to revisit and eventually rephrase the driving question. Typically, applying the following two rules when the problem domain still presents a high level of ambiguity may narrow down the possible options prematurely, making the driving question too directive.

Rule 3: The question could include the recipient and unfulfilled need. When the problem domain presents a relatively low level of ambiguity, specifying the recipients in demographical or behavioral terms, and indicating what they need, increases the effectiveness of the question:

// No

How might we leverage technology to help families?

// Yes

How might we leverage technology to help parents and children to communicate and feel safe?

Rule 4: The question could include the specific segment of the journey underpinning the problem. When the problem domain presents a relatively low level of ambiguity, specifying the particular moment that generates the unfulfilled need increases the effectiveness of the question:

// No

How might we make air travel more convenient for single parents traveling with children?

// Yes

How might we make boarding an aircraft more convenient for single parents traveling with children?

Step 3: Invite reflection. You consider the point of view statement and the different driving questions created, and you answer a set of questions aimed to expand your perspective and improve your decision-preparation and decision-making process:

// Consider the design perspective

Is the point of view still valid?

// Consider the organizational alignment

Is the driving question aligned with the current organization's priorities?

Is the driving question aligned with my line manager's priorities?

Is the driving question aligned with my team's priorities?

// Consider the desired outcome

Do we want to explore a new concept?

Do we want to scale or release an already validated concept?

// Consider the necessary resources

Does the design team possess the skills required to solve the problem?

Does the design team need input from other groups? If yes, when?

Describing a challenge and framing it into a problem that can be explored is crucial to cultivating creative collaboration. The more compelling is the result of this activity, the more extensive is the potential access to the brain's semantic network that the designers in your team can perform.[lx]

Efficacious problem framing allows individuals with adequate skills and knowledge to perform a vast number of neural connections in their brain, increasing the number of potentially original and appropriate solutions achievable by the group.[lxi] The act of framing problems is intrinsically connected with creating stories. Crafting a compelling narrative helps you connect information and emotions, provide meaning to problems, and mobilize individuals, engaging them in a collaborative effort.

Create Stories

Leading ahead with a strong viewpoint requires a supportive narrative to mobilize individuals rationally and emotionally. Storytelling allows you to engage the attention of other individuals and project a vision into which they can immerse and experience it directly. Listening to a compelling narrative creates a hormonal profile that, among other things, increases empathy, enhances information retention, and promotes collaboration.[lxii]

Creating stories also achieves another crucial objective that dramatically affects the social cognitive experience of your team members.[lxiii] From a cognitive perspective, human beings systematize experiences in a narrative form, where the mnemonic entities are organized into temporal narrative frames.[lxiv] This process is undeviatingly correlated with the development of our sense of self, defined as the feeling of identity, uniqueness, and self-direction.[lxv] We already explored this correlation in Chapter 7, introducing the practice of coaching and mentioning that one way with which we determine our identity is via the meanings we attribute to experiences and situations.[lxvi]

Creating narrative frames represents such a pivotal element of our cognitive existence that when a certain degree of inadequacy of communication and information generates narrative ambiguity within the team, individuals address this absence of data fabricating reasons to interpret that given situation.[lxvii] Creating stories support your team members in constructing a coherent narrative structure that serves as the cognitive foundation of their professional experience within the group. This experience fosters intrinsic motivation and drives the perception of "meaning of work" and "impact of work."[lxviii]

Notwithstanding, while being a crucial element of our cognitive existence, not all coherent narratives have the ability to compel the audience and promote a mental shift that triggers a form of behavioral engagement. When the objective is to mobilize individuals, a compelling story must develop at the intersection of an adequate level of rational and emotional information (Figure 8-10).

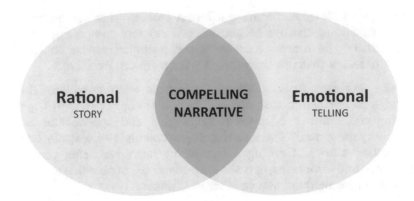

Figure 8-10. The two informational elements of storytelling: rational and emotional

When a narrative resonates with the audience, the brains of the speaker and the listener exhibit joint, temporally coupled, response patterns.[lxix] The speaker's brain activities generate a sound wave, the speech, that in turn influences the brain responses in the listener and brings them into alignment.[lxx]

This process is called *brain coupling*; the sharpest is the speaker-to-listener brain pattern alignment, and the strongest is the mutual sense of comprehension.[lxxi] This cognitive dynamic forms an interpersonal neural substrate that connects the individuals engaged in the conversation and conveys the emotion-eliciting context attached to the story from the speaker to the listener.[lxxii] Cognitively, the storytelling experience is a single act performed by two brains, represented by the brain-to-brain connection between the speaker and the audience.[lxxiii]

Narrative Structure

Storytelling, the activity of telling, writing, and visually representing stories, is composed of two foundational elements: the story and the telling.

The story: The beats of information composing the narrative. This element captures the content of the message. The story part defines the experiences of the protagonist instantiated into a set of episodes called *beats*. A compelling story always connects an objective to a challenge; you can capture this relationship using the following structure:

I need to tell a story about [objective], so that I can [challenge].

I need to tell a story about offering feedback, so that I can initiate a communication about performance.

Framing a story forces you to refine your thinking, simplify your message to its essence, and make the underlying need evident. After the first iteration, ask yourself the following questions to refine and eventually rephrase the statement:

Does my idea capture the problem that I'm trying to solve?

Is the idea clear and concise?

Is the idea compelling?

Once you have framed the story by connecting the objective to the challenge, you must consider how the narrative evolves. If the beats composing the episodes of the "story" represent the dots of your message, the "telling" part is the line that connects those dots into a meaningful pattern.

The telling: The sequence of the beats of information composing the narrative. This element captures the format of the message. The telling part defines the series of episodes punctuating the journey of the protagonist; this sequence is called the *story arc*. There are five primary types of story arcs.

Rising action. Use this story arc (Figure 8-11) if your presentation has an evident problem and solution. Beginning from an undesired state, during this story arc, the beats will raise the emotional level to a positive peak to subsequently decrease to a lower state that points to the successful conclusion of the story. This format represents the most commonly used story arc which is typically used, often unconsciously, to retrospectively present a design iteration introducing the challenge, showing the insights, explaining the solution, and then discussing the future evolution of the work.

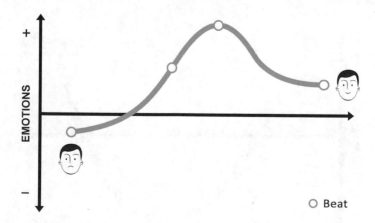

Figure 8-11. The rising action story arc

Rags to riches. Use this story arc (Figure 8-12) if you need to communicate a message of hope and optimism. Beginning from an undesired state, during this story arc, the beats will raise the emotional level, creating a positive tendency that identifies the successful conclusion of the narrative. This type of story arc can also be used reversely, beginning from a desirable state and creating a negative tendency that points to the unsuccessful conclusion of the narrative.

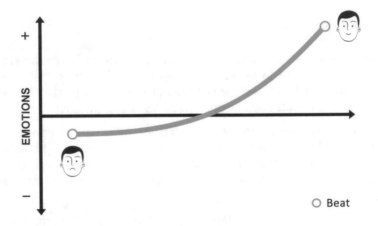

Figure 8-12. The rags to riches story arc

Hero's journey. Use this story arc (Figure 8-13) if you need to promote the adoption of a new idea. Beginning from a neutral state, during this story arc, the beats will decrease the emotional level to a negative peak to subsequently increase it, creating a positive tendency that points to the successful conclusion of the narrative.

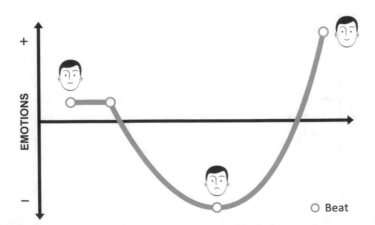

Figure 8-13. The hero's journey story arc

Cinderella story. Use this story arc (Figure 8-14) if you need to emphasize perseverance despite the presence of obstacles and experience failures during the journey. Beginning from an undesired state, during this story arc, the beats will increase the emotional level to a positive point to then fall into a lower state compared to the initial one and subsequently increase it, creating a positive tendency that identifies the successful conclusion of the narrative.

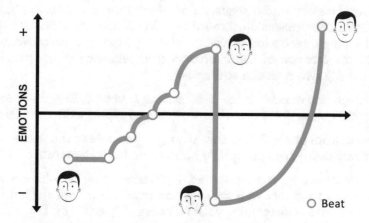

Figure 8-14. The cinderella story arc

In-medias res. Use this story arc (Figure 8-15) if you need to engage the audience from the earliest moments. Beginning from a sudden and unspecified state, during this story arc, the beats will decrease the emotional level articulating the backstory to a lower state to subsequently increase it, creating a positive tendency that points to the successful conclusion of the narrative.

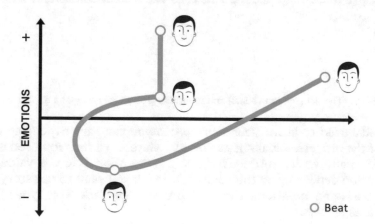

Figure 8-15. The in-media res story arc

Each story arc leverages a specific sequence of information that elicits a particular combination of emotions during the narrative. Counterintuitively, when emotions are experienced within a healthy range, they do not interfere with critical thinking.[lxxiv] Cognition and emotions represent two neurophysiologically and psychologically connected entities in our brain:[lxxv] cognition allows us to understand the world, while emotions enable us to interpret that knowledge and act accordingly. In that regard, storytelling with an effectual interconnected sequence of information and emotions enhances the listener's comprehension stimulating areas of the brain that are typically engaged during direct experiences.[lxxvi] When you craft a narrative, you can approach the creation of the interconnected sequence of information and emotions using two opposite strategies.

Bottom-up approach: With this strategy, you define the beats and subsequently select the appropriate story arc to support the narrative.

Top-down approach: With this strategy, you select the story arc and subsequently define the appropriate beats to deliver the narrative.

Irrespective of the approach adopted, a physical or digital whiteboard is an effective tool to support your thinking and manipulate both the information and the emotions underpinning your narrative (Figure 8-16). Once you have created the first draft of your story, you need to prototype it, test it, and reflect on its effectiveness to refine the message if necessary.

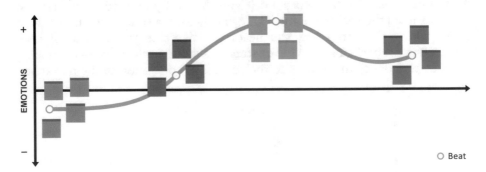

Figure 8-16. The information (beats) and emotions (story arc) story of a narrative

The media used to share your story prototype can vary depending on the stage of the process. Initially, it can be the whiteboard that you used to craft the first draft, while, subsequently, it can be your voice, a video, or a presentation depending on the context in which you want to use storytelling. There are some questions that can prompt feedback from your testing audience as follows:

In your opinion, what was the core message of the story?

Did you find this message compelling?

Did you find this story motivating?

If yes, what moved or motivated you?

If not, when did you lose interest?

While you iterate on the content of your message, remain open to the possibility to modify the current arc of your story or even create a novel one if that is what you need to optimize the interplay between the information and emotions underpinning your narrative. What is central is producing a positive impact on the audience, not the format used to deliver the message.

Framing challenges and creating stories is crucial to effectively lead ahead with an "explorer" mindset, especially at the beginning of a journey or every time the team traverses an uncertain situation that requires bold decisions. Once the team has developed a clear understanding of the direction of travel, you need to switch to a different modality to support the creative collaboration directly.

Lead Alongside

From this perspective, you lead in direct contact with your team's activities to remain in touch with the group's needs. In this role, you adopt the mindset of a "coach" to guide experimentation and assist the group and offer guidance during challenging moments to preserve the preconditions necessary to unlock experiential learning.

When: This modality of leadership is typically required when the team can benefit from your expertise or a novel perspective during the problem-solving activities.

The Creative Process

Cultivating creative collaboration assisting your team to go beyond incremental changes and unlock original and appropriate solutions requires a disciplined process. Counterintuitively, creative problem-solvers reach their maximum effectiveness when they have a structure in place to support their effort. Senior design practitioners interiorize this process, developing its elements into a mindset that allows them to operate in an adaptable way that may appear chaotic and unpredictable from outside, but that instead is highly disciplined. This behavior substantiates one of the core principles of human-centered design: be mindful of the process.

The problem-solving process is intentionally designed to leverage creative collaboration integrating diverse perspectives and eliminating barriers between individuals so that the group can create solutions that no individual contributor could have envisioned in isolation. One prerequisite to integrating diverse perspectives is to unite the group around two interconnected elements discussed in the previous section. The first is the *point of view statement* that describes the challenge to overcome, and the second is the *driving question* that frames that situation into an actionable form that can guide the collaboration through the two distinct phases of the creative process: divergence and convergence (Figure 8-17).

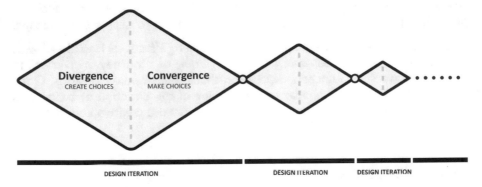

Figure 8-17. The two phases of the creative process

The divergent phase focuses on creating choices, while the convergent phase concentrates on making choices. As the group traverses these two reasoning phases iteratively during the creative problem-solving process, the designers switch their attitude embracing a divergent-oriented and convergent-oriented mindset to optimize the cooperation. Completing multiple iterations, the team reduces the ambiguity that permeates the problem domain and increases the level of confidence attached to the current solution.

Your responsibility as a design manager and leader is to initiate the mental shift from one thinking modality to the next and guide the collective energy through the moments of excitement, tension, disagreement, and discomfort that will inevitably arise during the process. In Chapter 9, you will learn how to optimize creative meetings and other types of discussions from an operational perspective.

Lead Divergence

Divergence aims to create choices, generating a set of possibilities that, within the problem domain, expand the focus of the problem-solving conversation and explore multiple directions. It is the moment when the group leverages analysis to navigate the ambiguity of a problem domain with a "what if" attitude exploring its aspects in different directions, making novel connections, reframing old ones, and taking informed risks to create a novel perspective.

Sustaining divergent thinking requires optimism and perseverance to manage the central characteristic of this phase: uncertainty. Divergence can include activities such as researching the desirability of a concept and creating investigational prototypes,[5] but the most typical practice is brainstorming.

During brainstorming, the team leverages the brain's semantic network to trace interesting leads, exploring new connections and taking informed risks to investigate novel ideas. Some of these explorations will bring the group to an impasse that is impossible to anticipate; therefore, it is crucial that you lead alongside encouraging the experimentation by nurturing curiosity and optimism.

The ideation phase that characterizes a brainstorming session has an approximate duration of 30–60 minutes, which represents an average estimation that accounts for all the different factors that affect sustained attention.[lxxvii] Figure 8-18 illustrates the typical ideation arc that distinguishes an ideation session.

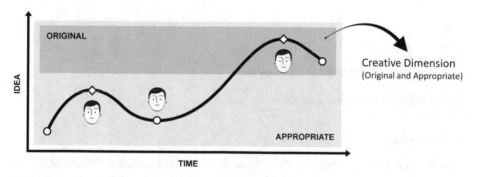

Figure 8-18. The ideation arc

The team can initially produce many "appropriate" ideas; these concepts typically do not possess an "original" component. After this initial peak, the production decreases. At this point, the group tends to feel that all the creative energy has been dissipated. Teams that are not familiar with the creative process typically conclude the session at this point, failing to produce any original idea. In this scenario, your role as a design manager and leader is to push the group beyond this drop point toward the second peak, generating "appropriate" and "original" concepts with a high *creative factor*.

---

[5] An investigational prototype materializes a concept with a high degree of abstraction, with the intent to promote reflection-in-action.

The most effective approach to support this crucial transition is encouraging the group to explore unreasonable ideas. Generating extremely original ideas with a low degree of appropriateness represents the most effective approach to craft creative concepts. Dictated by the excitatory and inhibitory nature of the connections of the semantic network, in practice, it is more effective generating an extremely original idea with unreasonable characteristics and bringing that concept down to the adequate balance between originality and appropriateness than trying to achieve the same result beginning from an extremely appropriate idea.[lxxviii] Creative ideas are more proximate to unreasonable concepts than reasonable ones (Figure 8-19). Provoking divergent thinking during a collaborative session demands you a meticulous approach across five elements.

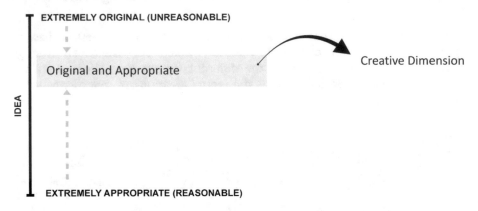

Figure 8-19. The creative dimension

Environment

The environment represents the physical or virtual space for the session. The environment needs to support the generation and circulation of ideas; this configuration enhances the team's cognitive ability to analyze the available information.[lxxix] A physical or digital whiteboard is a typical tool used to support the circulation of ideas without erasing or removing them. Additionally, it is also necessary to saturate the space with relevant visuals, research insights, and other data to help the team adopt the appropriate attitude. In a physical configuration, also consider sticky notes, blank paper, markers, and pens.

Constraints

The constraints represent the limitations or restrictions for the session. Constraints are crucial to guide divergence during a creative collaboration because they focus and drive the team's explorative thinking. Without restrictions, the group can disperse their cognitive resources by analyzing an excessive number of directions at once, resulting in an unproductive session. An efficacious constraint to guide the analysis of information is "time" because it prevents your team members from overanalyzing concepts switching to a convergent attitude that tends to judge ideas. Another effectual constraint is a "driving question" because it allows the team to frame a challenge into an explorable problem domain. As the exploration progress and knowledge increase, the team can refine or rephrase that given driving question to reflect the current level of comprehension of the problem.

Choreography

The choreography represents the objective, the desired outcome, and the potential obstacles of the session. Choreography is crucial to building energy, excitement, and momentum during the analysis of information. Always open the session with a creative warm-up that engages the brain and encourages participation. You can introduce an exercise like the *30 Circles* or ask provoking questions with a different degree of correlation with the session, such as the following examples:

// Generic

What are all the things you can do with a shoelace?

How would life be different if there were no electricity?

//Specific

Are other industries facing a similar challenge?

How do they approach it, and what can we learn from them?

After the warm-up, introduce the point of view statement and the driving question and connect it with the objective, the desired outcome, and the potential obstacles of the session. If the session needs to push the boundaries of creativity, you can also present some provocative ideas to encourage the group to release inhibitions and think radically.

Allow the team to spend the opening minutes silently to capture an initial set of ideas and then invite the group to share those concepts to initiate a collaborative conversation. A good session typically generates between 30 and 50 possibilities. Avoid the dispersion of ideas encouraging everyone to capture every consideration in a format that can be shared immediately within the digital or physical environment used for the session. Everyone in the team needs to be able to see those ideas and build on them. After the group has released all the creative energy, it's essential to provide closure to the session, grouping similar ideas into themes and identifying patterns that can be considered during the convergent phase.

Guidance

The guidance represents the management of the creative tension with the intent to activate, sustain, and direct that energy during the session to maintain an optimal level of engagement. Guidance is necessary to proactively maintain an adequate level of cognitive engagement to transform the generation and reception of ideas into a contagious experience that enhances creativity. When you sense that this optimal level drops below a certain point where it commences to affect productivity, intervene intentionally to nudge the creative energy. This situation is a typical dynamic that you need to consider before the session, preparing provocative questions that can be used to challenge the group in that circumstance; the following are four examples that can stimulate the group's curiosity and creativity:

What are five ideas that [competitor] would have?

What would you do if you had only $10 to make a solution?

What would you do if you, instead, had $10 million?

What if we were designing only for [extreme archetype]?

If the team trespass the problem domain veering their thinking in unproductive directions, guide them back by pointing the aspects of specific ideas that you consider potentially valuable by saying something similar to the following:

I love that this idea [characteristic].

What else can we think that [attribute]?

The deeper is your social connection with the team and your comprehension of the superpowers and shadow sides present within the group, the easier it is for you to manage the collective intelligence during these moments and maintain an adequate level of creative energy during the session.

Impediments

The impediments represent the obstacle that can interfere with divergent thinking during the session. Impediments can take different forms during a divergent session, from social to political and technical; the following are the three recurrent ones.

Obstacle 1: Safe ideas. If you sense that the team refrains from thinking radically and producing conservative concepts, stimulate their curiosity by introducing a radical view and asking the group to build on it:

> *If [constraint] was no issue, how would you solve the problem?*

> *What customer insights can we leverage to inspire our thinking?*

Obstacle 2: Hastened judgments. If you sense that the team evaluates an idea prematurely, whether with their body language or words, remind everyone that they must defer judgment during divergence. In this scenario, it is imperative for you to lead alongside using a neutral communication style and phrases such as "this is an interesting direction" instead of "this is a great idea."

Obstacle 3: Early convergence. If you sense that the team is unable to sustain divergent thinking long enough to explore the problem domain comprehensively, tending to manifest premature signs of convergent thinking, remind everyone that divergence concentrates on creating choices while making them is a prerogative of convergence. You can also force a mindset shift back to divergent thinking by asking questions similar to the following:

> *What are the solutions that our competitors are too scared to pursue?*

Proper divergence happens when you guide the team through the entire arc of ideation, reaching exciting, unreasonable, and unexpected places. At the end of the session, to evaluate the quality of the output, answer the following three questions:

> *Did we push the ideation into the creative dimensions?*

> *Did we produce surprising or unexpected ideas?*

> *Did we generate a significant variation of ideas?*

Sometimes, you may feel that the outcome of the session is incomplete and that the team needs to spend more time diverging. In those cases, use the output of the first brainstorming to plan and inform an additional divergent session. If you feel that the outcome of the divergent phase can be, instead, considered satisfactory, prepare the environment for the convergent phase.

Lead Convergence

Convergence aims to make choices, reducing the number of possibilities available to narrow the focus of the problem-solving conversation and give direction to the work. It is the moment when the group leverages synthesis to comprehend the essence of the ideas and reflectively decide to keep, and possibly integrate, the most promising elements among all the possibilities explored during the divergent phase.

Sustaining convergent thinking requires rigor and meticulousness to manage the central characteristic of this phase: complexity. Convergence is not a mere action of reduction, but a creative act that generates novel connections between the promising elements of the available possibilities (Figure 8-20); this creative trait distinguishes convergence from decision-making.

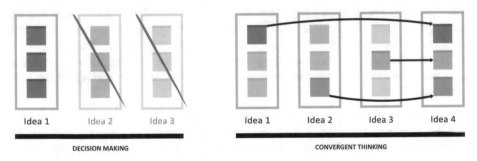

| Idea 1 | Idea 2 | Idea 3 | | Idea 1 | Idea 2 | Idea 3 | Idea 4 |

DECISION MAKING   CONVERGENT THINKING

Figure 8-20. The difference between decision-making and convergent thinking

Convergence can include activities such as conducting research on feasibility and viability, defining principles to guide the synthesis, and creating sacrificial prototypes.[6] Provoking convergent thinking during a collaborative session demands you a meticulous approach across five elements.

Environment

The environment represents the physical or virtual space for the session. The environment needs to support the visualization of all ideas simultaneously; this configuration enhances the team's cognitive ability to synthesize the available information.[lxxx] If you used a digital or physical whiteboard during the divergent phase, this tool needs to be present and central during the conversation. In a physical configuration, also consider sticky dots or other material used to vote ideas. In preparation for the session, saturate the space with relevant themes generated at the end of the divergent phase and the eventual principles defined to drive the decisional process.

---

[6] A sacrificial prototype materializes a concept with a low degree of confidence in originality and appropriateness, with the intent to promote reflection-in-action.

Constraints

The constraints represent the limitations or restrictions for the session. Constraints are crucial to guide convergent thinking during a creative collaboration because they enable the team to focus their decisions and drive their thinking. Without restrictions, the group can disperse their cognitive resources synthesizing the available directions using the erroneous criteria, resulting in an unproductive session. An efficacious constraint to guide the synthesis of information is "time" because it prevents your team members from preparing and making decisions considering the incorrect time frame; the following are the three criteria that can implement this limitation:

Now: Valuable immediately

Near: Valuable in the near future

Far: Valuable in the distant future

Another effectual constraint is "value" because it allows the team to prepare and make decisions considering the relevant perspective; the following are the three questions that can implement this limitation:

Does this generate tangible value for the user?

Does this fortify the relationship with the users?

Does this generate revenue for the company?

In Chapter 9, you will explore a framework that articulates the five main types of value that you can generate using human-centered design.

Choreography

The choreography represents the objective, the desired outcome, and the potential obstacles of the session. Choreography is crucial to create clarity, alignment, and momentum during the synthesis of information. Plan the session with the following three desired outcomes in mind.

Outcome 1: Generate a heat map. Use dot voting or poker chips to visually represent where the team's decisional direction is gravitating.

Outcome 2: Define desirable and undesirable concepts. Decide which idea to keep and which to discard.

Outcome 3: Understand the next steps. Decide the design direction, whether creating a prototype, conducting more research, or initiating another divergent thinking phase.

With these three objectives in mind, ensure to engage the appropriate stakeholders in the conversation and guide the session through three steps.

Step 1: Begin by considering the available ideas and themes.

Step 2: Elevate some concepts and deprioritize others.

Step 3: Conclude by evaluating the ideas and themes.

After the group has released all the creative energy, it's essential to provide closure to the session, assigning actions to the team members following the agreed design direction.

Guidance

The guidance represents the management of the creative tension with the intent to understand the options available adequately by assigning them sufficient airtime to uncover their desirable and undesirable elements. Guidance is necessary to intentionally allocate an equal amount of cognitive resources to every concept available to ensure that the synthesis of ideas is a comprehensive experience. When you sense that the creative tension begins to arise, intervene intentionally to guide the creative energy permeating the conversation using three strategies.

Strategy 1: Hold the conversation. Maintain a neutral approach by controlling your biases and emotions to preserve a psychologically safe climate and the willingness to share and collaborate. When you disagree, avoid expressions such as "I don't think we can do that" in favor of phrases like "tell me more about why we should do that."

Strategy 2: Be a provocateur. Stimulate the scrutiny of unconscious cognitive biases that can skew judgment and comprehension challenging assumptions. When you want to provoke reflection, avoid expressions such as "think deeper about that" in favor of phrases like "here is a provocation, think about the journey, what would make customers desire to use this service over and over again?".

Strategy 3: Focus on the "why." Investigate the core reasons behind individual and collective decisional orientations. When you want to investigate the rational or irrational factors that motivated a decision, avoid expressions such as "tell me which idea you prefer" in favor of phrases like "why do you like that idea?".

Impediments

The impediments represent the obstacle that can interfere with convergent thinking during the session. Impediments can take different forms during a convergent session, from social to political and technical; the following are the three recurrent ones.

Obstacle 1: Persistent divergence. If you feel that the team struggles to switch from analysis to synthesis, keep expanding certain concepts instead of narrowing the ideas available, and introduce a *parking slot* where the team can place concepts that cannot currently consider without losing them. This approach enhances the ability of the group to disengage cognitively and reset their mindset.[lxxxi]

Obstacle 2: Biased convergence. If you feel that the team is affected by the role power projected by specific individuals during decisional moments, remove the connection between ideas and their creators using numbers to categorize the different options: idea 1, idea 2, idea n. In a physically co-located session, you also can consider rewriting the sticky notes using your calligraphy style to anonymize them. You can also push the team beyond this tendency, inviting the group to reevaluate their ideas, saying something similar to the following statement:

For each idea, let's discuss desirability, viability, and feasibility.

Obstacle 3: Decisional conformity. If you feel that the team manifests a strong concurrence-seeking tendency that interferes with critical thinking, you can leverage selective vulnerability to introduce alternative viewpoints during the discussion that can guide the team to analyze the positive and negative aspects of all the available possibilities. The most effective approach to minimizing decisional conformity is to increase the levels of psychological safety,[lxxxii] but, additionally, you can also push the team beyond this tendency by asking questions that invite reflection-on-action:

Which idea feels most original and disruptive?

Which option would be more delightful for our consumers?

Which option would earn us the most recognition in five years?

Creating a solid structure that supports a disciplined process is necessary to effectively lead alongside your team with a "coach" mindset. In the next section, you will explore one of the crucial activities of this role, guiding experiments.

Guide Experiments

A crucial aspect of leading alongside is exploring problems purposefully; this act represents a core activity of the problem-solving process (Figure 8-21) and a pillar of a culture based on continuous learning. An experiment is the atomic unit of exploration: a formalization of the design intent. Integrating experimentation as a form of learning into the design operations requires the respect of a simple principle: given a problem to solve, never undertake a course of action if you do not expect to increase your comprehension.

Problem-Solving Process

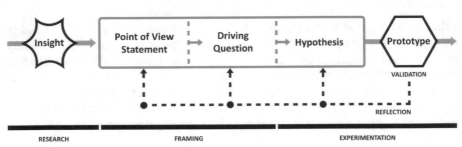

Figure 8-21. Experimentation within the broader context of the problem-solving process

Experiments can be part of the divergent phase as explorative prototypes to create choices, part of the convergent phase as a decisional prototype to make choices, or used as a testing prototype to validate hypotheses at the end of a design iteration. Irrespective of the details of its implementation, this form of cognitive exploration represents a step forward into the ambiguous dimension of the unknown, a necessary act of investigation aimed to understand the desirability, feasibility, and viability of an idea.

From the perspective of a design manager and leader, experiments are used in any circumstance where the team needs to generate, maintain, or recover momentum during the problem-solving process or when there is the need to decrease the level of risk attached to a given design decision. Guiding this form of exploration allows you to place small bets in the form of experiments intended to investigate concepts and decide where and how to allocate your resources.

Experimentation is cognitively demanding and operationally challenging for the team. For this reason, during the design iteration, the team requires your presence and active engagement at three specific moments that correspond to the three overarching phases of a formal experiential learning approach:

1. Create the hypothesis.
2. Test the hypothesis.
3. Integrate the learnings.

Moment 1: At the beginning. At this point of the process, you collaborate with the team to create the hypothesis statement that defines the supposition investigated with the experiment. A hypothesis statement captures the "why" and "what" of a given design perspective and must always derive a research insight:

// Insight

We have observed that [event] is causing [adverse effects] for [archetype].

// Hypothesis

We believe that [idea] will enable [archetype] to [goal].

The natural evolution of a hypothesis is a prototype that serves the purpose of materializing its design perspective. When you create a hypothesis to inform investigational and sacrificial prototypes during the divergent and convergent phases, the objective is to investigate the nature of an idea. When you create a hypothesis to inform a prototype that instead materializes a potential solution generated after the divergent and convergent phases, your objective is to investigate the application of an idea. In that case, you need to expand the structures of the hypothesis with an additional statement to specify which signal determines the acceptance or rejection of your thesis:

// Hypothesis

We believe that [idea] will enable [archetype] to [goal].

We will know we have succeeded when [measurable signal].

At the beginning of the experimentation, you have to help the team to create the hypothesis and even force the tempo to generate and maintain creative traction when necessary using different approaches. If you sense that the team is hesitant to approach a problem and struggle to analyze its aspects, you can help the group by asking questions aimed to identify crucial elements of the problem domain and increase focus:

What is most critical?

Which is the riskiest conjecture?

Which is the weaker design element?

If you sense that the team is hesitant to approach a problem and struggle to unlock learning, help the group by asking questions aimed to force the tempo. All these types of inquiry have the cognitive effect of optimizing the performance of the working memory by restricting the attention domain.[lxxxiii]

What is the essential information we need?

What can we learn in a week?

What can we do if we only focus on the customers/profits/easiest solution?

Moment 2: At regular intervals. At this point of the process, you diagnose the status of the investigation to ensure that the team preserves a bias toward action and allocates sufficient time to divergence and convergence. If you sense that the team is overwhelmed by uncertainty and struggle to maintain a bias toward action during convergence, and tend to invite reflection prematurely, you can help the group by asking to complete an exercise aimed to regain creative momentum:

Spend 15 minutes sketching alternative solutions.

Spend 30 minutes gathering inspiration from other team members' perspectives.

Switch roles and sketch solutions from other team members' perspectives.

If you sense that the team is overwhelmed by uncertainty and struggle to turn insights into actionable ideas during divergence, you can help the group by asking questions aimed to analyze concepts and reduce ambiguity:

If [constraint] was not an issue, what would we do?

Which customer insights can inspire our thinking?

Which solution are our competitors too scared to pursue?

What other industries have a similar challenge?

What can we learn from them?

If you sense that the team is overwhelmed by complexity and struggle to turn ideas into actionable possibilities during convergence, you can help the group by asking questions aimed to synthesize concepts and reduce ambiguity:

Which of these ideas would align more with our purpose?

Which of these ideas would generate more value for the user?

Which of these ideas feel more disruptive?

Which of these ideas feel more exciting?

Which of these ideas would earn us the most recognition?

Moment 3: At the end of a design iteration. At this point of the process, you create the necessary space to invite and support reflection to ensure that the team carefully considers any belief or supposed form of knowledge in the light of the ground that supports it. If you sense that the team is hesitant to assess the problem-solving experience and struggle to review the experiment, you can help the group by asking questions aimed to evaluate the outcome and examine the process:

Share considerations as "I like", "I wish", and "I wonder".

Describe how we tried to fulfill our purpose.

Describe how we tried to achieve our objectives.

Describe what we did not anticipate and its impact.

Describe what we have learned and its impact.

As we introduced in Chapter 4, the act of reflecting constitutes a foundational aspect of experiential learning, promoting a sense of exploration and discovery that represents a prerequisite to creative success.[lxxxiv]

Adapt Your Behavioral Strategy

Leading alongside adopting the mindset of a "coach" can be challenging because the cognitive impact of your leadership on the different individualities present in your team is not always directly observable. Furthermore, each team member can potentially perceive and respond to your actions differently and in ways that are not always instantly identifiable.

Assuming the presence of psychological safety and a trustworthy relationship with your team members, you can explore these perceptions using an active dialogue. During an individual conversation, ask your team members to candidly answer the following questions:

> *What is your sentiment on my contribution to the team?*
>
> *What do you think is working, and what can be improved?*
>
> *What is one thing that I can do to support you?*

As we introduced in Chapter 2, leadership is context dependent, and it exists in a specific dimension defined by the particular domain of a given role and organization.[lxxxv] Under this condition, your success as a person in charge of a team is a function of the needs of the group.[lxxxvi] These needs are likely to change with time as a consequence of several individual and collective factors, such as new designers joining the team and old members leaving the company, and as an organic consequence of the development of the collective intelligence. From your perspective, it is vital to maintain a constant connection with the group's perception and needs to continuously adapt your behavioral configuration to their current necessities.

Lead Behind

From this perspective, you lead through culture, engaging indirectly with your team's activities to influence what individuals believe and how they behave. In this role, you adopt the mindset of a "gardener" to cultivate the cognitive and social preconditions that support the behaviors necessary to sustain creative collaboration.

When: This modality of leadership is typically required when you need to define or preserve the social norms underlying specific desired behaviors at the beginning of a project or when you join a new team.

Moreover, leading from behind through culture must also be your default modality when leading from ahead and alongside is unnecessary. This strategy maps organically over the tendency of the brain to activate the social network for social cognition purposes during moments characterized by the absence of cognitive tasks that require the engagement of the analytical network.[lxxxvii]

Create Rituals

An essential component of leading through culture is creating rituals that cultivate the beliefs and behaviors that support creativity. Contrary to products and services, a creative and collaborative culture can be designed, but not directly. A ritual is a sequence of behaviors that an individual or a group implements following a pattern that symbolizes specific beliefs. Beliefs and behaviors represent the two foundational elements of any culture.[lxxxviii] Rituals can help you establish and reinforce descriptive and prescriptive norms, and by leveraging these socially determined consensual standards, you can codify critical behaviors aligned with specific beliefs that can unlock creative collaboration.

Beliefs and behaviors are interrelated: the evaluative associations of characteristics or attributes accepted as truth by an individual influence the observable or measurable actions of that person, and vice versa.[lxxxix] Beliefs cannot be observed directly; they can only be assessed indirectly via their behavioral manifestations. From that perspective, rituals make the intangible nature of those convictions tangible, visible, and interactive.[xc] The following list reports the four core observable behaviors and related signals that characterize the organizations with a high level of creative collaboration.

Observable behaviors:

- They are curious.
- They collaborate.
- They tangibly explore.
- They prioritize learning.

Observable signals:

- I feel inspired.
- I have ownership.
- I'm not afraid to fail.
- I'm growing.

Rituals establish pockets of creative collaboration that, when scaled purposefully, allow you to export those beliefs and behaviors across an entire organization. If a ritual is adequate to the demands and intentions of an organization, with time, those pockets of creative collaboration organically develop into practices becoming an integral part of the organizational culture. While this chapter concentrates explicitly on creative collaboration, you need to be aware that you can use this type of social event to promote beliefs and behaviors across five main organizational domains:

Creativity: Stimulate new ideas and visions.

Transition: Adapt to new circumstances.

Community: Form social connections.

Resilience: Manage tensions and frictions.

Performance: Support processes and flows.

Assuming the presence of trust and psychological safety, when you design a ritual, you have to consider four interconnected elements:[xci] the "belief" that you want to establish or reinforce, the operational "condition" within which you have to operate, the "behavior" that you desire to initiate or support, and the "outcome" that you intend to generate (Figure 8-22).

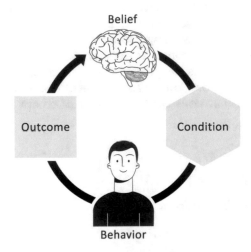

Figure 8-22. The four interconnected elements of a ritual

In some cultures, the beliefs required to trigger creative collaboration are already present, whether in a latent or apparent state, while in others, they need to be carefully cultivated to initiate the expected behaviors and generate the desired outcomes. Independently from the existing context, creating a ritual is a process that requires five steps.

Step 1: Identify the behavioral discrepancies. Begin considering the core observable behaviors and their related signals that characterize high levels of creative collaboration, and then analyze your organizational culture, answering the following questions:

Are those behaviors and signals present in our organization?

If not, what and where are the discrepancies?

Once you have answered these questions, select one crucial gap that needs to be filled and transform it into an objective using the following statement:

I want to initiate [behavior] in order to [desired outcome]

I want to initiate reflection-on-action in order to promote experiential learning.

Step 2: Capture a behavioral gap. Frame the objective into an attackable problem using a driving question. The following example prioritizes the need for candor as a way to unlock reflection-on-action and, therefore, experiential learning:

How might we promote candor during the design reviews?

Step 3: Brainstorm ideas. Use divergent thinking to explore the driving question and convergent thinking to decide which possibility you want to implement.

Step 4: Create the ritual. Define the practical details of the event to create a prototype. Navigate through this part of the process, answering the following questions:

What is the belief to promote?

What is the sequence of behaviors to initiate?

What is the outcome to generate?

Who needs to be involved?

How does the ritual integrate into the design operations?

How will we recognize behavioral progress?

How will we measure behavioral success?

Step 5: Try the ritual. Organize and run the event prototype. If the event is connected with an area of interest of a team member, you can consider delegating this responsibility to that person while retaining the accountability.

Step 6: Reflect on the ritual. Analyze how the team responded to the event prototype to understand how individuals behaved and which beliefs have been manifested. Summarize those responses answering the following questions:

Did the ritual promote the desired belief?

Did the ritual initiate the desired behaviors?

Did the ritual generate the desired outcome?

Similar to any other form of prototype, a new ritual typically requires some adjustments after the first iteration, and even in an optimal scenario, it still necessitates time to induce *behavioral momentum*. If the ceremony does not generate the desired outcome despite several refinements, consider the possibility of canceling the event and go back to step 3 to brainstorm additional ideas to explore. With diverse temporal cadences, your calendar needs to include rituals to initiate or reinforce all the beliefs and behaviors necessary to promote creative collaboration.

A crucial observation is that you should never attempt to blindly replicate the same social dynamics when creating the virtual counterpart of an already established in-person ceremony. In that scenario, consider the expected outcome within the digital media context and explore novel ways to enhance the strengths and mitigate the weaknesses of that given platform to promote the desired beliefs and initiate the appropriate behaviors. Establishing and reinforcing specific beliefs and behaviors is vital to effectively lead behind with a "gardener" mindset. The following section will explore a crucial aspect of this role, managing tension.

Manage Tension

The diversity of perspectives represents an enabler of creative collaboration but also an organic generator of tensions. Tension is an inevitable element of social groups, and it is defined as the feeling of psychological strain and uneasiness derived from a situation characterized by antagonist forces, including ideas, attitudes, behaviors, and emotions.[xcii] Within the context of a team, tension is present in three primary forms:[xciii] organic, episodic, and personal.

Organic tension. This feeling of psychological strain represents an intrinsic element of the creative process. This form of energy is generated by stressors associated with the uncertainty that permeates wicked problems and the multidisciplinarity that characterize modern teams and workgroups. Organic tension is an intrinsic part of a design team, and therefore it cannot be released but only maintained within healthy ranges. If you have created an environment permeated by trust, respect, and psychological safety, this type of tension rarely requires a direct intervention because it self-regulates.[xciv]

Episodic tension. This feeling of psychological strain represents an extrinsic element of the creative process. This form of energy is generated by stressors associated with collaborative inefficiencies derived from misaligned expectations and issues related to roles, goals, and procedures. Episodic tension is an extrinsic part of a design team and therefore can and must be released by preventing moments of irritation from escalating into conflicts. In more critical scenarios, this form of energy originates from frictions of ideas that escalate into frictions of personalities. If you have created an environment permeated by explicit descriptive and prescriptive norms, this type of tension can efficiently and productively be released and redirected into a creative direction.[xcv]

Personal tension. This feeling of psychological strain represents an underpinning of the creative process. This form of energy is generated by stressors associated with events characterizing the private life of an individual and can drastically alter the behavioral configuration of a person in some severe cases. Personal tension is typically concealed from the group and therefore out of your control. If you created an environment permeated by high levels of trust, respect, and psychological safety, you could consider offering the team the opportunity to discuss personal situations and provide support when requested and if possible.

As you can see in Figure 8-23, organic tension tends to be relatively stable, increasing and decreasing over a long period, typically oscillating when the team acquires or loses a team member or redirects its attention to a significantly more or less complex problem to solve.[xcvi] Episodic tension tends to be more unstable, increasing and decreasing with daily, weekly, or monthly oscillations typically based on the nature and quantity of the stressors originating from social interactions.[xcvii] Personal tension, from your viewpoint, is instead an unknown variable.[xcviii]

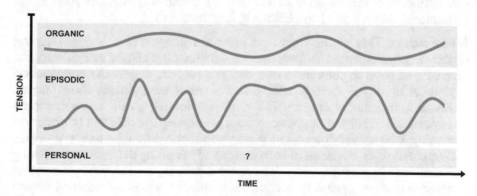

Figure 8-23. The three primary forms of tension experienced by a team

Given a situation, in order to optimize creative collaboration, you need to develop an understanding of the levels of tension present within the group and if they are fostering or hindering cooperation. For instance, if the team is demotivated but collaborates effectively, you need to increase organic tension by providing more meaningful challenges or more effective constraints. Instead, if the group is motivated but collaborates ineffectively, you need to decrease episodic tension by investigating frictions connected to expectations, roles, goals, and procedures.

If the total level of tension develops beyond a healthy threshold, the group's creative potential is reduced because the psychological strain experienced produces a disruptive response that impairs the working memory functions and limits the capacity of the semantic network in the brain to form neural connections.[xcix]

All forms of tension manifest themselves tangibly within three areas of the problem-solving process: decisions, collaboration, and reflection. Your role as a design manager and leader is to be aware of these three scenarios and manage the relative tension accordingly when necessary.

Decision

Tension can arise every time the team needs to make a commitment; convergence is often a common source of this form of energy. Synthesizing complex ideas is cognitively onerous, and the tension derived from the need to form insightful connections to prepare and make decisions can be perceived as oppressive. If this dynamic originates in the presence of low levels of trust and psychological safety, individuals also tend to devolve into a fixed mentality, assuming a self-preserving attitude that prevents creative collaboration.[c] You can maintain a healthy level of tension by channeling psychological strain productively using one of the following four methods.

Installation: This method aims to promote integrative thinking. Use the digital or physical space to present the various possibilities available to the team. If you need to commit to one design direction, display the concepts on a physical or digital surface, adding sticky notes to annotate their human, business, and technical aspects. This configuration allows the team members to experience the digital or physical room processing the different information and using it to establish a robust dialog about the advantages-disadvantages and opportunities-obstacles of each concept. Presenting the information using an installation improves the team's integrative thinking capabilities by enhancing the recall and recognition functionalities of the working memory, which expands the semantic network's ability to form meaningful neural connections.[ci] This approach allows the group to manage the tension originated by opposing concepts and, instead of choosing one at the expense of the others, generate a creative resolution of that tension in the form of a new possibility that contains elements of the opposing ideas but is superior to each of them individually.

Imagination: This method aims to mitigate cognitive biases. Assign one team member to each possibility available. Ask a person to argue in favor of a given concept against another individual that impersonates an inquisitorial role. Repeat the debate for each option available. You can assign multiple persons to the different ideas to create a diverse narrative about the same concept. Playing different roles forces a change in perspective that mitigates deviations from critical thinking during the decision-preparation and decision-making process.[cii] This method contributes to mitigating several biases, including, but not limited to, the authority,[7] the belief,[8] and the confirmation[9] bias.[ciii] Additionally, expanding the individual's perspective contributes to increasing the performance of the semantic network, which contributes to enhancing the efficacy of integrative thinking.[civ]

Construction: This method aims to overcome the decisional impasse. Encourage the team members to build to think, whether a simple sketch on the whiteboard, a mock-up, or a functional prototype. The team members can leverage concrete ways to synthesize possibilities to exceed the inherent perceptual limitations of dialogs. This approach creates a tangible representation of a concept that expands the cognitive substrate of reflection, which improves the ability to investigate the validity of an idea.

Examination: This method aims to increase inclusiveness. Invite the team members to write and share three aspirations and concerns related to the concepts under consideration. The team members write down silently and independently their six points before sharing them with the group using a digital or physical space. This exercise creates the preconditions for a robust dialog concentrated on the strengths and weaknesses of each concept. This approach lowers the cognitive threshold for participation across all personalities, from predominantly extrovert to introvert personalities.[cv] Additionally, it sends a clear message that all concerns are valid and appreciated during the conversation.

Managing tension is a situational ability that requires significant time and experience to develop, but it represents a necessary competence for a design manager and leader that aims to build a high-performing team. The preceding methods will help you to initiate this process, assisting you in regulating this crucial form of psychological energy. In Chapter 9, you will explore a similar challenge in regard to the management of stress and the definition of objectives for your team members.

---

[7] The authority bias is the tendency to accept the perspective of an individual based on the level of role power of that person.

[8] The belief bias is the tendency to evaluate the logical validity of an argument considering the believability of the conclusion.

[9] The confirmation bias is the tendency to process, interpret, and use information in a way that confirms a preconception.

Cooperation

Tension can arise every time the team needs to work jointly toward the same end; a project is often a common source of this form of energy. Synthesizing complex ideas is cognitively onerous, and the tension derived from the need to form insightful connections to prepare and make decisions can be perceived as oppressive. Creating a multidisciplinary group provides a diversity of perceptions that can have an ambivalent effect on the team. This condition enhances the creative potential of the group but also potentially introduces divergent interpretations of the social contract that intangibly inform the creative collaboration. If neglected, this situation can escalate into an unhealthy state that disrupts the social climate within the team.[cvi]

You can maintain a healthy level of tension by providing an opportunity to the team to surface frictions, discuss stressors, and establish a constructive dialogue that can prevent conflicts. One instantiation of this approach is to have a session where the team creates a representation of their emotional journey across the significant milestone of a project. During this type of session, initiate the conversation by visualizing the emotional journey of the group over a specific time frame. Use a digital or physical whiteboard to create a graph illustrating the relationship between time and emotions, asking the team members to represent their unique experiences (Figure 8-24).

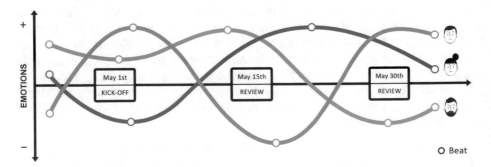

Figure 8-24. The emotional journey experienced by the team

When the entire team concluded the exercise, identify local and global points of positive and negative emotions to isolate patterns and pinpoint areas requiring investigation. The following three steps can guide you through this process every time you need to recognize, investigate, and resolve tensions.

Step 1: Recognize. Analyze the patterns delineated by the experiential journeys, assign them a name, and collectively decide from which moment of tension the group wants to begin the discussion.

Step 2: Investigate. Invite the team to leverage curiosity to explore the moment of tension curiously. Ask the group to answer the following questions candidly:

What was the context, and what happened?

What was your role, and how did you feel it?

What other factors could contribute to this situation?

Considering the answers to these questions allows you to surface and identify potential misalignments within the group. In some extreme situations, to gather comprehensive responses, you may need to adjourn the meeting to have one-on-one conversations with specific team members before resuming the investigation.

Step 3: Resolve. Decide how to address the moment of tension, individually with a one-to-one conversation, collectively as a group, or leveraging a combination of both types of communication. Explore possible approaches, asking the team the following questions:

How might we improve the way we collaborate?

What do we need to start doing?

What do we need to stop doing?

What do we need to continue doing?

At the end of the session, the group must use the answers to the preceding questions to formulate and collectively agree on a revised *team agreement* that can drive the group's future behavioral expectations. This conclusive action contributes to dissipating the psychological strain and releasing tension within the team by explicitly enshrining the revised social contract that intangibly informs the creative collaboration.

As we introduced in Chapter 5, explaining how to maintain a healthy social system, there is always an imperative need to rediscuss descriptive norms after a moment of tension to avoid a possible escalation into a conflict. Chapter 9 will continue to explore this subject, introducing how to define social norms and craft team agreements to support diversity and optimize the design operations.

Reflection

Tension can arise every time the team needs to unlock experiential learning; communicating about performance is often a common source of this form of energy. Unlocking experiential learning requires a questioning approach that inevitably produces a social exchange characterized by some degree of psychological strain. You can maintain a healthy level of tension by creating an environment permeated by trust, respect, and psychological safety that can invite and support reflection. As introduced in Chapter 6 and examined in Chapter 7, a crucial aspect of this effort is minimizing the perceived social threat and the experienced social pain associated with your communication across five specific dimensions: status, affiliation, fairness, certainty, and autonomy.

Endnotes

i. Foster, Richard N., and Sarah Kaplan. Creative Destruction: Why Companies That Are Built to Last Underperform the Market, and How to Successfully Transform Them. Currency, 2001.

ii. Feist, Gregory J. "A Meta-Analysis of Personality in Scientific and Artistic Creativity." Personality and Social Psychology Review, vol. 2, no. 4, 1998, pp. 290–309., doi:10.1207/s15327957pspr0204_5.

iii. Picchi, Andrea. "The 3 Dimensions of Design: A Model to Scale the Human-Centered Problem-Solving Practice across the Organization." ResearchGate, Jan. 2017, www.researchgate.net/publication/330634631_The_3_Dimensions_of_Design_A_Model_to_Scale_the_Human-Centered_Problem-Solving_practice_across_the_Organization.

iv. Runco, Mark A., and Garrett J. Jaeger. "The Standard Definition of Creativity." Creativity Research Journal, vol. 24, no. 1, 2012, pp. 92–96., doi:10.1080/10400419.2012.650092.

v. Runco, Mark A., and Garrett J. Jaeger. "The Standard Definition of Creativity." Creativity Research Journal, vol. 24, no. 1, 2012, pp. 92–96., doi:10.1080/10400419.2012.650092.

vi. Kaufman, James C., and Ronald A. Beghetto. "Beyond Big and Little: The Four C Model of Creativity." Review of General Psychology, vol. 13, no. 1, 2009, pp. 1–12., doi:10.1037/a0013688.

vii. Abraham, Anna. The Neuroscience of Creativity. Cambridge University Press, 2019.

viii. Abraham, Anna. The Neuroscience of Creativity. Cambridge University Press, 2019.

ix. Brandt, Anthony K., and David Eagleman. The Runaway Species: How Human Creativity Remakes the World. Canongate Books, 2017.

x. Berger, Peter Ludwig, and Thomas Luckmann. The Social Construction of Reality a Treatise in the Sociology of Knowledge. Penguin, 1991.

xi. Abraham, Anna. The Neuroscience of Creativity. Cambridge University Press, 2019.

xii. Brandt, Anthony K., and David Eagleman. The Runaway Species: How Human Creativity Remakes the World. Canongate Books, 2017.

xiii. Brandt, Anthony K., and David Eagleman. The Runaway Species: How Human Creativity Remakes the World. Canongate Books, 2017.

xix. Berger, Peter Ludwig, and Thomas Luckmann. The Social Construction of Reality a Treatise in the Sociology of Knowledge. Penguin, 1991.

xv. Fauconnier, Gilles, and Mark Turner. The Way We Think: Conceptual Blending and the Mind's Hidden Complexities. Basic Books, 2002.

xvi. Fuster, Joaquín M. The Prefrontal Cortex. Elsevier, AP, 2011.

xvii. Gazzaniga, Michael S., et al. Cognitive Neuroscience: The Biology of the Mind. Norton, 2014.

xviii. Binder, Jeffrey R., et al. "Where Is the Semantic System? A Critical Review and Meta-Analysis of 120 Functional Neuroimaging Studies." Cerebral Cortex, vol. 19, no. 12, 2009, pp. 2767–2796., doi:10.1093/cercor/bhp055.

xix. Vartanian, Oshin, et al. Neuroscience of Creativity. The MIT Press, 2016.

xx. Abraham, Anna. The Neuroscience of Creativity. Cambridge University Press, 2019.

xxi. Griffiths, Thomas L., et al. "Topics in Semantic Representation." Psychological Review, vol. 114, no. 2, 2007, pp. 211–244., doi:10.1037/0033-295x.114.2.211.

xxii. Abraham, Anna. The Neuroscience of Creativity. Cambridge University Press, 2019.

xxiii. Brandt, Anthony K., and David Eagleman. The Runaway Species: How Human Creativity Remakes the World. Canongate Books, 2017.

xxiv. Arends, Richard. Learning to Teach. McGraw-Hill, 2014.

xxv. Davidson, Richard J., and Bruce S. McEwen. "Social Influences on Neuroplasticity: Stress and Interventions to Promote Well-Being." Nature Neuroscience, vol. 15, no. 5, 2012, pp. 689–695., doi:10.1038/nn.3093.

xxvi. Fauconnier, Gilles, and Mark Turner. The Way We Think: Conceptual Blending and the Mind's Hidden Complexities. Basic Books, 2002.

xxvii. Hasson, Uri, et al. "Brain-to-Brain Coupling: A Mechanism for Creating and Sharing a Social World." Trends in Cognitive Sciences, vol. 16, no. 2, 2012, pp. 114–121., doi:10.1016/j.tics.2011.12.007.

xxviii. Fauconnier, Gilles, and Mark Turner. The Way We Think: Conceptual Blending and the Mind's Hidden Complexities. Basic Books, 2002.

xxix. Cowan, Nelson. Working Memory Capacity. Psychology Press, 2005.

xxx. Abraham, Anna. The Neuroscience of Creativity. Cambridge University Press, 2019.

xxxi. Cowan, Nelson. Working Memory Capacity. Psychology Press, 2005.

xxxi. Cowan, Nelson. Working Memory Capacity. Psychology Press, 2005.

xxxiii. Edmondson, Amy. Teaming: How Organizations Learn, Innovate, and Compete in the Knowledge Economy. Wiley, J., 2012.

xxxiv. Fauconnier, Gilles, and Mark Turner. The Way We Think: Conceptual Blending and the Mind's Hidden Complexities. Basic Books, 2002.

xxxv. Huettel, Scott A., et al. "Neural Signatures of Economic Preferences for Risk and Ambiguity." Neuron, vol. 49, no. 5, 2006, pp. 765–775., doi:10.1016/j.neuron.2006.01.024.

xxxvi. Huettel, Scott A., et al. "Neural Signatures of Economic Preferences for Risk and Ambiguity." Neuron, vol. 49, no. 5, 2006, pp. 765–775., doi:10.1016/j.neuron.2006.01.024.

xxxvii. Huettel, Scott A., et al. "Neural Signatures of Economic Preferences for Risk and Ambiguity." Neuron, vol. 49, no. 5, 2006, pp. 765–775., doi:10.1016/j.neuron.2006.01.024.

xxxviii. Abraham, Anna. The Neuroscience of Creativity. Cambridge University Press, 2019.

xxxix. Edmondson, Amy C. The Fearless Organization: Creating Psychological Safety in the Workplace for Learning, Innovation, and Growth. Wiley, 2018.

xl. Edmondson, Amy. Teaming: How Organizations Learn, Innovate, and Compete in the Knowledge Economy. Wiley, J., 2012.

xli. Yu, Lianchun, and Yuguo Yu. "Energy-Efficient Neural Information Processing in Individual Neurons and Neuronal Networks." Journal of Neuroscience Research, vol. 95, no. 11, 2017, pp. 2253–2266., doi:10.1002/jnr.24131.

xlii. Mink, J. W., et al. "Ratio of Central Nervous System to Body Metabolism in Vertebrates: Its Constancy and Functional Basis." American Journal of Physiology, vol. 241, no. 3, 1981, doi:10.1152/ajpregu.1981.241.3.r203.

xliii. Harris, Julia J., et al. "Synaptic Energy Use and Supply." Neuron, vol. 75, no. 5, 2012, pp. 762–777., doi:10.1016/j.neuron.2012.08.019.

xliv. Zhu, Xiao-Hong, et al. "Quantitative Imaging of Energy Expenditure in Human Brain." NeuroImage, vol. 60, no. 4, 2012, pp. 2107–2117., doi:10.1016/j.neuroimage.2012.02.013.

xlv. Bullmore, Ed, and Olaf Sporns. "The Economy of Brain Network Organization." Nature Reviews Neuroscience, vol. 13, no. 5, 2012, pp. 336–349., doi:10.1038/nrn3214.

xlvi.   Bullmore, Ed, and Olaf Sporns. "The Economy of Brain Network Organization." Nature Reviews Neuroscience, vol. 13, no. 5, 2012, pp. 336–349., doi:10.1038/nrn3214.

xlvii.   Hagura, Nobuhiro, et al. "Perceptual Decisions Are Biased by the Cost to Act." ELife, Feb. 2017, doi:10.7554/eLife.18422.001.

xlviii. Hagura, Nobuhiro, et al. "Perceptual Decisions Are Biased by the Cost to Act." ELife, Feb. 2017, doi:10.7554/eLife.18422.001.

xlix.   Edmondson, Amy. Teaming: How Organizations Learn, Innovate, and Compete in the Knowledge Economy. Wiley, J., 2012.

l.   Hardy, Jay H., et al. "Outside the Box: Epistemic Curiosity as a Predictor of Creative Problem Solving and Creative Performance." Personality and Individual Differences, vol. 104, 2017, pp. 230–237., doi:10.1016/j.paid.2016.08.004.

li.   Loewenstein, George. "The Psychology of Curiosity: A Review and Reinterpretation." Psychological Bulletin, vol. 116, no. 1, 1994, pp. 75–98., doi:10.1037/0033-2909.116.1.75.

lii.   Loewenstein, George. "The Psychology of Curiosity: A Review and Reinterpretation." Psychological Bulletin, vol. 116, no. 1, 1994, pp. 75–98., doi:10.1037/0033-2909.116.1.75.

liii.   Loewenstein, George. "The Psychology of Curiosity: A Review and Reinterpretation." Psychological Bulletin, vol. 116, no. 1, 1994, pp. 75–98., doi:10.1037/0033-2909.116.1.75.

liv.   Gilson, Lucy L., and Christina E. Shalley. "A Little Creativity Goes a Long Way: An Examination of Teams' Engagement in Creative Processes." Journal of Management, vol. 30, no. 4, 2004, pp. 453–470., doi:10.1016/j.jm.2003.07.001.

lv.   Edmondson, Amy. Teaming: How Organizations Learn, Innovate, and Compete in the Knowledge Economy. Wiley, J., 2012.

lvi.   Hardy, Jay H., et al. "Outside the Box: Epistemic Curiosity as a Predictor of Creative Problem Solving and Creative Performance." Personality and Individual Differences, vol. 104, 2017, pp. 230–237., doi:10.1016/j.paid.2016.08.004.

lvii.   Edmondson, Amy. "Psychological Safety and Learning Behavior in Work Teams." Administrative Science Quarterly, vol. 44, no. 2, 1999, p. 350., doi:10.2307/2666999.

lviii. Edmondson, Amy C. The Fearless Organization: Creating Psychological Safety in the Workplace for Learning, Innovation, and Growth. Wiley, 2018.

lix. Ryan, Richard M., and Edward L. Deci. Self-Determination Theory: Basic Psychological Needs in Motivation, Development, and Wellness. The Guilford Press, 2017.

lx. Vartanian, Oshin, et al. Neuroscience of Creativity. The MIT Press, 2016.

lxi. Vartanian, Oshin, et al. Neuroscience of Creativity. The MIT Press, 2016.

lxii. Dahlstrom, Michael F. "Using Narratives and Storytelling to Communicate Science with Non-Expert Audiences." National Academy of Sciences, vol. 111, no. 4, 2014, pp. 13614–13620., doi:10.1073/pnas.1320645111.

lxiii. Lieberman, Matthew D. "Social Cognitive Neuroscience: A Review of Core Processes." Annual Review of Psychology, vol. 58, no. 1, 2007, pp. 259–289., doi:10.1146/annurev.psych.58.110405.085654.

lxiv. Bruner, Jerome. "The Narrative Construction of Reality." Critical Inquiry, vol. 18, no. 1, 1991, pp. 1–21., doi:10.1086/448619.

lxv. Bruner, Jerome. "The Narrative Construction of Reality." Critical Inquiry, vol. 18, no. 1, 1991, pp. 1–21., doi:10.1086/448619.

lxvi. Adler, Alfred. Social Interest: A Challenge to Mankind. Martino Fine Books, 1938.

lxvii. Gottschall, Jonathan. The Storytelling Animal: How Stories Make Us Human. Houghton Mifflin Harcourt, 2012.

lxviii. Bruner, Jerome. "The Narrative Construction of Reality." Critical Inquiry, vol. 18, no. 1, 1991, pp. 1–21., doi:10.1086/448619.

lxix. Hasson, Uri, et al. "Brain-to-Brain Coupling: A Mechanism for Creating and Sharing a Social World." Trends in Cognitive Sciences, vol. 16, no. 2, 2012, pp. 114–121., doi:10.1016/j.tics.2011.12.007.

lxx. Hasson, Uri, et al. "Brain-to-Brain Coupling: A Mechanism for Creating and Sharing a Social World." Trends in Cognitive Sciences, vol. 16, no. 2, 2012, pp. 114–121., doi:10.1016/j.tics.2011.12.007.

lxxi.   Hasson, Uri, et al. "Brain-to-Brain Coupling: A Mechanism for Creating and Sharing a Social World." Trends in Cognitive Sciences, vol. 16, no. 2, 2012, pp. 114–121., doi:10.1016/j.tics.2011.12.007.

lxxii.   Hasson, Uri, et al. "Brain-to-Brain Coupling: A Mechanism for Creating and Sharing a Social World." Trends in Cognitive Sciences, vol. 16, no. 2, 2012, pp. 114–121., doi:10.1016/j.tics.2011.12.007.

lxxiii.   Hasson, Uri, et al. "Brain-to-Brain Coupling: A Mechanism for Creating and Sharing a Social World." Trends in Cognitive Sciences, vol. 16, no. 2, 2012, pp. 114–121., doi:10.1016/j.tics.2011.12.007.

lxxiv.   Immordino-Yang, Mary Helen. "The Role of Emotion and Skilled Intuition in Learning." Mind, Brain, and Education, 2009, pp. 66–81.

lxxv.   Gazzaniga, Michael S., et al. Cognitive Neuroscience: The Biology of the Mind. Norton, 2014.

lxxvi.   Hasson, Uri, et al. "Brain-to-Brain Coupling: A Mechanism for Creating and Sharing a Social World." Trends in Cognitive Sciences, vol. 16, no. 2, 2012, pp. 114–121., doi:10.1016/j.tics.2011.12.007.

lxxvii.   Farley, James, et al. "Everyday Attention and Lecture Retention: The Effects of Time, Fidgeting, and Mind Wandering." Frontiers in Psychology, vol. 4, 2013, doi:10.3389/fpsyg.2013.00619.

lxxviii. Griffiths, Thomas L., et al. "Topics in Semantic Representation." Psychological Review, vol. 114, no. 2, 2007, pp. 211–244., doi:10.1037/0033-295x.114.2.211.

lxxix.   Kozhevnikov, Maria, et al. "Creativity, Visualization Abilities, and Visual Cognitive Style." British Journal of Educational Psychology, vol. 83, no. 2, 2013, pp. 196–209., doi:10.1111/bjep.12013.

lxxx.   Kozhevnikov, Maria, et al. "Creativity, Visualization Abilities, and Visual Cognitive Style." British Journal of Educational Psychology, vol. 83, no. 2, 2013, pp. 196–209., doi:10.1111/bjep.12013.

lxxxi.   Monsell, Stephen, and Jon Driver. Control of Cognitive Processes. MIT, 2000.

lxxxii. Edmondson, Amy. Teaming: How Organizations Learn, Innovate, and Compete in the Knowledge Economy. Wiley, J., 2012.

lxxxiii. Baddeley, Alan. "Working Memory." Current Biology, vol. 20, no. 4, 2010, doi:10.1016/j.cub.2009.12.014.

lxxxiv. Fogg, Brian Jeffrey. Tiny Habits: The Small Changes That Change Everything. Virgin Books, 2020.

lxxxv. Day, David V. "Leadership Development: A Review in Context." The Leadership Quarterly, vol. 11, no. 4, 2000, pp. 581–613., doi:10.1016/s1048-9843(00)00061-8.

lxxxvi. Day, David V. "Leadership Development: A Review in Context." The Leadership Quarterly, vol. 11, no. 4, 2000, pp. 581–613., doi:10.1016/s1048-9843(00)00061-8.

lxxxvii. Lieberman, Matthew D. Social: Why Our Brains Are Wired to Connect. Oxford University Press, 2015.

lxxxviii. Alvard, Michael S. "The Adaptive Nature of Culture." Evolutionary Anthropology: Issues, News, and Reviews, vol. 12, no. 3, 2003, pp. 136–149., doi:10.1002/evan.10109.

lxxxix. Lange, Paul A. M. van, et al. Social Psychology. The Guilford Press, 2021.

xc. Durkheim Émile, et al. The Elementary Forms of Religious Life. Oxford University Press, 2008.

xci. Aronson, Elliot, et al. Social Psychology. Pearson, 2016.

xcii. Aronson, Elliot, et al. Social Psychology. Pearson, 2016.

xciii. Arnold, John, et al. Work Psychology: Understanding Human Behaviour in the Workplace. Pearson, 2020.

xciv. Arnold, John, et al. Work Psychology: Understanding Human Behaviour in the Workplace. Pearson, 2020.

xcv. Arnold, John, et al. Work Psychology: Understanding Human Behaviour in the Workplace. Pearson, 2020.

xcvi. Razinskas, Stefan, and Martin Hoegl. "A Multilevel Review of Stressor Research in Teams." Journal of Organizational Behavior, vol. 41, no. 2, 2019, pp. 185–209., doi:10.1002/job.2420.

xcvii. Razinskas, Stefan, and Martin Hoegl. "A Multilevel Review of Stressor Research in Teams." Journal of Organizational Behavior, vol. 41, no. 2, 2019, pp. 185–209., doi:10.1002/job.2420.

xcviii. Razinskas, Stefan, and Martin Hoegl. "A Multilevel Review of Stressor Research in Teams." Journal of Organizational Behavior, vol. 41, no. 2, 2019, pp. 185–209., doi:10.1002/job.2420.

xcix. Sandi, Carmen. "Stress and Cognition." Wiley Interdisciplinary Reviews: Cognitive Science, vol. 4, no. 3, 2013, pp. 245–261., doi:10.1002/wcs.1222.

c. Dweck, Carol S. Mindset: The New Psychology of Success. Ballantine Books, 2016.

ci. Goolkasian, Paula, and Paul W. Foos. "Presentation Format and Its Effect on Working Memory." Memory & Cognition, vol. 30, no. 7, 2002, pp. 1096–1105., doi:10.3758/bf03194327.

cii. Thomas, Oliver. "Two Decades of Cognitive Bias Research in Entrepreneurship: What Do We Know and Where Do We Go From Here?" Management Review Quarterly, vol. 68, no. 2, 2018, pp. 107–143., doi:10.1007/s11301-018-0135-9.

ciii. Friedman, Hershey H. "Cognitive Biases That Interfere with Critical Thinking and Scientific Reasoning." SSRN Electronic Journal, 2017, doi:10.2139/ssrn.2958800.

civ. Abraham, Anna. The Neuroscience of Creativity. Cambridge University Press, 2019.

cv. Gundogdu, Didem, et al. "Investigating the Association between Social Interactions and Personality States Dynamics." Royal Society Open Science, vol. 4, no. 9, 2017, p. 170194., doi:10.1098/rsos.170194.

cvi. Jehn, Karen A., et al. "Why Differences Make a Difference: A Field Study of Diversity, Conflict, and Performance in Workgroups." Administrative Science Quarterly, vol. 44, no. 4, 1999, p. 741., doi:10.2307/2667054.

Optimize Design Operations

Bill Moggridge, a co-founder of IDEO, brilliantly indicated that a few persons are aware of it, but there is nothing made by human beings that does not involve a design decision somewhere.[i] Design decisions are crucial moments during the problem-solving process that require a clear intent and a human-oriented attitude. Under this premise, your role as a design manager and leader is to systematize human centricity by establishing the necessary preconditions to prepare and make design decisions effectually and efficiently. These preconditions encompass processual but, more importantly, social aspects of the problem-solving process that support the mindset and conduct required by that methodology. Without that cultural element defined by the social norms that guide the team's attitudinal and behavioral patterns, the process alone is inadequate to produce any significant change.[ii]

Attribution Process

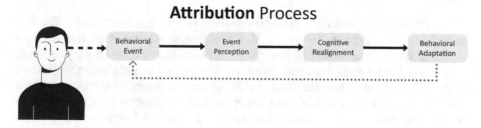

Figure 9-1. The attribution process

© Andrea Picchi 2022
A. Picchi, *Design Management*, https://doi.org/10.1007/978-1-4842-6954-1_9

The way in which human beings perceive and interpret events around them affects their future attitude and behaviors; this dynamic is called the *attribution process* (Figure 9-1). Human beings continuously evolve their cognitive structure based on their interpretations of experienced events, forming the *perceptual history* that defines their empirical significance of the information.[iii] Specifically, individuals analyze a behavioral event to determine its positive or negative connotation from a survival perspective, and then they revise their cognitive frames accordingly, realigning their *reflective assumptions* regarding that specific causal relationship.[iv] For instance, if you take a risk to explore a problem without success, and while you have learned from that experiment, you nonetheless receive criticisms as a consequence of that behavior, you realize that taking risks is not a productive behavioral strategy in your organization and that it can jeopardize your career progression. This dynamic defines the trial and error, action and reaction, "response cycle" that in design is symbolized by the word "feedback," an essential evolutionary process that extends across all the three horizons of psychological and, consequently, behavioral adaptation: *learn*, *develop*, and *evolve*.

As the person in charge of the team, your objective is to establish these social and processual elements of the human-centered design methodology with the intent to align the necessary mindset with the descriptive and prescriptive norms required to promote the desired behaviors.

In this book, you have already explored the intentional management of this social and processual dynamic under different circumstances. If you want to establish a physiologically safe environment, you must invite participation and respond productively to it. If you want to increase the team's learning capacity, you must support "intelligent failure" and respond constructively to it. If you want to promote creative problem-solving, you must foster curiosity and respond enthusiastically to it. Beyond an effective process, the social element is always the determining factor in any context of permanent behavioral change.[v]

In this chapter, you will learn how to optimize the design operations. You will learn how to establish the social and processual elements of the human-centered design methodology with the intent to align the necessary mindset with the descriptive and prescriptive norms required to promote the desired behaviors. You will analyze how to build and support the infrastructure required to establish the design operations. You will examine how to create a fertile environment for productive diversity, find and activate the team purpose, institute social norms, illuminate the types of design value, frame the design effort, and unlock high levels of performance. You will also explore how to create a design strategy, communicate about design, nurture collaboration with individual accountability, and scale design management and leadership within your organization.

Build the Team Infrastructure

In this section, you will learn how to build the necessary infrastructure to enable the design operations. You will explore how to create a healthy, sustainable, and productive environment for diversity to maximize the team's creative potential. You will also examine how to illuminate the different types of design value and how to find and activate the team purpose. This section also presents how to institute social norms to promote cohesiveness and establish objectives for the team and goals for the team members calibrating exogenous stress to increase their zone of proximal development.

Create a Fertile Environment for Productive Diversity

The creative output of a team is strongly correlated with the degree of diversity present within the group.[vi] The degree of diversity of a group is determined by its cognitive, social, and technical variety derived from the difference in team members' perceptual history that defines the amount of *empirical significance of information*[1] that the group can process during the problem-solving process.[vii]

This variety of assets expands the ability of the group to engage empirically in perceptions of the world and, consequently, the capacity of the collective brain to create a diversified set of *reflective assumptions* constructing meanings processing the meaningless information from the physical world.[viii]

Despite its enhancing effect on the collective brain, the strong emphasis that diversity places on the need for variety and collaboration can negatively affect the individual team members' self-efficacy and perception of their respective creative potential.[ix] As the person in charge of the team, you can mitigate this undesired side of that dichotomous relation by leveraging appreciative feedback delivered individually and privately.

Promote Group Cohesiveness

Creative collaboration is positively influenced by diversity exclusively when the team presents a sufficient level of group cohesiveness.[2] Group cohesion is indicated by the strength of the psychological and behavioral bonds that link

---

[1] Revisit Chapter 3 to review the concept of the empirical significance of information and how the brain constructs meanings and creates reflective assumptions.
[2] Group cohesiveness refers to the unity and solidarity of a set of individuals, including their integration for both social-related and task-related intentions around a common purpose.

members to the group as a whole.[x] Without that level of unity and solidarity around a common purpose, that degree of social, cultural, and technical variety becomes profoundly divisive.[xi]

Group cohesion is an antecedent of group performance, not a consequence.[xii] When that social condition is fulfilled, the group operates in a state of *productive diversity* where, among other dynamics, the potential ability of the team to solve problems creatively increases as a consequence of the group's capacity to regulate the creative tension allowing frictions of ideas while preventing conflicts of personalities.[xiii] Contrarily, as introduced previously, that degree of social, cultural, and technical variety can instead become profoundly divisive, evolving into barriers that hinder creative collaboration and decrease team performance.[xiv]

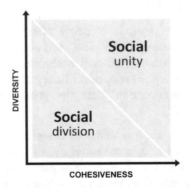

Figure 9-2. The relationship between team diversity and cohesiveness

As illustrated in Figure 9-2, when diversity is high but cohesiveness is low, the group experiences a significant degree of "social division" that inhibits creative collaboration.[xv] When diversity and cohesiveness are equally high, the team instead operates in a climate of "social unity" that sustains productive problem-solving interactions that increase creative problem-solving.[xvi]

Frame Physiological and Behavioral Synchrony

A state of productive diversity requires group cohesiveness, which originates from a state of physiological and behavioral synchrony (Figure 9-3). Interpersonal synchrony represents a profound evolutionary-based mechanism at the foundation of human ultrasociality that facilitates social and informational exchanges for survival purposes.[xvii]

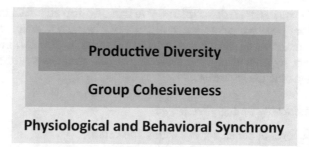

Figure 9-3. The relationship between psychological and behavioral synchrony, group cohesiveness, and productive diversity

Physiological and behavioral synchrony can be deconstructed into three foundational elements (Figure 9-4) in relation to their potential level of resonance or dissonance.[xviii]

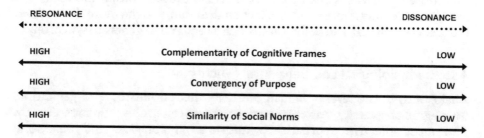

Figure 9-4. The three foundational elements of physiological and behavioral synchrony

The most cost-effective strategy to increase the group's complementarity of cognitive frames is to hire and release team members accordingly. As we mentioned in Chapter 7, discussing the needs for a growth mindset, while changing a person's attitude is possible by investing a significant amount of time and effort, this approach is often not feasible nor desirable within a business context. Hiring and releasing team members remains the most practical and cost-effective tool at your disposal to ensure the presence of any cognitive traits.

Under the assumption that the team members intend to integrate into a given working environment, regulating divergent purposes and dissimilar social norms instead represents a feasible and desirable objective to achieve. You can increase the similarity of social norms by instituting appropriate cultural standards, and you can decrease the variance of purpose by articulating the team's reason to exist beyond making profits.

If you quantify these three descriptors, you can develop an initial understanding of the potential positive or negative impact of diversity within your team and the potential managerial effort required to regulate it. You can investigate this potential effect by answering the following three questions and scoring them using a three-point Likert scale:

// *Complementarity of cognitive frames*

What is the level of social variety within the team?

// *Convergency of purpose*

What is the level of technical variety within the team?

// *Similarity of social norms*

What is the level of cultural variety within the team?

The more significant is the degree of diversity present within the team, the more considerable is its potential effect on the team, and the more demarcated is the social, cultural, and technical structure required to generate synchrony.[xix]

Assess Physiological and Behavioral Synchrony

Analogously to the level of human centricity that permeates an organization, the degree of diversity that characterizes a team cannot be estimated directly. The most effective approach to measure physiological and behavioral synchrony is to investigate this construct indirectly across social-oriented and results-oriented situations and measure it using behavioral and attitudinal operationalizations.[xx]

The questionnaire in Table 9-1 implements that strategy and quantifies the perception of eight associated states positively correlated with a productive environment characterized by high degrees of social, cultural, and technical variety.[xxi] You can assess physiological and behavioral synchrony annually, semiannually, or every time the team or workgroup is subject to a significant structural change.

Table 9-1. Physiological and behavioral synchrony questionnaire

Physiological and Behavioral Synchrony

Belonging

- In this team, individuals care about me.

Trustworthiness

- In this team, I can manifest vulnerability.

Psychological safety

- In this team, I can communicate candidly.

Purposefulness

- In this team, I know our reason to exist beyond making profits.

Fairness

- In this organization, individuals are recognized and rewarded.

Integration

- In this organization, different opinions are valued and respected.

Execution

- In this organization, divergent perspectives are actively considered.

Heterogeneity

- In this organization, cognitive, social, and technical variation is actively pursued.

7-Point scale: Strongly disagree, Disagree, Slightly disagree, Neutral, Slightly agree, Agree, Strongly agree.

A state of physiological and behavioral synchrony is temporal and adaptive in nature and tends to fluctuate based on several social-related factors, such as the level of commitment to a given purpose and the healthiness of interpersonal interactions.[xxii] When present, this underlying state is also positively correlated with numerous elements that affect organizational behavior, like the openness to embrace social standards, the tendency to develop healthy and productive relationships, and the capacity to experience job satisfaction.[xxiii]

A substrate of physiological and behavioral synchrony constitutes the foundation of every strong design team by representing a social, cultural, and technical meta-layer that enables the team members to integrate into the working environment by expressing their uniqueness productively and sustainably.[xxiv]

Foster Collaborative Practices

Developing group cohesiveness requires implementing physiological and behavioral synchrony in a way that increases clarity around collaborative practices to overcome the natural inclination of a socially, culturally, and technically heterogeneous group to interact in dissimilar manners.[xxv] In order to produce performance benefits, collaborative practices must be conceptualized and introduced as a behavior instead of an outcome.[xxvi] The following example demonstrates this distinction:

> // No: Practice as an outcome
>
> *Collaborate creatively.*
>
> // Yes: Practice as a behavior
>
> *Defer judgment.*
>
> *Build on the idea of others.*
>
> ... ...

Diversity is subsidiary to collaboration and represents an asset solely when used in service of the collective intelligence, under clear and precise social norms. A diverse team within a socially unregulated environment tends to underperform a homogenous one by a significant margin.[xxvii]

For this reason, the team's degree of social, cultural, and technical variety must be considered subsidiary to collaboration and, more profoundly, the willingness of an individual to integrate into a given working environment. Healthy and productive diversity cannot exist without the intention to integrate into a collective entity in respect of a given set of predefined norms; this attitude requires proactiveness and adaptability and constitutes a requirement that affects social groups at any scale.[xxviii]

This form of social, cultural, and technical multifariousness must be manifested within the boundaries of a healthy culture developed on effective problem-solving practices. This cultural strategy is founded on the principle that the team, intended as a collective entity, is always more valuable than its individual members because, as you learned discussing the social component of creativity in Chapter 8, "all of us are smarter than any of us."

Find and Activate the Team Purpose

Any organization has a purpose. Your role as a design manager and leader is to consider that intent and support your team in finding and activating their reason to exist beyond making profits in alignment with it: the team's "why," the contribution that the group wants to provide to that intention. Based on the process introduced in Chapter 4 to connect with your professional purpose, the following three steps can help to find and activate the team purpose.

Step 1: See it. The team looks back at what the group has done in the past and identifies moments of fulfillment. Ask the team to answer the following questions to drive this part of the process:

Who are our team's primary users?

What product/service do we offer to meet their needs?

Why is it essential that we provide this product/service?

What are our team's strengths and capabilities?

What would happen if our team didn't exist?

Step 2: Name it. The team discusses the answers to the previous questions and looks forward to the team's future to craft the first iteration of the team purpose statement. Capture the team purpose statement using the following structure:

We exist to [impact] in order to serve [audience].

Step 3: Do it. The team activates and substantiates its purpose and defines priorities connected to objectives, goals, and actions. While the team's purpose never changes, its implementation requires adaptation to remain aligned with possible shifts of the organizational strategy. At the end of the session, set a yearly reflection meeting to review the purpose statement collectively and, if necessary, discuss changes to the team objectives, goals, and actions.

Institute Social Norms

Social norms, both descriptive and prescriptive, represent socially determined consensual standards that regulate how the members of a social group[3] perceive and interact with one another and approach decisions. Instituting social norms unlocks the fulfillment of all the five social needs of your team, in particular, "dependability" and "structure and clarity."[xxix]

In the form of availability standards, they affect "dependability," regulating an individual's confidence level associated with the possibility of counting on other persons to do excellent work on time. Availability norms typically specify social aspects such as meeting hours for synchronous and asynchronous sessions. When a team adopts a hybrid or a fully remote configuration, they

---

[3] A social group refers to a collection of interdependent individuals who influence one another through social interactions characterized by a given degree of cohesiveness and underpinned by a shared purpose and a set of norms and roles.

also clarify the daily time frame that identifies when everyone is available on the messaging application of choice or the structure of the working hours for specific groups.

In the form of collaborative standards, they affect "structure and clarity," regulating an individual's confidence level associated with the capacity to understand roles, responsibilities, and decisional processes. Collaborative norms typically specify how each person's social, cultural, and technical contribution integrates into the collective effort with the intent to leverage diversity healthily and productively.

Social norms can be instituted to regulate individual and collective behaviors. Besides what was already introduced in Chapter 8, other examples of rules aimed to cultivate creative collaboration can be speaking to one person at a time, not allowing interruptions, and blocking contributions that stray the conversation outside its area of focus without adding tangible value.

Certain social norms are essential to instantiate and substantiate the core human-centered beliefs of the group and will never change with time. In contrast, other noncore socially determined consensual standards can be revised if a significant change modifies the working conditions.

Define Descriptive Norms

Descriptive norms are defined by our perception of whether persons considered important already perform or not perform a given behavior.[xxx] Within the context of a design team, they delineate the traits of its design culture that intangibly permeate the creative collaboration. Descriptive norms allow you to promote "dependability," define individual and collective accountability, and develop "social unity."[xxxi] When the team leverages divergent thinking, for instance, among other things, it implicitly agrees on creating choices exploring possibilities optimistically, deferring judgment, and building on the ideas of others. When the team leverages convergent thinking, instead, among other things, it implicitly agrees on making choices assigning sufficient airtime to each possibility to understand them adequately, embracing tension to encourage ideas to collide in unexpected ways, and defining criteria to drive decision-making. In the case of creativity, these types of socially determined consensual standards optimize the collaboration reducing the likelihood of unproductive interactions, and when creative tension generates a friction that produces a moment of irritations, the group can use them to self-regulate their behavior and prevent conflicts.[xxxii]

Social norms also create the precondition for distributed cultural accountability, a necessary condition to maintain the homeostasis of the design culture. Furthermore, this form of social agreement offers a model of reference that every team member can use to monitor and preserve the behaviors that delineate the design culture by providing appreciative and constructive feedback.[xxxiii] As you implicitly explored in Chapter 8, learning how to create rituals, from a behavioral perspective, social norms can be examined considering four interconnected elements:[xxxiv] the underpinning belief, the operational condition, the expected behavior, and the desired outcome (Figure 9-5).

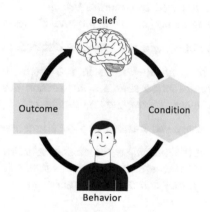

Figure 9-5. The four interconnected elements of a ritual

The following four steps can be used to structure a session where you can explore the current design-related beliefs present within the team and purposefully connect the operational conditions, the expected behaviors, and the desired operational outcomes.

Step 1: Surface beliefs. Begin the conversation by asking the team to articulate the purpose of their collaboration and describe their personal and professional objectives. At this point, reconcile unrealistic or misaligned expectations if necessary.

Step 2: Connect conditions, behaviors, and outcomes. Continue the conversation by asking the team to create a *hero shot* where they envision what a fictional newspaper headline might say about their contribution to their current project. Considering the hero shot, ask the group to work backward and articulate the detail of the design practice that unlocked that contribution by answering the following question:

What would have to be true to [contribution]?

Investigate the details of that answer by exploring the different aspects of the contribution that every team member envisioned for the team, considering areas such as, but not limited to, working styles, collaborative practices, and communicative patterns. At this point, reconcile unrealistic or misaligned expectations if necessary.

Step 3: Define descriptive norms. Deepen the conversation by asking the team to capture the "beliefs," "conditions," "behaviors," and "outcome" using concise and meaningful sentences; the following are three examples:

// *Belief: To promote creativity, you must defer judgment*

If a team believes that refraining from evaluating ideas during a convergent session (condition) is required to generate original and appropriate concepts (outcome), they can agree on "deferring judgment" (behavior).

// *To facilitate collaboration, you must remove distractions*

If a team believes that having phones in the room during a collaborative session (condition) represents a source of distraction that impedes creativity (outcome), they can agree on having a "no-phone zone" (behavior).

// *To increase productivity, you must have focused sessions*

If a team believes that the morning is the most productive time of the day and that they need a moment where they can focus intensely (condition) on their design work (outcome), they can agree on one or more "no-meeting morning" (behavior).

Step 4: Radiate descriptive norms. Conclude the conversation by integrating these forms of commitment into your environment to use them as a *behavioral prompt*: a source of passive influence. As introduced in Chapter 5, a prompt is a retrieval cue that can trigger an intender to perform a behavior by fortifying the relationship between an intention and an action and making a given comportment salient.[xxxv] You can make the descriptive norms visually tangible using physical and digital tools depending on the working configuration of the group: from beautifully framed posters on a wall to digital notes on your digital collaboration tool of choice. Defining the descriptive norms of the team represents merely the initial part of a more strategic and sustained effort oriented to establish a creative and efficient design culture within your team or workgroup. Within the context of this effort, rituals represent an influential tool at your disposal that can facilitate the process of nurturing and reinforcing the belief-condition-behavior-outcome behavioral chain that delineates the culture of your team.

Clarify Prescriptive Norms

Prescriptive norms, also called injunctive norms, are defined by our perception of whether persons considered important desire the performance or nonperformance of a given behavior.[xxxvi] Within the context of a design team, they delineate the traits of its design culture that intangibly permeate the design operations. Prescriptive norms allow you to promote "structure and clarity," define individual and collective responsibilities, and facilitate communication about performance.[xxxvii] The team needs to know how to collaborate creatively; this means that, as the person in charge of the group, you have to clarify the individual and collective expected behaviors corresponding to each role present within the team at any point of the problem-solving process. Each team member needs to comprehend how to complement the collective intelligence. This social element also enhances the promotive effect of descriptive norms on "dependability."[xxxviii]

Beyond understanding roles and responsibilities, prescriptive norms that instantiate human-centered behaviors can benefit junior designers who still need to acquire all the mental models underpinning the design process. Additionally, specifying individual and collective expected behaviors also facilitates the creation of a shared definition of success that supports the communication about performance, providing a frame of reference that can be used to offer appreciative and constructive feedback. In Chapter 7, you learned how to assist the team members in identifying their superpowers and developing a sensibility around their unique contribution. In this section, you will learn how to orchestrate those abilities with the intent to clarify the prescriptive norms underpinning the design operations. This process can be divided into two steps.

Step 1: Clarify expected individual behaviors. These are the activities that a person needs to complete to fulfill the personal responsibilities in relation to a specific role. You can investigate these requirements by answering questions similar to the following ones:

> *Is a researcher expected to create a list of actionable insights after a research session?*

> *Is a designer expected to create a prototype to discuss a design decision?*

> *Is a team member expected to provide feedback to collaborators?*

Step 2: Clarify expected collective behaviors. These are the activities that a person needs to complete to fulfill the collective responsibilities in relation to a specific role. You can investigate these requirements by answering questions similar to the following ones:

> *Is a researcher expected to collaborate with designers to uncover and frame insights?*

> *Is a designer expected to collaborate with developers to identify challenges and frame problems?*

> *Is a team expected to demonstrate a bias toward action when solving problems?*

Discussing and collectively agreeing on individual and collective expected behaviors may sound redundant and unnecessary. However, the higher is the degree of diversity present within the team, the higher is the likelihood of encountering divergent behaviors, and therefore the higher is the need for "structure and clarity" around roles and responsibilities. Additionally, communicating clear behavioral expectations reinforces the perception of a psychologically safe climate because it reduces the level of uncertainty and, consequently, risk associated with taking action.[xxxix] While restrictive from a superficial viewpoint, this level of contextual awareness counterintuitively equips team members with an empowering perspective that increases their proactiveness and enables them to trespass those boundaries when necessary to support the collective effort.[xl] Creativity requires discipline and management; to break the rules, you must first master them.

Promote Accountability

Once you have introduced descriptive and prescriptive norms, you must promote individual and collective accountability toward these social standards. You can concurrently approach this challenge from two opposite angles: use constructive feedback to modify undesired behaviors and appreciative feedback to reinforce desired ones. Beyond increasing accountability toward social norms, offering feedback also decreases performance ambiguity, which directly increases the perception of "structure and clarity" and, consequently, group cohesiveness.[xli]

As you learned in Chapter 7, feedback can be offered using individual conversations and collective discussions; one-on-one meetings and moments of group reflection represent two typical examples of this form of communication. When you intend to provide constructive feedback to modify an undesired behavior, you must first understand the context of the related situation to ensure that your communication addresses the underlying causes and not the symptomatic[4] effects. Answering the following questions can help you to explore these conditions:

Does that person/team have the required skills?

Does that person/team have the required resources?

Does that person/team lack intrinsic or extrinsic motivation?

---

[4] A symptom can be a cause (e.g., fatigue as a cause of long hours), can be an effect (e.g., depression as an effect of being made redundant), and can be both (e.g., depression as an effect of losing the job and cause of suicide).

Concurrently with constructive conversations, you also have to utilize appreciative feedback to reinforce the desired behaviors already present within the team. Expressing appreciation leverages the need for social recognition that remains one of the deepest and intrinsic human motivators. Beyond impromptu exchanges, you can institutionalize offering appreciative feedback using different strategies, from supporting casual interaction creating a #kudos channel on your messaging tool of choice to formal ones such as instituting an award assigned during an official ceremony.

Irrespective of the communicative strategy adopted, you must always link these celebrations to the group's process, product, and professional objectives to reinforce the sense of accountability. Public celebrations must always prioritize collective entities over individuals to increase group cohesiveness, promote prosocial focus, and maximize collective performance.[xlii] This celebration strategy is oriented to create a reward system that associates high social status to prosocial behaviors that generate value for the collective intelligence and prevent the development of unhealthy individualities.[xliii]

Despite the need to prioritize collective entities, celebrating individuals is also necessary to preserve the social economy of the team. You can address this dichotomy by celebrating the group in public and commemorating individuals in private.

Illuminate the Design Value

Design is the practice of generating value through problem-solving.[xliv] The more value a given solution generates, the more it is considered "appropriate." From the perspective of a human being, life is permeated by several different classes of value that can all be reconducted to five core types. The following list ranks these five types, from top to bottom, in ascending order of perceptual and cognitive intensity:[xlv]

Functional: Does this have the required capability?

Financial: Does this have an affordable cost?

Emotional: Does this trigger my desired psychological state?

Identity: Does this reflect my qualities and attributes?

Meaningful: Does this align with my belief system?

The first two, functional and financial, are quantitative and tangible types of value; they typically account for a counterintuitively small part of the total value exchanged within the context of a given relationship.[xlvi] The last three, emotional, identity, and meaningful, are qualitative and intangible types of value; they typically account for the significant majority of the total value experienced exchanged within the context of a given relationship.[xlvii] Figure 9-6 illustrates the cognitive relationship between quantitative and qualitative types of values.[xlviii]

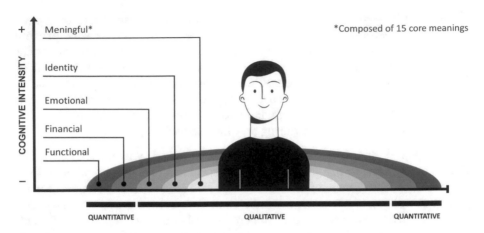

Figure 9-6. The five core types of human value

The most significant and intense type of value, meaningful, is composed of 15 core meanings.[xlix] Louis Cheskin, clinical psychologist and scientific researcher who pioneered marketing research and customer-driven product design, found that human beings experience products and services independently from their social traits via a set of 15 core universal meanings:

Accomplishment: The sense of fulfilling an intention

Beauty: The sense of appreciating a quality that provides pleasure

Community: The sense of connecting with other human beings

Creation: The sense of producing something original and appropriate

Duty: The sense of devoting to a responsibility

Enlightenment: The sense of understanding through practice and reflection

Freedom: The sense of living without undesired constraints

Harmony: The sense of balancing the relationship of parts to a whole

Justice: The sense of ensuring an equitable treatment

Oneness: The sense of unifying with the surrounding environment

Redemption: The sense of making amend

Security: The sense of living without undesired uncertainty

Truth: The sense of committing to integrity

Validation: The sense of receiving respect for a contribution

Wonder: The sense of experiencing beyond comprehension

Meaningful, identity, and emotional are qualitative types of values characterized by an invisible nature. One rare occasion where this type of business value is distinctly visible is when a company is acquired by another organization, and the balance sheet of the company purchased is distinctly different before and after the sale.

On April 9, 2012, Facebook acquired Instagram. One week prior to the acquisition, the *book value* of Instagram was $36 million; a few days later, it increased to $86 million with an additional investment of capital.[i] On that day, Facebook acquired a company with a book value of $86 million for $1.1 billion. Why did Facebook accept to pay 13 times more than the functional and financial value of Instagram? Why did Facebook pay an additional $1 billion to acquire a simple social photo feed application? That monetary portion of that transaction, which accounted for 92% of the total company value, was the estimated worth of the emotional, identity, and meaningful value that Instagram generated within the context of a given relationship with its customers before the acquisition.

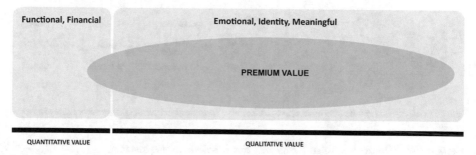

Figure 9-7. The relationship between quantitative, qualitative, and a premium value

The qualitative value is typically added to the book value using the label *good-will*, and it almost entirely accounts for the *premium* value (Figure 9-7) generated by a given organization within the context of a relationship with its customers.[ii] The premium value identifies the additional cost that a customer would pay for a product or service over an equivalent replacement, and it represents the most effective form of competitive advantage because it cannot be copied.[iii] Specifically, all five types of value, but more pronouncedly qualitative and premium ones, can only be exchanged within the context of a relationship, a social connection between a human being and another entity, whether another person or an organization. The peculiar aspect of a relationship is that it cannot be reproduced by merely copying its functional element.[iiii] Once more, Facebook made this crucial characteristic of value evident when it unsuccessfully made a millionaire offer to Snapchat to acquire the company, a business based on an ordinary camera application. Only when Snapchat declined the offer, Facebook decided to copy its functionalities.

Introducing this concept to the team and expanding their perception and awareness of the type of value that they can generate with their work deepens the level of their design conversations and enhances their ability to prepare and make decisions.

Frame the Design Effort

The value that the design team can generate needs to be illuminated but also framed within the context of the company's business model. This intent requires connecting the decisions made during the problem-solving process to the impact produced by those actions. Framing the design effort means attaching significance to every problem-solving initiative by elucidating the connection between six interrelated elements: the purpose, mission, and vision of the organization, the challenge faced, the problem identified, and the solution proposed (Figure 9-8).

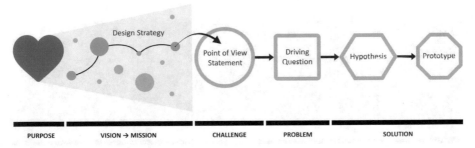

Figure 9-8. The six interrelated elements of a problem-solving instance

Purpose: The aspiration that defines the reason to exist beyond making profits. This element represents the company's "why," and its primary intent is to manifest the fundamental intention of the organization:

// Organizational purpose statement

To encourage exploration by inspiring curiosity for the good of humanity.

Vision: The unattainable dream that embodies the purpose. This element represents the company's "how," and its primary intent is to create a shared sense of direction across the organization:

// Vision statement

To discover and expand knowledge of the solar system for the benefit of humanity.

Mission: The attainable overarching objective that instantiates the vision. This element represents the company's "what," and its primary intent is to define a measurable, inspirational, and shared achievement across the organization:

// *Mission statement*

To lead innovative and sustainable programs of exploration to enable human expansion across the solar system and bring new knowledge and opportunities back to Earth.

Challenge: A situation to overcome to achieve the mission. This element represents a potential barrier toward the achievement of the overarching objective, and its primary intent is to describe the needs and constraints of a specific condition:

// *Point of view statement*

A mars inhabitant needs an artificial habitat in order to survive the adverse conditions.

Problem: A condition that requires a resolution. This element represents a practical and instrumental step forward toward the achievement of the mission. This element represents an actionable representation of that challenge, and its primary intent is to identify a creative direction:

// *Driving question*

How might we build a pressurized environment that can protect mars inhabitants from the toxic atmosphere and enable socialization?

Hypothesis: A possible resolution to a problem. This element represents a formalization of a potential step forward toward the achievement of the mission, and its primary intent is to create the precondition to prove or disprove a creative direction:

// *Hypothesis statement*

We believe that placing silica aerogel shields over sufficiently ice-rich regions of the Martian surface will enable photosynthetic life to survive.

Prototype: A possible resolution to a problem. This element represents a tangible manifestation of a potential step forward toward the achievement of the mission, and its primary intent is to prove or disprove a creative direction:

// *Prototype*

Any tangible embodiment of the hypothesis.

Fulfilling a purpose, realizing a vision, and accomplishing a mission require overcoming a sequence of obstacles. Each barrier is identified by a "point of view" framed into a problem with a "driving question" that is used to formulate a hypothesis proved or disproved using a prototype. As a design manager and leader, communicating all these informational elements is essential to engage collaborators and stakeholders across the organization:

> ... we observed that ... (point of view), and we asked ourselves how might we ... (driving question)? We believe that ... (hypothesis).

The knowledge acquired overcoming a given obstacle informs the decisional process responsible for identifying the most appropriate next step. In retrospect, the sequence of challenges that a team decided to face determines the specific design strategy deployed by that group.

Later in this chapter, you will learn how to create a design strategy that can drive your and your team's actions, allowing you to craft an active participation in your organization's future.

Unlock High Levels of Performance

Achieving high levels of performance cannot prescind from a certain degree of discretionary effort: the additional attitudinal and behavioral effort demonstrated by a person to complete a task beyond its minimal functional requirements. Discretionary effort is a derivate of *personal engagement*, which is promoted by a sense of psychological safety, meaningfulness of the work produced, and availability of the resources necessary to fulfill the job requirements.[liv] Assuming an understanding of psychological safety introduced in Chapter 6, in this section, you will explore how to promote "meaning of work" and foster engagement to create the environment necessary to establish challenging objectives as a way to unlock high performance levels.

Promote Meaning of Work

When you establish a challenging objective for a team member, your intention must be to define that request at the intersection of the three dimensions defined by the NPR model: developmental needs (N), passions (P), and role requirements (R). When the match between an individual and a goal is improper, everything that follows, no matter how diligently pursued or fervently desired, inevitably suffers. The following questions can help you to reflect on the needs of a specific person at the intersection of these three dimensions:

What matters the most to [person]?

How [objective] can fulfill [person]'s developmental needs?

How [objective] can fulfill [person]'s role requirements?

In order to establish meaningful objectives for a given team member, you need to invest time and effort to investigate and surface the cardinal elements that define that person's nuances of "meaning of work." The weekly one-on-one meetings and dedicated coaching sessions represent your best opportunities to explore the cognitive and social aspects of these elements. During those conversations, you can adopt two different viewpoints to learn more about your interlocutor and understand which activities that person perceives as meaningful and impactful.

Design is the practice of generating value through problem-solving.[iv] The first perspective allows you to understand the desired cognitive layer from which a team member prefers to approach specific problems. That unique perspective correlates to the *aspect* of the problem-solving process perceived as engaging and is commonly associated with the area of the contribution that identifies the person's *superpower*; the following are the three possible levels:

- **Strategic:** Proposition layer
- **Systemic:** Interaction layer
- **Graphic:** Presentation layer

The second perspective allows you to understand the desired social layer from which they prefer to approach specific problems. That unique perspective correlates to the *role* in the problem-solving process perceived as engaging and is commonly associated with the form of contribution that identifies that person's *superpower*; the following are the three possible examples:

- **Creator:** Productional contribution
- **Exemplar:** Inspirational contribution
- **Enhancer:** Supportive contribution

Different team members are likely to experience fulfillment contributing to the problem-solving process from different cognitive and social layers, often even concurrently. Some individuals experience satisfaction predominantly providing a contribution from a "strategic" layer with an "exemplar" or "enhancer" role; persons demonstrating this personality[5] trait tend to develop a career following a managerial track. Other individuals experience satisfaction predominantly providing a contribution from a "systemic" or "graphic" layer with a "creator" role; persons demonstrating this personality trait tend to develop a career following an individual contributor track.

Beyond all the possible shades of personality, the pivotal concept behind this attitudinal and behavioral dynamic is that every team member is unique, and when possible, you must establish objectives considering this discriminator. Depending on the level of emotional intelligence maturity and, consequently, the degree of self-awareness of your interlocutor during these conversations, this dialogue can require several sessions to explore all these layers. When completed collaboratively, the following six steps can help you to introduce priorities, objectives, and goals, preserving a sense of empowerment and promoting individual accountability:

Step 1: Introduce the organizational purpose.

Step 2: Articulate the connection between the purpose and one priority.

Step 3: Derive one to three objectives from that priority.

Step 4: Define one to three goals for each objective.

Step 5: Define the actions necessary to accomplish each goal.

Step 6: Formalize those actions into SMART[6] goals.

Similarly to what you have learned concerning communication about performance, this process represents an activity that can trigger a threat response that induces a certain level of social pain and stress. This dynamic[lvi] is more likely to be observed among personalities characterized by low levels of conscientiousness.[7] You can mitigate this tendency by projecting emotional stability and demonstrating the ability to calibrate exogenous stress during the process.[lvii]

During this conversation, it is also necessary to clarify that introducing priorities, objectives, and goals that can be perceived as meaningful, impactful, and desirable is not always possible, and when necessary, every team member must demonstrate the capacity to navigate those less motivating moments

---

[5] A personality is intended as the enduring attitudinal and behavioral configuration that delineates an individual's unique adjustment to life.

[6] SMART stands for specific, measurable, achievable, relevant, and time-bounded.

[7] Conscientiousness expresses the tendency to be organized, responsible, and diligent, construed as one end of a dimension of individual differences in the Big Five personality model.

effectively. Framing the design effort articulating the connection between the organizational purpose and the design priorities, establishing objectives, and collaboratively setting goals is typically sufficient to preserve an adequate level of motivation during these demanding situations.

Foster Engagement

According to William A. Kahn, Professor of Management and Organizations at Boston University's Questrom School of Business, who originated the concept of personal engagement at work in 1990, this construct refers to the simultaneous employment and expression of a person's true self in task behaviors that promote significant connections to the work produced and to other individuals.[lviii]

This condition identifies an active and positive work-related state characterized by vigor, dedication, and absorption.[lix] Within the context of a specific request, vigor refers to a high level of mental resilience, dedication refers to a high level of enthusiasm, and absorption refers to a high level of concentration.[lx]

Engagement is temporal in nature and tends to fluctuate based on the sequence of events punctuating a given period of time.[lxi] Engagement is also multifactorial and consists of attitudinal, emotional, and behavioral components associated with the performance.[lxii] This attitudinal, emotional, and behavioral state is influenced by three antecedent conditions:[lxiii]

Experienced psychological safety: A sense of being able to manifest and employ the self[8] in the role performances without fear of negative consequences to self-image, social status, or career progression

Experienced meaningfulness of work: A sense of return on investment of the self in the role performances

Experienced resource availability: A sense of possessing the physical, psychological, and emotional resources necessary to invest the self in the role performances

Within the context of a role performance, these three conditions are directly influenced by three experiential components:[lxiv] the social norms that regulate the working environment, the responsibilities that define the nature of the job, and the resources allocated to that specific role. These experiential components contribute to fostering engagement that exerts discretionary effort, a crucial driver of high levels of performance.[lxv] Figure 9-9 illustrates the elements of personal engagement with its antecedents and consequences.[lxvi]

---

[8] The self is intended as the totality of the individual, consisting of all characteristic attributes, conscious and unconscious, psychological and physical.

Engagement Elements

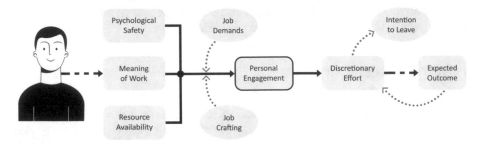

Figure 9-9. The elements of personal engagement

You have learned how to create psychological safety in Chapter 6 and promote "meaning of work" in the previous section of this chapter. In order to foster a sense of resource availability, you have to leverage the three modalities of leadership introduced in Chapter 8 to provide active and passive support to products, processes, and persons by removing any form of obstacle that impedes creative collaborations. The questionnaire in Table 9-2 can help you to assess the current level of engagement of your team.

Table 9-2. Survey questionnaire to assess personal engagement

Psychological Safety$^{\alpha}$

- At work, I can express my personality.
- At work, I can bring up problems and issues without repercussions.
- At work, I know what is expected of me every day.

Meaning of Work$^{\alpha}$

- The work I do on this job is important to me.
- The work I do on this job is meaningful to me.
- The work I do on this job is worthwhile.

Resource Availability$^{\alpha}$

- At work, I have the support I need to complete my job.
- At work, I have the resources I need to complete my job.
- At work, I am confident that the organization in the future will provide me with the support and resources I need to complete my job.

Person-Organization Alignment$^{\alpha}$

- I feel my values are aligned with the ones of my colleagues.
- I feel my values are aligned with the ones of my organization.
- I feel my personality is reflected by the values of my organization.

(continued)

Table 9-2. (continued)

Affective Commitment[α]
- I feel personally attached to my organization.
- I feel personally attached to my team.
- I feel personally attached to the problems we are trying to solve.

Discretionary Effort[α]
- I exert myself when I find it necessary.
- I persist when I need to overcome obstacles and complete crucial tasks.
- I exceed expectations to achieve personal and organizational success.

α 5+1-Point scale: Strongly disagree, Disagree, Neutral, Agree, Strongly agree.

In a state of engagement, a team member is more likely to commit to an objective that requires leaving the comfort zone and demonstrating sufficient discretionary effort to fulfill that demand.[lxvii] Psychological safety, "meaning of work," and access to resources are precursors of attitudinal, emotional, and behavioral engagement, which in turn exert discretionary effort and lead to high levels of performance.[lxviii]

Establish Challenging Objectives

Perception is a function of the locus and attention.[lxix] Human beings can only sustain one locus of attention and one thought at a time.[lxx] The capacity to direct team members' locus of attention and consequently influence the executive functions of those persons' brains by establishing clear objectives is essential to develop the "ability to drive for results," one of the five clusters of capabilities constituting the *Leadership Tent model*.[lxxi] As introduced in Chapter 4, the prefrontal cortex serves a superordinate functionality that underpins eight cognitive abilities known as executive functions:[lxxii] attention, memory, working memory, planning, decision-making, temporal integration, monitoring, and inhibitory control.

Aligning on a precise target also promotes "dependability" and "structure and clarity" within the group and allows you to run the design operations efficiently by coordinating efforts and, when necessary, mitigating frictions and resolving conflicts. As a design manager and leader, it is crucial to learn how to challenge the right persons with an adequate set of objectives that can maximize learning, elicit high performance, and concurrently avoid exhaustion.

Define Objectives for the Team

There are three main types of objectives that, directly or indirectly, you can establish for the team: process, product, and persons. The scope of your role determines your accountability and responsibility for these types of objectives.

Process objective: An achievement aimed to improve the efficacy of the problem-solving process. If you are in charge of a team, you typically have the accountability for these types of objectives. These are the achievements that model the behaviors oriented to nurture creative collaboration and optimize design operations. This type of outcome can be individual or collective depending on the behavioral domain of the objective, but it is always framed as a behavior to elicit cooperation.[lxxiii]

Product objective: An achievement aimed to improve the outcome of the problem-solving process. If you operate in a multidisciplinary environment, you typically share the accountability for these types of objectives with teams or workgroups, depending on the structure of your organization. These are the achievements that are connected with the purpose, vision, and mission of the organization. This type of outcome is always collective to foster collectivism and prevent individualism, and it is framed as a specific result to stimulate goal commitment and job satisfaction.[lxxiv]

Person objective: An achievement aimed to improve an ability required by the problem-solving process. If you are in charge of a team, you do not have the accountability and responsibility for these types of objectives. The social contract that underlies *coaching for development* specifies that you have a supportive role in fulfilling aspirations at the intersection of that person's needs (N), passions (P), and the functional requirements (R) of the organization. As mentioned in Chapter 7, the accountability and responsibility of these types of intentions belong to the coachee. This type of outcome can be individual or collective depending on the behavioral domain of the objective, but it is always framed as a behavior to promote self-efficacy and learning.[lxxv]

Besides developing the team professionally, investing time up front in defining process and product objectives also provides two significant benefits.

Decision-making. A clear objective provides a point of reference that facilitates collaboration and maintains social frictions within a healthy range. This benefit helps the team to answer questions similar to the following one:

Does [idea] help us to achieve our objective?

Distributed accountability. A clear objective provides a level of strategic alignment that facilitates synchronous and asynchronous contribution and communication about performance. This benefit helps the team to answer questions similar to the following one:

Does [behavior] help us to achieve our objective?

Depending on the scope of your role and your stage of contribution, you may need to define a different combination of strategic and tactical objectives spanning different time horizons:

- **Circa five years:** Typically referred to as visionary
- **Circa one year:** Typically referred to as strategic
- **Circa one quarter:** Typically referred to as tactical

An objective must always derive from an organizational priority to ensure its alignment with the company's purpose, vision, and mission. A priority can be identified by answering the following question:

What is most important for the next [time horizon]?

The steps required to fulfill a priority identify the concatenation of objectives that needs to be achieved within that specific time horizon. An objective can be identified by answering the following question:

What do we need to do to [priority]?

Once you have identified one or a sequence of objectives, you have to decompose each one of them into goals and actions. During this part of the process, you must minimize the number of elements allocated to a specific time frame to maximize the team's level of potential sustained attention and operational efficiency.[lxxvi] The more distant is the time horizon of an objective, the more ambiguous is the process of defining goals and actions is, and the greater is the level of abstraction of these forms of achievements. Within a given time frame, the degree of abstraction decreases from a priority to an objective, goal, and action. The following example demonstrates this relationship based on a process objective:

// Priority

Create a human-centered organization.

// Objective

Create a learning-oriented culture.

... ...

// Goal

Reach 100% of the problem framed using a formal hypothesis structure by Q2.

... ...

// Action

Run 1 workshop to demonstrate how to frame challenges into problems by May.

... ...

Under favorable social conditions, identifying and connecting priorities, objectives, and goals form an operational structure that enables high performances.[lxxvii] This practice also reduces the level of uncertainty and risk associated with taking actions, promoting creativity, and reinforcing the perception of a psychologically safe climate.[lxxviii] Furthermore, framing the design effort and establishing clear objectives provides visibility across the organization on what individuals and groups within the design team are committed to achieving; this condition increases team members' accountability and elevates the group's reputation.[lxxix]

Define Objectives for the Team Members

As a design manager and leader, you need to push individuals outside their comfort zone in an area where their competencies can be expanded healthily, supporting their personal development and increasing their potential contribution to the team and organization. This capacity predominantly necessitates two complementary competencies: a social-oriented one, the "ability to inspire and motivate others to high performance," and a results-oriented one, "the ability to establish challenging objectives."

This intertwined interplay represents another substantiation of the fact that social-oriented skills magnify the expression of results-oriented skills and vice versa.[lxxx] In this specific example, they both assume the role of a *competency companion* for the other ability. Even in this circumstance, a psychologically safe social substrate is essential to promote the candid conversations necessary to facilitate the process of establishing challenging objectives and calibrate the inevitable exogenous stress associated with it.[lxxxi]

Calibrate Exogenous Stress

Optimizing the design operations cannot prescind from considering the social component of collaboration, which, as you learned in Chapter 8, enhances the ability to confront two neurologically distinct peculiarities of uncertainty: risk and ambiguity. A central regulator of that social dynamic is stress, intended as an adaptive response triggered by a *stressor* that generates a perception interpreted as a threat to existence or well-being.[lxxxii] Stress can be endogenous when the trigger originates from internal sources and exogenous when the stimulus that elicits that reaction derives from the external environment.[lxxxiii]

Targets, whether in the form of an objective or a goal, represent a form of exogenous stress that originates from a situation characterized by an onerous demand associated with a condition of scarce resources that exceeds the person's coping ability. When prolonged under unhealthy conditions, this situation tends to develop a state of psychological and physical exhaustion accompanied by decreased motivation and a negative attitude toward oneself and others.[lxxxiv] Specific symptoms that are considered typical for this form of fatigue also occur in depression; the distinguishing factor for depression is that it is also characterized by low self-esteem and suicidal tendencies.[lxxxv]

Ultimately, managing exogenous stress means calibrating the psychological and physical fatigue potentially generated by objectives, with the intent of maximizing learning and eliciting high performance without reaching a state of exhaustion. In order to find this balance, you have to understand stress from a broader perspective. Hans Selye, the researcher who brought the term *stress* into our everyday vocabulary, divided this form of physiological response into two categories: "eustress," also referred to as positive, healthy stress, and "distress," also called negative, unhealthy stress.[lxxxvi]

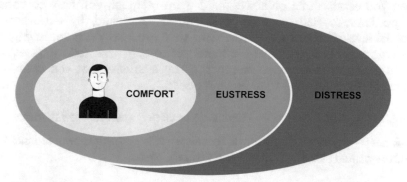

Figure 9-10. The connection between the zone of comfort, eustress, and distress

As illustrated in Figure 9-10, starting from the inner part of the Venn diagram, the most proximal dimension of a person is represented by the "comfort zone," where the level of psychological safety is high, but objectives are not sufficiently challenging to generate any level of statistically significant stress. In this dimension, the person feels safe but cognitively disengaged.[lxxxvii] Under these conditions, if psychological safety decreases in combination with a low level of stress, the experience degrades further into an "anxiety zone," an unhealthy dimension where the team feels unsafe and disengaged.[lxxxviii]

The next dimension is represented by the "learning zone," where the level of psychological safety is high, and the objectives are sufficiently challenging to generate a healthy level of stress or "eustress."[9] In this dimension, the person feels safe and cognitively engaged because the demand and resources required to achieve a given objective go beyond the comfort zone but remain within

---

[9] Eustress constitutes the positive stress response, involving optimal levels of stimulation, a type of cognitive tension that results from challenging but attainable tasks.

the person's coping ability.[lxxxix] This social condition maximizes the individual's sustainably attainable performance and represents a precondition for building a continuous learning culture.

The most distant dimension is represented by the "anxiety zone," where the level of psychological safety is low, and the objectives are excessively challenging, generating an unhealthy level of stress called "distress."[10] In this dimension, the person feels unsafe and cognitively strained because the demand and resources required to achieve a given objective go beyond the comfort zone and the person's coping ability.[xc] This social condition inhibits critical and creative thinking and can be highly destructive, affecting the brain functionalities to a point where they develop into a systemic malfunction.[xci]

When you establish an objective for a team member, you have to consider that person's capacity to cope with the demand and, by extension, the potential level of stress associated with that request. With some degree of abstraction related to "environmental factors,"[11] the following equation illustrates the relationship between the key elements that the brain, often unconsciously, processes to assess a request:

Stress = f (available abilities - demanded abilities) / environmental factors

Abilities = personal character + technical skills + (results-oriented skills x social-oriented skills) + organizational change skills.

The environmental factors include, among other constituents, the antecedents of engagement such as the perceived level of psychological safety, meaning of work, and resources available and, additionally, the stressors derived from the personal life of the team members. Once the brain has processed a request, the result of that assessment shapes the person's attitudinal and behavioral tendency toward that given demand: the person's salient beliefs concerning the outcome of the required behavior and its implications.[xcii] The following question summarizes the rationality of this process:

Am I able to complete this task successfully without jeopardizing my social status within the group?

---

[10] Distress constitutes the negative stress response, involving pessimal levels of stimulation, a type of cognitive tension that results from challenging but unattainable tasks.

[11] Beyond the stressors derived from personal life, the environmental factors also include the antecedents of engagement such as the perceived level of psychological safety, "meaning of work," and resources available.

Individuals manifest an automatic and relatively consistent physiological response to eustressing and distressing situations delineating what is defined as the *general adaptation syndrome*.[xciii] The physiological reaction to these situations traverses three phases: alarm, resistance, and adaptation or exhaustion (Figure 9-11).

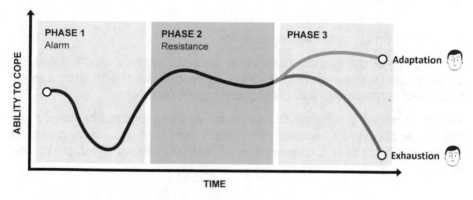

Figure 9-11. The general adaptation syndrome

The "alarm phase" occurs when an individual processes a behavioral request that triggers a threat response.[xciv] The magnitude of this reaction, described by the aforementioned equation, is a function of the intensity of the stressor that, as introduced in Chapter 7, originates from five dimensions: status, affiliation, fairness, certainty, and autonomy.[xcv] In this phase, the threat response generates a temporary physiological and psychological shock that negatively affects the brain's executive functions reducing the ability to cope with the request.[xcvi]

The "resistance phase" occurs when the brain activates various physiological responses that, among other substances, produce cortisol to favor the engagement of coping mechanisms initiated to overcome the effects of the stressor.[xcvii] In this phase, to support this mechanism, there is a vast increase in energy production and utilization of nutrients in the body that increase the ability to cope with the request.[xcviii]

The initial two phases characterize both eustressing and distressing situations with minimal variations, while the third and concluding phase is distinct and specific to these two types of adaptive responses.[xcix]

The "adaptation phase" in a eustressing situation occurs when, after a protracted exposure to a stressor that trespasses the comfort zone but remains inside the person's coping ability, the body consolidates the physiological response into a specific adaptation that results in a functional enhancement aimed to restore the homeostasis.[c]

The "exhaustion phase" in a distressing situation occurs when, after a prolonged exposure to a stressor that exceeds the comfort zone and the person's coping ability, the body fails to consolidate the physiological response into a specific adaptation and gradually depletes its energy resources in the attempt to cope with the request to a point where the situation generates systemic damage.[ci]

Exogenous Stress and Neuroplasticity

The intensity of the stressors applied to the team members and the capacity of the group to cope with these forces represent two crucial determinants that affect development and, consequently, performance.[cii]

When teams and workgroups solve problems under eustressing conditions, they operate in an optimal neural-rich environment that promotes neuroplasticity and, as a consequence of that, continuous development and learning.[ciii]

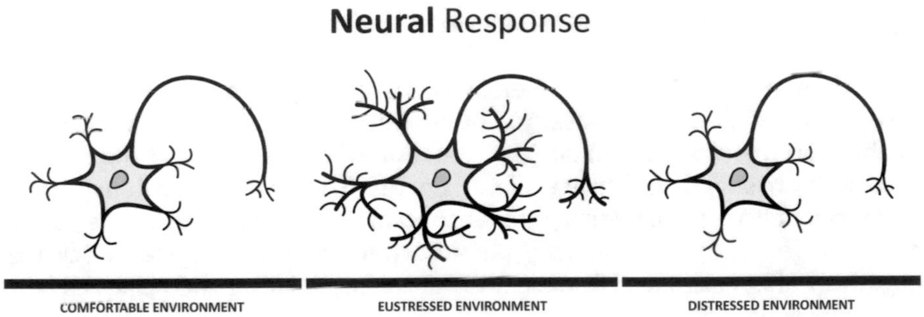

Neural Response

COMFORTABLE ENVIRONMENT   EUSTRESSED ENVIRONMENT   DISTRESSED ENVIRONMENT

Figure 9-12. A multipolar neuron in comfortable, eustressed, and distressed environments

The *pro-synaptogenesis effect* of a eustressing condition that characterizes a neurally rich environment can be mapped directly on the level of exogenous stress experienced by each individual within that specific context.[civ] As illustrated in Figure 9-12, in a state of eustress, the brain cells produce a response characterized by a higher rate of synaptogenesis[12] and a more complex arbor of dendrites that leads to an enhanced brain activity and an increased overall potential.[cv]

---

[12] Synaptogenesis refers to the formation of synapses between neurons as axons and dendrites grow.

Monitor Objectives

Once the team or the team member has accepted the objective, it is necessary to monitor the cognitive impact of that request to ensure that it promotes the optimal environment for neuroplasticity, development, and learning; the following three steps capture this process:

1. Establish the objective.

2. Observe the behavioral response.

3. If necessary, adapt the objective.

After you have established the objective, you need to observe the behavioral response of the receiving part.[13] If you set a target that does not produce an initial "alarm phase," you can assume that your request did not push the person or group outside the comfort zone.[cvi]

Suppose that, after the expected initial reduction in performance triggered by the "alarm phase," the person or group fails to adapt to the behavioral requirement of the demand, gradually and constantly decreasing the coping abilities and consequently performance. In that case, your request generated a distressing condition that requires you to adapt the objective to reduce its complexity or intensity to match the coping capacity of that person or group.[cvii]

Suppose instead that, after the expected initial reduction in performance triggered by the "alarm phase," the person or group seems capable of adapting sustainably to the behavioral requirement of the objectives. In that case, your request generated a eustressing condition that possesses the potential to promote neuroplasticity, development, and learning.[cviii]

Establishing objectives is not solely an essential tool to promote "dependability" and increase "structure and clarity," but when those requests continuously push the team outside the "comfort zone" into the "eustress" area, they create a cognitively stimulating environment that provides a sense of professional advancement.[cix] This correlation also represents a crucial element that you must consider when you identify the activities that punctuate your developmental plan.

---

[13] This concept also applies to yourself when you identify the activities that punctuate your developmental plan.

Increase the Zone of Proximal Development

The zone of proximal development is defined as the area between an individual's current and potential developmental levels.[cx] This cognitive distance is determined by the gap between the problem-solving capacity of that person working independently and under the guidance of a more capable collaborator.[cxi]

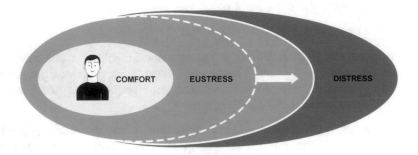

Figure 9-13. The expansion of the eustress domain

Increasing the zone of proximal development of an individual means expanding the eustress domain within healthy ranges (Figure 9-13), increasing that person's coping ability and, consequently, the level of attainable and sustainable performance.[cxii] A consequence of this expansion is the redefinition of the situations that can cause distress. You can increase the zone of proximal development of individuals and groups in two primary ways.

Assemble a heterogeneous duo: Pair a person with an individual with higher levels of proficiency.

Assemble a heterogeneous team: Form a group composed of individuals with different levels of proficiency.

In both scenarios, the collaboration between more and less capable persons promotes a form of cooperative learning, also known as *scaffolding*. Scaffolding is defined as the process that enables individuals to solve a problem that would otherwise be beyond their problem-solving capacity.[cxiii] Scaffolding requires a more knowledgeable individual controlling those elements of the task that are initially beyond the learner's coping ability, allowing that person to concentrate on completing only the parts within a range of competence.[cxiv] A designer collaborating with a more senior practitioner, a design review, and a coaching session can be three typical examples of scaffolding.

When the assisted person struggles, the more senior person reinforces the scaffolding, increasing the specificity of the instructions provided to unlock progress and promote development; the instructions can be offered verbally or behaviorally, modeling the required conduct.[cxv]

When the assisted person demonstrates signs of efficiency, the more senior person instead provides less specific support to promote confidence and autonomy.[cxvi]

The instructions received during this form of social interaction are processed and internalized by the assisted person in the form of a mental model that, once consolidated, can guide that person's behavior independently.[cxvii] When that individual has developed a mental model of the required behavior, specific instructions are no longer necessary, and any form of additional support must be provided via coaching, leveraging active listening and proactive inquiry.

Support the Team Activities

In this section, you will learn how to support the necessary infrastructure to optimize the design operations. You will explore the creative nature of strategy and learn how to make strategic decisions while navigating ambiguous problems. You will also explore how to communicate about design to promote social exchanges that can initiate candid and productive discussions without generating psychological strain and being perceived as a social threat. This section also presents how to choreograph meetings, conduct design critiques, report effectively to your line manager, nurture collaboration with individual accountability, and integrate your intuition in your decision-preparation and decision-making process.

Create a Strategy

Thinking and behaving strategically while navigating ambiguous situations are essential competencies for a design manager who manages like a leader. Strategy is the act of making choices under uncertainty and competition to increase the likelihood of being successful in the present and the future.[cxviii] Depending on your current stage of contribution, you can have different degrees of control over the present, but even if you are a Chief Design Officer, the future remains, de facto, uncontrollable. Despite this fact, by driving your and your team's actions, a strategy allows you to craft a proactive participation in the future.

The Creative Nature of Strategy

Strategy is often approached as an analytical exercise where logical reasoning is leveraged to develop a plan of action. Logical reasoning is at the foundation of the scientific method that, given a context, aims to build a logical and self-consistent model that describes that specific situation and discerns the activities by which success can be achieved.

Aristotle is accredited as the author of the earliest systematic treatise on scientific inquiry in the western tradition. With the *Prior and Posterior Analytics* written around 400 B.C., the Greek philosopher and polymath provided the first formal conception of cause and effect by articulating his seminal scientific method defining boundary conditions for his theory.[cxix] In that treatise, Aristotle peremptorily specified that the analytical approach must only be used in "the part of the world in which things cannot be other than they are."[cxx] From a design perspective, this admonition means that the scientific method must only be used to solve *tame* problems.[14]

In his treatise, Aristotle also cautioned that there is also another "part of the world in which things can be other than they are," and that under these circumstances, the analytical approach is utterly inappropriate.[cxxi] In that context, predominantly characterized by relationships and interactions between human beings, Aristotle peremptorily specified that the method used to develop our understanding of the situation must be the *rhetoric*: a dialogue between parties that builds comprehension and contributes to shape and alter that part of the world. In a world in which things can be other than they are, in opposition to the scientific method, rigor is instantiated, leveraging creativity to imagine possibilities and choose the one for which the most compelling argument can be made.[cxxii] From a design perspective, this admonition means that a *wicked* problem[15] can only be solved using the creative method.

A strategy is a creative exercise where imaginative reasoning is leveraged to identify possibilities and test hypotheses to prove or disprove those successful conditions.[cxxiii] A strategy possesses three distinct characteristics.[cxxix]

Characteristic 1: Human-centered and business-focused. A strategy leverages problem-solving to generate value for a particular group of individuals within the context of a specific business model.

Characteristic 2: Rigorously imaginative. A strategy concurrently observes the world analytically to understand how it is and creatively to envision how it could be different.

Characteristic 3: Socially enhanced. A strategy invites engagement across all the relevant stakeholders and final users.

---

[14] As introduced in Chapter 7, a tame problem is not unique and can be standardized, and the precise definition of the problem also unveils the solution.

[15] As introduced in Chapter 7, a wicked problem is unique and cannot be standardized, and the ambiguous frame of the problem does not unveil the solution.

When we work on something unique with novel or innovative attributes, it is not possible to prove analytically that a new idea is "good" in advance because there is no data about how that concept will interact with the world.[cxxv] Strategic decisions must be rigorously creative, not purely analytical, because it is impossible to predict the future.[cxxvi]

In the following two sections, you will explore a strategy blueprint created by Roger L. Martin, Professor Emeritus at the Rotman School of Management at the University of Toronto, where he served as Dean and a precursor that championed the use of creativity to create a strategy. The blueprint comprises a decisional framework to destructure possibilities and a process map to navigate the problem-solving journey.[cxxvii]

The Strategy Framework

A strategy is an integrated set of choices that uniquely positions a business in a specific market; specifically, it provides answers to five interrelated questions.[cxxviii] The framework in Figure 9-14 presents these five choices as part of a reinforcing cascade process that you can iteratively follow to deconstruct a strategic possibility.[cxxix] The choice at step X sets the context for the decision at step X+1, and the answer at step X+1 refines the question at step X.

Figure 9-14. The strategy framework

Decision I: Strategic aspiration. This choice defines success for your team, workgroup, or organization. At this point, you decide how to describe the desired future state. The strategic aspiration instantiates the purpose and vision of the organization considering the type of value generated for the customers. Typically, it tends to connect with specific benchmarks that measure the progress toward its achievement; the following is the Nike strategic aspiration:

To bring inspiration and innovation to every athlete in the world.

If you create a strategy for your team, you have to purposefully derive the team's strategic aspiration from the organizational one to connect the design effort to the business impact and promote "meaning of work."

Decision 2: Where to compete. This choice defines the competitive field. At this point, you decide how to describe the context of the desired future state. Where to compete identifies the specific activities that need to be undertaken to achieve the strategic aspiration and the business domain within which the team, workgroup, or organization can find its competitive advantage. This domain is comprised of five dimensions that can be identified by answering the following five questions. The significance of each dimension varies based on the business context of the company.

Geography: The countries or regions of the competition. Are you going to compete on local or global geography?

Customer: The primary segment of individuals to serve. What are the needs and pain points of your core customers?

Offer: The primary product or service to sell. What product or service will represent your core offering?

Channel: The business connection to use. Are you going to reach the customers directly or via external channels?

Production: The accountability and responsibility of the product. Are you going to develop the offering internally or via external production?

Decision 3: How to succeed. This choice defines the competitive advantage. At this point, you decide how to produce the value that characterizes the desired future state. How to succeed identifies the specific decisions necessary to generate quantitative and qualitative value at the intersection of the aforementioned five business dimensions. These choices can follow one of the two distinct approaches.

Focus on quantitative value: A strategy based on a low-cost, high-volume value proposition that generates predominantly functional and financial value.

Focus on qualitative value: A strategy based on a high-price, low-volume value proposition that generates predominantly emotional, identity, and meaningful value.

Decision 4: Specific capabilities. This choice defines the activities required to build your competitive advantage. At this point, you decide how to develop or acquire the abilities necessary to achieve the desired future state. The specific capabilities underpin elements such as, but not limited to, design, development, and marketing.

Decision 5: Management infrastructure. This choice defines the systems required to support, foster, and quantify the strategy. At this point, you decide how to build and maintain the specific capabilities necessary to achieve the desired future state. The management infrastructure underpins elements such as, but not limited to, the product front end and the back end, the organizational structure, and key performance indicators.

With an understanding of how to deconstruct and articulate a strategic possibility, in the next section, you will explore how to navigate the entire problem-solving process, from identifying the challenge to making the final strategic choice.

The Strategy Map

The creation of a strategy is often interpreted as a *thinking* activity, but in reality, it possesses the same distinct *doing* component that characterizes human-centered design. The map in Figure 9-15 presents a sequence of seven steps as part of an exploratory process that you can iteratively follow to navigate uncertainty and ambiguity, maintaining a bias toward action.[cxxx]

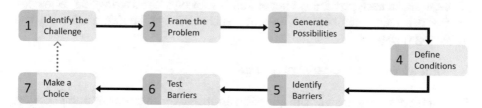

Figure 9-15. The strategy map

Step 1: Identify the challenge. Depending on the scope of your role, this step analyzes your team, workgroup, or organization to discover the most critical challenge to overcome. The following questions can drive this investigation:

> *What are our purpose and vision?*
>
> *What are our critical challenges?*
>
> *What is preventing us from achieving our strategic aspirations?*

The answers to the preceding questions can help you to identify a challenge that is instrumental to the success of your company.

Step 2: Frame the problem. This step explores the challenge and turns it into a problem using a driving question:

> *How might we [challenge]?*

The driving question builds the foundation for generative thinking; use it as a prompt to preserve the appropriateness of your underlying idea while pushing its originality to enhance its creative factor. If needed, you can refine the driving question by answering the following questions:

Geography: What if we compete in [different market]?

Customer: What if we serve [different primary segment]?

Offer: What if we provide [novel product or service]?

Channel: What if we reach customers using [different connection]?

Production: What if we change the accountabilities and responsibilities?

The answers to the preceding questions can help you to extremize the constraints of the "where to play" dimensions.

Step 3: Generate possibilities. This step brainstorms strategic choices that could answer the driving question compellingly and solve its correlated problem.

Step 4: Define successful conditions. This step identifies the circumstances that can make each possibility successful. Adopting a "lead alongside" behavior invites the team members to evaluate each strategic choice answering the following question:

> *What would have to be true to [possibility]?*

For each possibility, answer the question adopting three different perspectives:

Customers: The needs and pain points of your core customer

Company: The capabilities necessary to design, deliver, and maintain the possibility

Competition: The direct, adjacent, and emerging players in your industry

Step 5: Identify barriers. This step identifies the most critical condition for each of the possibilities generated, the ones with a high level of uncertainty and ambiguity. These conditions represent the hypothetical barriers to success. Explore which *what would have to be true* condition can potentially prevent you from choosing a given possibility by answering the following questions:

> *Which condition is the most uncertain?*
>
> *Which condition is the most ambiguous?*

For example, if it would have to be true that you have to acquire one million customers by Q4 and that requires some specific capabilities, ask the team: To what extent do we believe we could make it true? If the group is confident that they can make it true, that condition isn't a barrier.

Step 6: Test barriers. This step develops a hypothesis for each critical condition identified and then conducts experiments to prove or disprove those assumptions.

Step 7: Make a choice. This step reviews the test results using a barrier-test-learning logical structure. With the knowledge acquired, the team decides which possibility is willing to support and pursue.

The Strategy Workshop

Deciding what to do and, consequently, what not to do embodies the essence of a strategy; a workshop represents an effective method to come to these conclusions collaboratively. The workshop structure presented in this section offers you a seven-step plan that you can follow to implement the "strategy framework" and the "strategy map," preparing and making decisions that are both human-centered and business-focused:

Session one (1-2 days)

Step 1: Define the problem.

Step 2: Frame a question.

Step 3: Generate possibilities.

Between meetings (1-2 weeks)

Refine possibilities.

Session two (1-2 days)

Step 4: What would have to be true.

Step 5: Identify barriers.

Step 6: Test to learn: design and assign tests.

Between meetings (2-12 weeks)

Test to learn: conduct tests.

Session three (1 day)

Step 7: Make a choice.

The timeline presented in this example is optimized to develop a strategy at an organizational level. If your intent is to create a strategy for your team, which is derived from the organizational one and has a reduced scope, you can map this multisectional workshop on the five-day *design strategy sprint* structure. Irrespective of the decisional domain of your workshop, using the "strategy map" and the "strategy framework" allows you to combine rigor and creativity to analyze the world as it is and imagine how it could be better.

Communicate About Design

Social exchanges where information can be leveraged to initiate candid conversations about design without generating psychological strain and being perceived as a social threat are essential to achieving and sustaining high levels of performance.[cxxxi] Communicating about design constitutes a pivotal element of several structured and unstructured design activities such as critiques and moments of reflection that can unlock experiential learning and increase self-awareness.[cxxxii] These forms of communication can happen during one-on-one interaction or, more commonly for a design team, during a group activity. During these occasions, developing the ability to structure and choreograph these information exchanges is as crucial as the content of the message communicated.

Choreograph Meetings

Meetings are actively designed conversations where a design team sets directions, clarifies intentions, and advances problem-solving. When meetings are adequately managed, they raise individuals, groups, and organizations to tremendous levels of achievement, increasing, among other things, problem-solving capabilities but also engagement, satisfaction, and productivity.[cxxxiii]

An effective meeting increases the perception of "structure and clarity," leaving the attendees with a strong sense of purpose, direction, and control that provide them with a cognitive and behavioral understanding of the next objectives, goals, and actions to complete. When instead a proactive direction of meetings is not present, they reduce individual and collective productivity, and by increasing the levels of uncertainty, they raise the levels of stress to a point where it can negatively impact the well-being of the team members.[cxxxiv] A meeting represents a social event that underpins three temporal windows, the moments before, during, and after the gathering. In the following three sections, you will explore the aspects and requirements of these phases.

Before the Meeting

A meeting begins before its formal social gathering where an adequate amount of preparatory time needs to be invested initially to understand if it is necessary and, in that case, prepare it appropriately. This amount of time varies based on the type and purpose of the event. In Table 9-3, you can see the four primary types of meetings that typically characterize the day-to-day operations of a design manager and leader.

Table 9-3. The four primary types of meetings

Purpose	Intent	Example
Discuss	Present information to inform processes and practices.	Daily standup and research presentation.
Solve	Frame problems, explore solutions, and make decisions.	Convergent and divergent sessions.
Implement	Supervise the execution and effectiveness of strategic decisions.	Board and leadership meetings.
Reflect	Evaluate beliefs and behaviors against priorities, objectives, and goals.	Design reviews and retrospectives.

As mentioned in Chapter 4, one of your underlying priorities is to avoid unnecessary meetings by setting a demanding scheduling threshold because when numerous, especially if unstructured, they decrease productivity and increase the levels of perceived stress, fatigue, and workload.[cxxxv] Every time you feel that a meeting is necessary, validate that perception by answering the following three questions:

Do I need additional input to make progress?

If not, do not schedule the meeting.

Does this progress require a real-time conversation?

If not, consider the use of digital asynchronous communication.

Does this conversation necessitate a face-to-face interaction?

If not, consider the use of digital synchronous communication.

If the answers to the preceding questions are "yes" and you decide to schedule a meeting, you have to design the structure of the conversation and its participants intentionally.

Step 1: Create the agenda. This document defines a plan that structures the conversation. The meeting agenda must always include an introduction to present the purpose of the discussion, whether to discuss, solve, implement, or reflect, describe its objective, and clarify its expected outcome. Each item on the agenda must have a start time and an owner responsible for that specific segment of the conversation. A clear agenda circulated in advance via the meeting invitation also provides the attendees with the information necessary to prepare for the discussion.

Step 2: Define the participants. This list identifies the persons required for the conversation. That number of participants must be tailored around the purpose of the conversation, and it must include only the persons with the expertise and knowledge necessary to actively contribute to the discussion and achievement of the expected objectives. Irrespective of the meeting "purpose," you must always keep that number as low as possible to optimize communication and prevent social loafing[16] and the degradation of group performance;[cxxxvi] this social dynamic is more pronounced in teams with low levels of cohesiveness.[cxxxvii] Every additional individual added to a given meeting adds an extra layer of complexity that needs to be evaluated against the potential value added to the discussion by that specific person. As a general rule, the meeting must only include the minimum number of persons necessary to move the conversation forward.

Step 3: Prewire the meeting. This practice allows you to reduce the degree of potential misalignment during the conversation. Prewiring a meeting, whether oriented to discuss, solve, implement, or reflect, means introducing the relevant elements of your message to each attendee individually ahead of the formal gathering to receive feedback and have the opportunity to refine aspects of your content accordingly. This practice is essential to understand every attendee's sentiment and perspective concerning the points on the agenda. Every time you invest enough time in prewiring a meeting, you significantly increase your opportunities to achieve a win-win agreement at the end of the conversation; the ideal number of prewire discussions necessary is defined by the number of interactions required to achieve cognitive alignment ahead of the formal social gathering.

If you and the team have to present a new design, for instance, it means scheduling a series of conversations with every attendee to understand the implication of that given work on their activities, capture their sentiment, process their feedback, and, when appropriate, even make changes. During a prewire conversation, maintain an informal and friendly tone, leverage *relationship power* to connect to your interlocutor, and gather specific expectations concerning the meeting purpose and expected outcome; the following are two typical questions to ask in this case:

What is your view on [fact connected to the expected outcome]?

How does [expected outcome] affect [team or business]?

Asking these questions can help you to understand the primary decisional driver of that specific attendee and react accordingly. During those moments, approach the conversation with curiosity and a collaborative mentality to navigate the intrinsic ambiguity and complexity that characterize this type of social interaction.[cxxxviii]

---

[16] Social loafing identifies the reduction of individual effort that occurs when persons work in groups compared to when they operate alone.

Even if you diligently prewire meetings, you cannot eliminate social unpredictability from complex conversations with a high number of *points of agreement*. Despite that, if you invest the necessary time in this preparatory phase, creating a clear agenda, inviting an appropriate number of persons, and following the guidelines presented in the subsequent sections, you can decrease social unpredictability to a level where it does not represent a decisive factor.

During the Meeting

The value of a meeting is primarily defined by the quality of the conversation that originated and the extent to which that discussion concurred to the achievement of the objectives on the agenda. At this point, your responsibility is to regulate the exchange of information. The following guidelines can help you to manage this part of the meeting.

Remind social norms. When you open the conversation, introduce the social standards that need to be adopted to support the purpose and intent of that specific conversation. These standards create behavioral expectations that attendees need to consider during the discussion. The higher is the degree of diversity present within the group, the higher is the need for psychological and behavioral synchrony.[cxxxix]

Start on time. When you open the conversation, do it punctually. Meeting lateness is pervasive and potentially highly consequential for individuals, teams, and organizations. When a person breaks this elemental social norm, it triggers adverse social reactions that negatively influence individual and collective effectiveness, decreasing the perceived job satisfaction.[cxl] Starting a meeting late also negatively affects the perception of your ability to lead and manage the group,[cxli] demonstrates a lack of respect for the punctual attendees, and weakens their intention to be punctual in the future. You must always start a meeting on time, even if someone is missing. If your team or workgroup follows the good practice of having a buffer between consecutive events on the calendar, the likelihood of having this issue reduces significantly.

Delegate the agenda. When you need to "lead alongside" in direct contact with the team, consider using a team member to ensure the respect of the time allocated to each item on the agenda so that you can concentrate on two crucial tasks: monitor the social dynamics of the conversation and moderate the discussion. As you learned in Chapter 4, achieving peak performance requires you to operate in a state of deep concentration, free from distraction and concurrent activities.[cxlii]

Preserve focus. When you monitor the social dynamics of the conversation and moderate the discussion, you have one primary responsibility, maintain high levels of concentration to maximize the problem-solving capabilities of the collective intelligence. During a meeting, a conversation diverges and converges organically as a natural consequence of the brain attempting to construct meanings and create *reflective assumptions* from the inherently meaningless information shared in the physical world.[cxliii] When necessary, you need to intervene to refocus the exchange of information and realign the discussion under the purpose and objective of the agenda. The following two examples illustrate a possible response to a relevant and an irrelevant thinking direction:

> // *Relevant thinking response*
>
> *This is a great point that we haven't considered, [name].*
>
> *Let's add it to the parking lot to explore it at the end of the meeting properly and refocus on [objective].*
>
> // *Irrelevant thinking response*
>
> *This is an interesting observation, [name].*
>
> *For the purpose of this meeting, let's remain focused on [objective].*

The first part expresses gratitude for the contribution offered to preserve a trustworthy and psychologically safe climate, while the second section realigns the conversation, correcting the direction of the discussion toward the expected outcome.

Use a parking lot. When someone offers a perspective that is relevant to the purpose of the meeting but not appropriate for the specific objective on the agenda, you need to capture that thinking for future considerations without derailing the current discussion. A parking lot increases the subsequent participation and performance of individuals that offered a relevant but not appropriate viewpoint by facilitating the process of disengaging cognitively from the idea shared.[cxliv] With this approach, attendees can see that their contribution on that distinct occasion is not lost and will receive attention at the appropriate moment.

Create action items. When you conclude the discussion on a specific item on the agenda, assign accountabilities and responsibilities to the appropriate individuals to ensure that the conversation will generate progress and create value; the following is an example of that exchange:

> *In regard to [objective], [name] is going to [action], by [date].*
>
> *[name], how does this sound to you? Is there anything you need to complete this task?*

If the conversation necessitates another meeting, discuss and agree on the date and time of that session. Close a given discussion by deciding "who" does "what" by "when"; asking for explicit acceptance is crucial to increase commitment, engagement, and consequently, the likelihood of achieving the

Finish on time. When you close the conversation, do it punctually. In addition to what concerns lateness at the beginning of the meeting, not finishing on time also generates additional moments of friction due to the potential disruption that this socially disrespectful behavior can cause to the attendees' calendar. Starting punctual is necessary but not sufficient to demonstrate the ability to lead and manage meetings effectively; you must also finish on time.[cxlv]

After the Meeting

A meeting does not conclude after its formal social gathering. The information exchanged during the conversation, irrespective of its form, has value only if converted into actions that generate progress and create value. The following guidelines can help you to manage this part of the meeting.

Circulate notes. When the meeting is concluded, you have to formalize the content and result of the conversation. Meeting notes summarize the salient points of the discussion and list the action items assigned to the appropriate person. An efficient follow-up begins with taking notes during the session annotating the agreed action: "who" does "what" by "when." Circulate the notes remarking your availability to provide support to ensure the removal of potential blocker.

Control progress. When the meeting is concluded, you have to remain aware of the status of the work assigned. Create a reminder on your calendar for each action agreed, and use the one-on-one meetings to explore their status with the appropriate person. If applicable, request a report in terms of green, amber, or red.

Conduct Design Critique

Given that we do not see the world as it is, but we only perceive it through our reflective assumptions,[cxlvi] the act of evaluating the originality and appropriateness of a design solution represents an indispensable step of the problem-solving process. Conducting healthy and productive design critiques relies on two interconnected actors: the designers composing the group and you, the person in charge of the team.

The designers must be mature enough to comprehend that criticizing a design solution does not mean questioning a person's capabilities and, as a consequence of that belief, demonstrate the ability to discuss their work without perceiving those conversations as personal attacks to their social status within the group. Cognitively, each designer must be able to produce healthy and productive perceptions of the world across the three time frames that characterize short-, medium-, and long-term psychological adaptations: learning, developing, and evolving.[cxlvii] While coaching for development can have a long-term impact on a designer's mindset, the only realistic approach for you to control this element is via hiring or releasing team members

The person in charge of the group must be able to regulate the exchange of information in a way that promotes creative collaboration and minimizes socially adverse reactions to preserve the team's ability to think critically and reflect.[cxlviii] The presence of an environment permeated by trust and psychological safety is necessary but not sufficient to maintain healthy levels of social interaction; the communication also needs to be structured accordingly.[cxlix]

Structure Design Critiques

Your responsibility as a design manager and leader is to structure the design critiques with the intent to promote a candid, specific, and comprehensive analysis of the work. During a design critique, like any other moment during the problem-solving process, designers reach their maximum effectiveness when they can leverage a structure that supports their cognitive and behavioral effort.

The amount of structure required by a given team is directly related to two specific factors.

Ability: The ratio between junior and senior members. The higher is the level of seniority present within the team, the lower is the level of structure required to support a productive critique.[cl] The more senior an individual is, the more likely that person will operate autonomously, driven by a mature design mindset instead of needing a predefined procedure to follow.[cli]

Personality: The ratio between introvert and extrovert members. The higher is the level of predominant extroversion present within the team, the lower is the level of structure necessary to support a comprehensive critique.[clii] The more predominantly extrovert an individual is, the more likely that person will collaborate spontaneously, driven by a tendency toward the outer world instead of needing a predefined structure to openly express ideas.[cliii]

Personality Continuum Scale

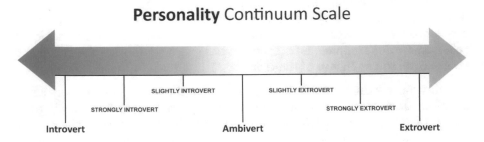

Figure 9-16. The personality continuum scale

Human beings are not entirely extroverts[17] or introverts,[18] and that orientation can vary on a continuum scale (Figure 9-16) based on personal and environmental factors.[cliv] A team member's personality can manifest predominantly extrovert characteristics in a specific situation while demonstrating predominantly introvert attitudes in others.[clv] Within the context of a design team, the higher is the level of seniority and emotional intelligence of an individual, the lower is the degree of fluctuation exhibited by that given personality.[clvi]

Fostering productive social interaction between different personality types during a design critique requires a clear set of prescriptive and descriptive norms to regulate the social exchanges of information.[clvii] The following three universal principles can help you define an initial set of social norms that you can use to drive any design conversation, including those during the design critiques. Once you have crafted the first set of social standards, radiate them into your digital and physical environment to use them as an attitudinal and behavioral prompt.

Be candid: Consistently share your opinion regarding a design solution; do not remain silent. During a video call, always have your camera on when you share your feedback to engage actively and sustain an empathetic climate.

Be specific: Consistently articulate your opinion connected with the particular design problem; do not express personal views that are irrelevant to the objective discussed.

Be comprehensive: Consistently discuss positive and negative aspects of solutions to preserve elements that are working, and improve or abandon the ones that are not satisfactory.

Information without context is meaningless.[clviii] During a design critique, a designer needs to frame the work presented by providing an adequate context to facilitate the cognitive process of elaborating the information and extracting meanings from it. Ensure that the conversation addresses the following four questions.

Question 1: Which problem is this design trying to solve? This part introduces the business objective, the *point of view* used to attack the design problem, and the *driving question* leveraged to generate the proposed solution:

This quarter our primary objective is to [objective].

… we observed that [user] needs to [need] because [insight].

… we thought how might we [question]?

---

[17] Extroversion corresponds to the orientation of one's interests and energies toward the outer world of people and things rather than the inner world of subjective experience.
[18] Introversion corresponds to the orientation of one's interests and energies toward the inner world of thoughts and feelings rather than the outer world of subjective experience.

Question 2: What are the design constraints? This part introduces the restrictions that need to be considered to critique the work, typically temporal, technical, and economical.

Question 3: At which stage is the design process? This part introduces the current phase of the problem-solving process and its expected outcome. This information also helps to set the expectation around the level of efficacy and fidelity adopted to present the solution:

> *The work is currently at the stage of [phase]. We are now working on [deliverable], and our objective is to [outcome].*

Question 4: Which type of feedback are you looking to receive? This part introduces the adequate comment required to improve the current state of the solution:

> *At this stage, we are interested in your view on the approach used and your feedback on the [design aspect].*

Before initiating the design critique, it is also essential to ask the participants if they feel that some critical element for the conversation is missing or needs to be clarified further:

> *Am I missing anything here? Does anyone have something to share before we start the critique?*

Proactively defining the structure of the design critique establishes the social preconditions necessary to manage the communication constructively. Depending on the problem-solving process phase, you may need to conduct a design critique internally with the design team or externally with multidisciplinary stakeholders.

Internal Design Critiques

This type of critique is necessary during the entire problem-solving process as a deliberate practice that is oriented to promote reflection-on-action. During these moments of reflection, your intention must be to offer feedback with the intent to protect embryonal concepts by ensuring that ideas are explored with sufficient cognitive resources without undermining their nature or dismissing them prematurely. Assuming that a solution does not present foundational incongruences, besides being candid, specific, and comprehensive, your feedback needs to possess three essential characteristics.

Processable: Feedback needs to be offered using objective communication. This approach reduces the likelihood of triggering a threat response that increases stress and impairs the ability of the brain to analyze thoughts critically, reflect on-action, and envision new possibilities.[clix]

Actionable: Feedback needs to be offered at the intersection of a specific business objective, user need, and driving question domain. This approach provides context to your critique and facilitates the process of elaborating the information and extracting meanings from it.[clx]

Empowering: Feedback needs to be offered as a suggestion. This approach demonstrates trust in the ability of the designer to iterate a solution and promotes engagement.[clxi]

The following tripartite template, created at the Design School at Stanford University, implements these characteristics and can effectively structure your message:

> // Use case: Book accommodation online
>
> [processable] I like that the user can search nearby stays.
>
> [actionable] I wish it would also be possible to get inspired by different types of locations.
>
> [empowering] What if we create a section of curated experiences based on user preferences?

The first part makes the critique processable; opening the conversation with appreciative feedback reduces the likelihood of triggering a threat response and consequently increases cognitive engagement.[clxii] The second part makes the critique actionable; offering constructive feedback connected to a specific user need and business objective triggers reflection. The third part makes the critique empowering; introducing a possible design direction as a suggestion rather than a mandate promotes commitment.

External Design Critiques

This type of critique is necessary during the entire problem-solving process as a deliberate practice oriented to promote cross-disciplinary inclusion and alignment. During these moments of collaboration, your primary intention must be connecting the design effort to the business impact to introduce promising concepts, ensuring that ideas are analyzed with a human-centric, value-oriented mindset by the workgroup without undermining their nature or dismissing them erroneously. If necessary, govern certain parts of these conversations leveraging the counterargument management strategy introduced in Chapter 5.

A distinctive ability of a designer is the ability to process feedback. Given the broad number of attendees and the diversity of backgrounds typically present during these types of meetings, the use of a digital or physical whiteboard to support the conversation can be beneficial to implement the tripartite template "I like, I wish, what if" to facilitate the management of information shared during the session. After the presentation and an initial moment of reflection, allocate ten minutes to structure the feedback asking each participant to create a minimum of one to three sticky notes for each of the "I like," "I wish," and "what if" sections (Figure 9-17).

Figure 9-17. The "I like, I wish, what if" template implemented on a whiteboard

At this point, you can manage the social interaction like a divergent session: encourage the exploration of feedback using "yes, and ..." conversations, form *themes* to surface possibilities, and create *actions* to initiate the subsequent design iteration. You can also use this approach to information management during internal design critiques every time the number of attendees surpasses the single-digit mark.

Report to Your Line Manager

Your line manager is the single most influential individual in your career because that person can promote or impede your professional progression in several ways. Despite the broad adoption of 360-degree feedback as the de facto method to employee evaluation, developing a harmonious relationship with your line manager remains a crucial competence to increase your likelihood of being successful in your role. While an individual can be open to feedback and improvements, it is unlikely that your line manager will change their working style to suit your expectations. In practical terms, if you dislike your supervisor, it is your problem, and if your supervisor dislikes you, it is also your problem. If there are no conflicts of personalities, this type of relationship can be effectively managed by building relationship power, trust, and adaptability, if necessary.

On the foundation of what was presented in Chapters 5 and 6 regarding relationship power and trust, learning how to report work effectively can significantly facilitate the development of a transparent, productive, and low-stress relationship with your line manager, and at the appropriate moment, demonstrate to that person that you are ready to increase the scope of your role and receive a promotion. When you report to your line manager, you must structure your communication based on three mental models.

Inform: These are the things that your line manager needs to know to remain in contact with your work.

Discuss: These are the things that your line manager needs to consider in the view of future decisions.

Decide: These are the things that your line manager needs to understand to make a decision that unblocks your workflow.

If the conversation is part of a meeting, use these three mental models to structure the agenda accordingly.

Share the agenda with your line manager one day before the session, including any eventual attachment required to provide context and help that person to join the conversation prepared. Take notes during the conversation and after the meeting, and share a summary that includes the topic discussed and the decisions made. When you report on the design impact, connect the outcomes to the generation of the five types of design values, and quantify the business impact they produce using key performance indicators to provide a qualitative view of the work.

Nurture Collaboration with Individual Accountability

Creative collaboration and, more broadly, design operations necessitate access to a diverse set of information from a multidisciplinary array of contributors. Designers at all levels of seniority that operate in a creative and collaborative culture possess a sense of collective responsibility that brings them to view seeking a contribution from colleagues as natural, regardless of whether a formal job description defines that interaction. Nevertheless, collaboration must not be confused with consensus, which represents an obstacle to rapid and effective decision-making, especially in situations characterized by a high degree of complexity and ambiguity. Creative collaboration is not in antithesis with individual accountability. Every design decision during the problem-solving process is an individual decision, informed by a diverse set of perspectives but still a resolution made by a single person.

Optimizing the design operations also means developing a creative and collaborative culture where designated individuals are expected to make decisions and own the consequences of their resolutions, while others are expected to contribute to the preparation of the decision and, if necessary, disagree and commit after the decision is made. These decisional functions can and must be fluid based on the context and attributes of the specific decision. Depending on the scope of your roles, you can be the designated decider of certain types of decisions while having a contributory role in other situations.

Accountability and collaboration can be complementary, and in a trustworthy and psychologically safe environment, the former can drive the latter. You can nurture collaboration with individual accountability, instituting social norms that identify the specific decider in a particular context and clarify behavioral expectations concerning the decision-preparation and decision-making process. Furthermore, holding yourself accountable publicly for your choices, even when that can potentially impact your social status, demonstrates high integrity and reinforces the sense of individual accountability within the team.[clxiii]

Regulate Decisions

Preparing and making decisions represent crucial moments within the economy of a team that can reinforce or weaken the group cohesion and everything that interconnects with it, such as the psychological and behavioral synchrony and, consequently, creative collaboration.[clxiv] In respect of the creative problem-solving process introduced in Chapter 8, decisions extend across three primary phases.

Phase 1: Frame the problem. This is the preliminary moment where a shared cognitive frame is introduced to structure the exchange of information.

Phase 2: Prepare the decision. This is the preparatory moment where the thinking diverges to explore the available design possibilities.

Phase 3: Make the decision. This is the moment where the thinking converges to commit to one of the available design possibilities. If you are the designated decider, at this point, you must consciously process information from three distinct sources:

- **Indirect experience:** Another person's conscious perception of a design possibility
- **Direct experience:** Your conscious perception of a design possibility
- **Intuitive experience:** Your nonconscious[19] reaction to a design possibility

In order to maximize your capacity to make effective decisions, you have to learn how to effectively process both rational and irrational information, which implies developing the sensibility necessary to integrate intuition into your decisional process.

---

[19] A nonconscious event describes any mental process that is not available to introspection or report.

Integrate Intuition

In opposition to the notion of direct and indirect experience as a form of knowledge related to the domain of conscious, serial, and analytical information processing, intuitive knowledge can be seen as a nonsequential information processing modality comprised of both cognitive and emotional elements that construct comprehension without the use of conscious reasoning.[clxv] Conscious and unconscious evaluations complement one another during the decisional process. However, the role of intuition is particularly relevant when the information that needs to be processed presents a low level of structure, and the decision is characterized by a high level of ambiguity (Figure 9-18).

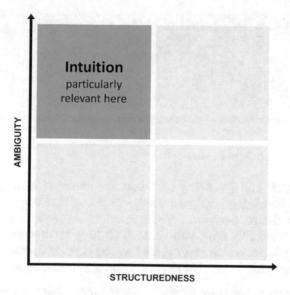

Figure 9-18. Intuition within the context of informational structuredness and decisional ambiguity

The quality of intuition is directly correlated to the comprehensiveness of the experiential learning of an individual, defined by the interrelatedness of the *reflective assumptions* that construct the *empirical significance of information* and perceptual history of that person.[20]

Intuition constructs knowledge using a holistic process that identifies a cue that allows the brain to recognize a pattern that generates a reaction that activates a specific adaptive behavioral response.[clxvi] The deeper and more

---

[20] Revisit Chapter 3 to review how the brain extracts meanings and constructs the empirical significance of information.

comprehensive is the experiential learning of an individual, the vaster is the set of patterns available to that person, and, consequently, the more accurate are the potential intuitive responses processed.[clxvii]

Intuition is your gateway to any remote location of your design experience that can help you translate that knowledge into judgment and decisional capacity. Intuition as a decision-making tool, it's not an antagonist to analysis. Based on the specific requirements of a decisional context, analysis can and must be complemented by intuition to enhance your decisional potential. Figure 9-19 illustrates the decisional continuum scale that extends across intuition and analysis.

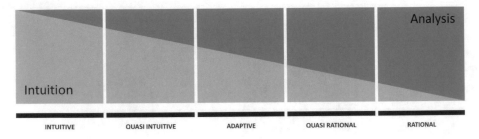

| INTUITIVE | QUASI INTUITIVE | ADAPTIVE | QUASI RATIONAL | RATIONAL |

Figure 9-19. The decisional continuum scale

Given a situation and assuming that you accumulated enough experiential learning to do it reliably, you must always begin by listening to your intuition when facing a design decision. When you initiate by rationally analyzing a decision, you inevitably limit your access to long-term memory to what is consciously available, which represents a vastly smaller dataset than the total amount of information available in your brain accessible by your intuition.[clxviii]

If you struggle to sense your intuitive preference toward a design possibility, you can use your imagination to envision the outcome of the decision or your words to describe it expressively; this approach often triggers an emotional response that can indicate your favorite direction. Sensing your intuitive preferences is a skill that takes time and practice to develop and, as you will learn later in this section, leverages self-awareness, one of the four dimensions of emotional intelligence. When you integrate conscious and unconscious evaluation, you must discern the influences of four forms of visceral forces that can affect your decisional processes:[clxix]

Incidental emotions: Your psychological states that are triggered by the decision

Situational emotions: Your psychological states that are not triggered by the decision

Personal characteristics: Your personality traits and individual preferences

Context characteristics: Your perception of the success probability of an option and the interpersonal consequences of a decision

As presented in Figure 9-20, these four visceral influences permeate the entire decisional process, including the unconscious nonsequential part that constructs intuitive knowledge.[clxx]

Decisional Process

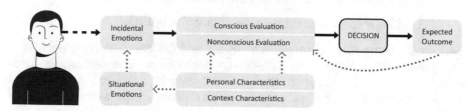

Figure 9-20. The principal influences on the decisional process (simplified)

"Incidental emotions" can be interconnected to different working-related and nonworking-related situations:[clxxi] the implications of the option available, but also a conversation that happened during the day, the social relationship with an individual present at the time of the decision, and even the weather. This type of psychological state can also affect the "situational emotions" triggered at the time of the decision.[clxxii] Your "personal characteristics," or those of the designated decider, and the "context characteristics" can affect "situational emotions" and influence conscious and nonconscious information processing.[clxxiii]

Mitigating the impact of these four visceral influences is crucial to maximizing the reliability of your intuition and the overall efficacy of your decisions.[clxxiv] While it is not possible to eliminate their psychological impact, there are a few precautions you can take to mitigate their effect on your cognitive process.

Frame decisions comprehensively. Allocate more time to the preparation than the deliberation to enrich your experiential learning and consequently increase the likelihood to identify a design cue that can access and retrieve intuitive knowledge.[clxxv] The previous section introduced decision-preparation and decision-making as two separated phases to support this scenario.

Discern intuitions from desires. Inspect the situation concentrating on the purpose, objective, and, if present, principles of the decision to mitigate any *characteristic* involved, whether personal or contextual.[clxxvi]

Listen to initial intuitions. Concentrate on the first psychological state triggered by the decision that, assuming that you developed sufficient experiential learning, statistically tends to be the most accurate.[clxxvii] Your subsequent reactions are more likely to be affected by critical thinking.[clxxviii]

Complement intuitions with analysis. Explore the first psychological state triggered by the decision using critical thinking. During this process, leverage your experiential learning and examine your intuition contextually, imagining the scenario unlocked by that specific decision. If you struggle to connect to a particular reaction, you can also use the following process to explore and compare multiple intuitions. This process can be divided into two steps:

1. Map strengths and weaknesses of the option.

2. Use mental simulations to evaluate the option.

Ultimately, being able to integrate intuition and critical thinking is necessary to evolve from making blind data-driven decisions to comprehensive data-informed ones. The capacity to complement experiential learning with formal analysis represents a differentiating quality that distinguishes every exceptional design manager and leader.

Refine Intuition

As a design manager and leader, you must refine your intuition as part of your developmental program to continuously increase its reliability and prevent significant biases in judgments that can negatively affect your decision-making performance.[clxxix] You can refine your intuition by developing self-awareness and self-management, two of the four dimensions of emotional intelligence.[clxxx] Within the context of decision-making, self-awareness enables your emotional consciousness, while self-management enables your emotional control. The following two homeostatic cognitive strategies can help you to regulate your emotions, refine your intuition, and enhance your decision-making performances.

Expand emotional knowledge. Examine your emotional states by identifying and describing their distinctive characteristics in relation to a specific context. Describe how a given emotion manifests itself differently under different circumstances, and then answer the following question to investigate the feelings[21] associated with that experience:

How experiencing [feeling 1] is similar to and different from feeling [feeling 2]?

This analysis increases your *emotional consciousness*, expanding your categorical knowledge of your emotional states surfacing their origin, manifestation, and impact in specific contexts.[clxxxi]

---

[21] Feelings are mental experiences of body states, which arise as the brain interprets emotions, which themselves are physical states arising from the body's responses to external stimuli. For example, I am threatened, experience fear, and feel horror.

Increase emotional sensibility. Investigate your experience during a determinate decisional moment by conducting an emotional audit. Describe how you experience a given condition in relation to a specific decision characterizing the different feelings associated with it. Ask yourself the following question to initiate this investigation:

How am I feeling now?

This exercise increases your *emotional control* by increasing your ability to articulate your different emotional states at the time when you experience them and, therefore, increasing your capacity to discern the four visceral influences that can affect your intuition and, more broadly, decisional processes.[clxxxii]

Assess Self-Awareness

In addition to the questionnaire introduced in Chapter 2 to assess your emotional intelligence, the questions in Table 9-4 can specifically evaluate your self-awareness level.[clxxxiii]

Table 9-4. Self-awareness questionnaire

Self-Awareness

- I have clear values that outline what is essential to me.
- I approach my life driven by my values.
- I can describe my ideal personal environment.
- I can describe my ideal professional environment.
- I can describe my personal goals.
- I can describe my professional goals.
- I understand what I find personally rewarding.
- I understand what I find professionally rewarding.
- I know what activities bring me fulfillment.
- I know what activities bring me joy.
- In any given situation, I can predict how I will behave.
- In any given situation, I can objectively assess my performance.
- In any given situation, I can see themes in my behavior.
- When I interact with people, I'm conscious of the impact of my actions.
- When I interact with people, I examine how they respond to my actions.

6-Point scale: Very untrue of me, Untrue of me, Somewhat untrue of me, Somewhat true of me, True of me, Very true of me.

Assessing your levels of self-awareness investigating similarities and differences between your past experiences and their underlying emotions and interrelated feelings can assist you to develop clarity concerning your present condition and inform your developmental program.

Scale Design Management and Leadership

The ultimate challenge for every design manager and leader is to scale design management and leadership. Scaling, per se, is never the objective; creating value at scale is always the intention. In this context, the managerial and leadership aspect of design only exists as a function of the individual contribution produced by each team member. Design management and leadership are unnecessary without the human-centered activities that characterize the problem-solving process.

As a consequence of this nonobvious relationship, scaling design as a human-centered problem-solving function translates into scaling design management and leadership. In some situations, the demand for design cascades from the C-level with a top-down mandate; under other unfavorable circumstances, that demand needs to be created with a bottom-up process.

Create the Demand for Design

When you are in charge of a team, irrespective of your current level of contribution, you need to ensure that the demand for design within your organization is always present as a prerequisite for scaling its managerial and leadership aspect. Without a request for human-centered problem-solving, what design management and leadership have to offer is irrelevant. If necessary, the following three questions can help you to investigate your current organizational context:

Is your organization already committed to

the creation of a design team?

the allocation of budget to the design team?

the use of the design team as a strategic partner?

Assuming that there is the need to create the demand for design, Eric Quint, former Chief Brand and Design Officer at 3M, developed a three-step approach (Figure 9-21) that you can use to inspire relevant stakeholders to embrace design as a strategic partner.[clxxxiv]

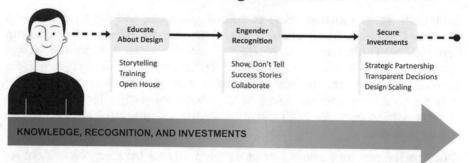

Create Design Demand

Figure 9-21. The three-step process for creating the design demand

Step 1: Educate about design. You introduce the design activities and tools that the organization can leverage to produce value for the users and sustainably propel its business model. During this initial moment, storytelling is crucial to connect design effort and business impact, demonstrating how design can derisk the exploration of new ideas and communicate concepts in ways that promote multidisciplinary collaboration. At this stage, you also need to share your vision for the design team articulating its future development within the current organizational structure.

Step 2: Engender recognition for design. You investigate the current organizational landscape to find projects through which the design team can prove its business-oriented effectiveness. During this moment, you need to focus on building trust, respect, and credibility, demonstrating the ability to collaborate, make judicious decisions, and achieve desirable results. This effort also includes creating spaces where design outcomes can be radiated and defining human-centered KPIs to complement existing business ones and measure the design impact effectively. At this stage, sharing success stories from other companies that have already decided to support human-centered problem-solving effectively is also beneficial.

Step 3: Secure investments in design. You discuss the financial and operational effort necessary to adequately support the human-centered problem-solving activities inside and outside the design team. During this moment, stakeholders voluntarily make the required changes to their processes and practices to include design as a strategic partner and benefit from its contribution. At this stage, stakeholders become ambassadors of human-centered problem-solving, enabling you to elevate the design function at the highest ranks within your organization.

Once the demand for design is present and your organization is committed to financially and operationally supporting the design function, you can begin to concentrate on scaling design leadership and management.

Quantitative and Qualitative Scale

Scaling the design function refers to adjusting the size and the type of design resources available in accordance with the current nature of that given demand.[clxxxv] The objective is to elevate creative collaboration and design operations at a level where strategic collaboration with design leaders becomes structural and the budget for hiring designers becomes available.[clxxxvi] In order to achieve and maintain a balance between design demand and supply, you have to consider both the quantitative and qualitative aspects of the scaling process.

Qualitative scaling: The effort to develop an adequate degree of competency sophistication to fulfill the needs of the design demand.[clxxxvii] Qualitative scaling aims to advance the role of the design function by elevating it from being a tactical supporter that solves predefined problems or implements predefined solutions to a strategic partner that frames challenges into problems and enables the business to explore new possible directions. Indirectly, qualitative scaling enables the participation of design managers and leaders in critical conversations that set the strategic direction of the organization and offer them an opportunity to provide a crucial contribution to it.

Quantitative scaling: The effort to develop the adequate amount and range of competencies necessary to fulfill the needs of the design demand.[clxxxviii] Quantitative scaling aims to hire an adequate number of in-house designers to reach a critical mass that allows the design function to counterbalance the social power projected by the other department within the company. Indirectly, quantitative scaling enables the presence of design managers and leaders in critical business areas that critically contribute to the strategic direction of the organization.

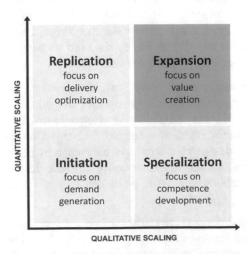

Figure 9-22. The quantitative and qualitative scaling objectives and activities

Figure 9-22 illustrates the scaling objectives and their related quantitative and qualitative scaling activities.[clxxxix]

Quantitative and qualitative scaling never happen in isolation, and you always need to balance purposefully. Realistically, you will often prioritize one activity over the other based on the current most pressing need of the team.

Quantitative scaling must be prioritized when the immediate need is to develop the current operational capacity of the design function, typically to increase the throughput at which recurring problems are solved: for instance, if the design team needs to increase its production speed against a deadline or a multidisciplinary group does not include a designer.

Qualitative scaling should be prioritized when the immediate need is to expand the operational capacity of the design function, typically to solve novel types of problems: for instance, if the team needs to include service design as part of its core activities.

Reflect on the Scaling State

Every company is different, and scaling design management and leadership is a wicked problem that cannot be solved with a predefined solution; it requires an exploratory and iterative conduct. Reflection is an integral part of this process. In order to investigate the impact of your work, there are six dimensions that you can use to examine the effectiveness of your current approach:

Strategy: Is the team or workgroup aligned with the purpose of the organization?

Structure: Is the team or workgroup adequate to fulfill the purpose of the organization?

Process: Is the flow of information and activities adequate to fulfill the purpose of the organization?

Infrastructure: Is the environment adequate to establish the desired beliefs and behaviors?

Incentives: Is the environment adequate to reward the desired beliefs and behaviors?

Development: Is the environment adequate to grow the desired beliefs and behaviors?

Adopting these six lenses, you can begin to diagnose the state of creative collaboration and design operations within your organization, assess the application of everything you have learned in this book, and identify eventual blockers requiring your attention. Your primary role as a design manager who manages like a leader is to shape the conditions that can unlock high levels of performance and value creation in your team or workgroup. Your group will continuously evolve along their professional journey, and your responsibility is to ensure that these conditions develop with them and continue to remain relevant.

Endnotes

i.   Kelley, Tom, and David Kelley. Creative Confidence: Unleashing the Creative Potential within Us All. William Collins, 2015.

ii. O'Donnell, Clifford R., and Lois Yamauchi. Culture & Context in Human Behavior Change: Theory, Research, and Applications. International Academic Publishers, 2005.

iii. Kelley, Harold H. "The Processes of Causal Attribution." American Psychologist, vol. 28, no. 2, 1973, pp. 107–128., doi:10.1037/h0034225.

iv. Kelley, Harold H. "The Processes of Causal Attribution." American Psychologist, vol. 28, no. 2, 1973, pp. 107–128., doi:10.1037/h0034225.

v.   O'Donnell, Clifford R., and Lois Yamauchi. Culture & Context in Human Behavior Change: Theory, Research, and Applications. International Academic Publishers, 2005.

vi. Bouncken, Ricarda. "Creativity in Cross-Cultural Innovation Teams: Diversity and Its Implications for Leadership." Milieus of Creativity, 2009, pp. 189–200., doi:10.1007/978-1-4020-9877-2_10.

vii. Bouncken, Ricarda. "Creativity in Cross-Cultural Innovation Teams: Diversity and Its Implications for Leadership." Milieus of Creativity, 2009, pp. 189–200., doi:10.10077/978-1-4020-9877-2_10.

viii. Lotto, Beau. Deviate: The Creative Power of Transforming Your Perception. Weidenfeld & Nicolson, 2018.

ix. Kurtzberg, Terri R. "Feeling Creative, Being Creative: An Empirical Study of Diversity and Creativity in Teams." Creativity Research Journal, vol. 17, no. 1, 2005, pp. 51–65., doi:10.1207/s15326934crj1701_5.

x.   VandenBos, Gary R. APA Dictionary of Psychology. American Psychological Association, 2015.

xi.   Jehn, Karen A., et al. "Why Differences Make a Difference: A Field Study of Diversity, Conflict, and Performance in Workgroups." Administrative Science Quarterly, vol. 44, no. 4, 1999, p. 741., doi:10.2307/2667054.

xii. Chang, Artemis, and Prashant Bordia. "A Multidimensional Approach to the Group Cohesion-Group Performance Relationship." Small Group Research, vol. 32, no. 4, 2001, pp. 379–405., doi:10.1177/104649640103200401.

xiii. Bradley, Bret H., et al. "Reaping the Benefits of Task Conflict in Teams: The Critical Role of Team Psychological Safety Climate." Journal of Applied Psychology, vol. 97, no. 1, 2012, pp. 151–158., doi:10.1037/a0024200.

xiv. Jehn, Karen A., et al. "Why Differences Make a Difference: A Field Study of Diversity, Conflict, and Performance in Workgroups." Administrative Science Quarterly, vol. 44, no. 4, 1999, p. 741., doi:10.2307/2667054.

xv.   Beal, Daniel J., et al. "Cohesion and Performance in Groups: A Meta-Analytic Clarification of Construct Relations." Journal of Applied Psychology, vol. 88, no. 6, 2003, pp. 989–1004., doi:10.1037/0021-9010.88.6.989.

xvii. Beal, Daniel J., et al. "Cohesion and Performance in Groups: A Meta-Analytic Clarification of Construct Relations." Journal of Applied Psychology, vol. 88, no. 6, 2003, pp. 989–1004., doi:10.1037/0021-9010.88.6.989.

xvii. Gordon, Ilanit, et al. "Physiological and Behavioral Synchrony Predict Group Cohesion and Performance." Scientific Reports, vol. 10, no. 1, 2020, doi:10.1038/s41598-020-65670-1.

xviii. Barile, Sergio, et al. "Modelling and Measuring Group Cohesiveness with Consonance: Intertwining the Sociometric Test with the Picture Apperception Value Test." Systems Research and Behavioral Science, vol. 35, no. 1, 2016, pp. 1–21., doi:10.1002/sres.2418.

xix. Bouncken, Ricarda. "Creativity in Cross-Cultural Innovation Teams: Diversity and Its Implications for Leadership." Milieus of Creativity, 2009, pp. 189–200., doi:10.1007/978-1-4020-9877-2_10.

xx.   Chang, Artemis, and Prashant Bordia. "A Multidimensional Approach to the Group Cohesion-Group Performance Relationship." Small Group Research, vol. 32, no. 4, 2001, pp. 379–405., doi:10.1177/104649640103200401.

xxi. Gordon, Ilanit, et al. "Physiological and Behavioral Synchrony Predict Group Cohesion and Performance." Scientific Reports, vol. 10, no. 1, 2020, doi:10.1038/s41598-020-65670-1.

xxii. Salas, Eduardo, et al. "Measuring Team Cohesion: Observations from the Science." Human Factors: The Journal of the Human Factors and Ergonomics Society, vol. 57, no. 3, 2015, pp. 365–374., doi:10.1177/0018720815578267.

xxiii. Hellriegel, Don, and John W. Slocum. Organizational Behavior. South-Western College Pub, 2011.

xxiv. Hoehl, Stefanie, et al. "Interactional Synchrony: Signals, Mechanisms and Benefits." Social Cognitive and Affective Neuroscience, vol. 16, no. 1-2, 2020, pp. 5–18., doi:10.1093/scan/nsaa024.

xxv. Beal, Daniel J., et al. "Cohesion and Performance in Groups: A Meta-Analytic Clarification of Construct Relations." Journal of Applied Psychology, vol. 88, no. 6, 2003, pp. 989–1004., doi:10.1037/0021-9010.88.6.989.

xxvi. Beal, Daniel J., et al. "Cohesion and Performance in Groups: A Meta-Analytic Clarification of Construct Relations." Journal of Applied Psychology, vol. 88, no. 6, 2003, pp. 989–1004., doi:10.1037/0021-9010.88.6.989.

xxvii. Bouncken, Ricarda. "Creativity in Cross-Cultural Innovation Teams: Diversity and Its Implications for Leadership." Milieus of Creativity, 2009, pp. 189–200., doi:10.1007/978-1-4020-9877-2_10.

xxviii. Baron, Robert A., and Nyla R. Branscombe. Social Psychology. Pearson, 2017.

xxix. Bettenhausen, Kenneth L., and J. Keith Murnighan. "The Development of an Intragroup Norm and the Effects of Interpersonal and Structural Challenges." Administrative Science Quarterly, vol. 36, no. 1, 1991, p. 20., doi:10.2307/2393428

xxx. Greifeneder, Rainer, et al. Social Cognition: How Individuals Construct Social Reality. Routledge, 2018.

xxxi. Fiske, Susan T., et al., editors. The Handbook of Social Psychology. Wiley, 2010.

xxxii. Fiske, Susan T., et al., editors. The Handbook of Social Psychology. Wiley, 2010.

xxxiii. Fiske, Susan T., et al., editors. The Handbook of Social Psychology. Wiley, 2010.

xxxiv. Aronson, Elliot, et al. Social Psychology. Pearson, 2016.

xxxv. O'Keefe, Daniel J. Persuasion: Theory and Research. SAGE, 2016.

xxxvi. Greifeneder, Rainer, et al. Social Cognition: How Individuals Construct Social Reality. Routledge, 2018.

xxxvii. Fiske, Susan T., et al., editors. The Handbook of Social Psychology. Wiley, 2010.

xxxviii. Fiske, Susan T., et al., editors. The Handbook of Social Psychology. Wiley, 2010.

xxxix. Frazier, M. Lance, et al. "Psychological Safety: A Meta-Analytic Review and Extension." Personnel Psychology, vol. 70, no. 1, 2016, pp. 113–165., doi:10.1111/peps.12183.

xl. Fiske, Susan T., et al., editors. The Handbook of Social Psychology. Wiley, 2010.

xli. Kaymaza, Kurtuluş. "The Interactive Effect of Feedback Sign and Task Type on Motivation and Performance Kluger and Van-Dijk - PsycEXTRA Dataset - 2005." Business and Economics Research Journal, vol. 2, no. 4, 2011, pp. 115–134., doi:1309-2448.

xlii. Gordon, Ilanit, et al. "Physiological and Behavioral Synchrony Predict Group Cohesion and Performance." Scientific Reports, vol. 10, no. 1, 2020, doi:10.1038/s41598-020-65670-1.

xliii. Kerrin, Máire, and Nick Oliver. "Collective and Individual Improvement Activities: The Role of Reward Systems." Personnel Review, vol. 31, no. 3, 2002, pp. 320–337., doi:10.1108/00483480210422732.

xliv. Picchi, Andrea. "The 3 Dimensions of Design: A Model to Scale the Human-Centered Problem-Solving Practice across the Organization." ResearchGate, Jan. 2017, www.researchgate.net/publication/330634631_The_3_Dimensions_of_Design_A_Model_to_Scale_the_Human-Centered_Problem-Solving_practice_across_the_Organization.

xlv. Diller, Steve, et al. Blind Spot: Illuminating the Hidden Value of Business. Two Waves Books, 2016.

xlvi. Diller, Steve, et al. Blind Spot: Illuminating the Hidden Value of Business. Two Waves Books, 2016.

xlvii. Diller, Steve, et al. Blind Spot: Illuminating the Hidden Value of Business. Two Waves Books, 2016.

xlviii. Diller, Steve, et al. Blind Spot: Illuminating the Hidden Value of Business. Two Waves Books, 2016.

xlix. Diller, Steve, et al. Blind Spot: Illuminating the Hidden Value of Business. Two Waves Books, 2016.

l. Diller, Steve, et al. Blind Spot: Illuminating the Hidden Value of Business. Two Waves Books, 2016.

li. Susan Fournier, Susan, et al., editors. Consumer-Brand Relationships: Insights for Theory and Practice. Routledge, 2012.

lii. Susan Fournier, Susan, et al., editors. Consumer-Brand Relationships: Insights for Theory and Practice. Routledge, 2012.

liii. Baron, Robert A., and Nyla R. Branscombe. Social Psychology. Pearson, 2017.

liv. Kahn, William A. "Psychological Conditions of Personal Engagement and Disengagement at Work." Academy of Management Journal, vol. 33, no. 4, 1990, pp. 692–724., doi:10.5465/256287.

lv. Picchi, Andrea. "The 3 Dimensions of Design: A Model to Scale the Human-Centered Problem-Solving Practice across the Organization." ResearchGate, Jan. 2017, www.researchgate.net/publication/330634631_The_3_Dimensions_of_Design_A_Model_to_Scale_the_Human-Centered_Problem-Solving_practice_across_the_Organization.

lvi. Perry, Sara Jansen, et al. "The Downside of Goal-Focused Leadership: The Role of Personality in Subordinate Exhaustion." Journal of Applied Psychology, vol. 95, no. 6, 2010, pp. 1145–1153., doi:10.1037/a0020538.

lvii. Perry, Sara Jansen, et al. "The Downside of Goal-Focused Leadership: The Role of Personality in Subordinate Exhaustion." Journal of Applied Psychology, vol. 95, no. 6, 2010, pp. 1145–1153., doi:10.1037/a0020538.

lviii. Kahn, William A. "Psychological Conditions of Personal Engagement and Disengagement at Work." Academy of Management Journal, vol. 33, no. 4, 1990, pp. 692–724., doi:10.5465/256287.

lix. Schaufeli, Wilmar B., and Arnold B. Bakker. "Job Demands, Job Resources, and Their Relationship with Burnout and Engagement: A Multi-Sample Study." Journal of Organizational Behavior, vol. 25, no. 3, 2004, pp. 293–315., doi:10.1002/job.248.

lx. Schaufeli, Wilmar B., and Arnold B. Bakker. "Job Demands, Job Resources, and Their Relationship with Burnout and Engagement: A Multi-Sample Study." Journal of Organizational Behavior, vol. 25, no. 3, 2004, pp. 293–315., doi:10.1002/job.248.

lxi. Saks, Alan M. "Antecedents and Consequences of Employee Engagement." Journal of Managerial Psychology, vol. 21, no. 7, 2006, pp. 600–619., doi:10.1108/026839406 10690169.

lxii. Saks, Alan M. "Antecedents and Consequences of Employee Engagement." Journal of Managerial Psychology, vol. 21, no. 7, 2006, pp. 600–619., doi:10.1108/026839406 10690169.

lxiii. Kahn, William A. "Psychological Conditions of Personal Engagement and Disengagement at Work." Academy of Management Journal, vol. 33, no. 4, 1990, pp. 692–724., doi:10.5465/256287.

lxiv. Kahn, William A. "Psychological Conditions of Personal Engagement and Disengagement at Work." Academy of Management Journal, vol. 33, no. 4, 1990, pp. 692–724., doi:10.5465/256287.

lxv. Shuck, Brad, et al. "Employee Engagement: An Examination of Antecedent and Outcome Variables." Human Resource Development International, vol. 14, no. 4, 2011, pp. 427–445., doi:10.1080/13678868.2011.601587.

lxvi. Saks, Alan M. "Antecedents and Consequences of Employee Engagement." Journal of Managerial Psychology, vol. 21, no. 7, 2006, pp. 600–619., doi:10.1108/0268394061 0690169.

lxvii. Macey, William H., and Benjamin Schneider. "The Meaning of Employee Engagement." Industrial and Organizational Psychology, vol. 1, no. 1, 2008, pp. 3–30., doi:10.1111/j.1754-9434.2007.0002.x.

lxviii.   Macey, William H., and Benjamin Schneider. "The Meaning of Employee Engagement." Industrial and Organizational Psychology, vol. 1, no. 1, 2008, pp. 3–30., doi:10.1111/j.1754-9434.2007.0002.x.

lxix.   REF: Fuster Joaquín M. The Prefrontal Cortex. Elsevier, AP, 2011.

lxx.   REF: Fuster Joaquín M. The Prefrontal Cortex. Elsevier, AP, 2011.

lxxi.   Zenger, John H., and Joseph R. Folkman. The Extraordinary Leader. McGraw-Hill, 2009.

lxxii.   REF: Fuster Joaquín M. The Prefrontal Cortex. Elsevier, AP, 2011.

lxxiii.   Beal, Daniel J., et al. "Cohesion and Performance in Groups: A Meta-Analytic Clarification of Construct Relations." Journal of Applied Psychology, vol. 88, no. 6, 2003, pp. 989–1004., doi:10.1037/0021-9010.88.6.989.

lxxiv.   Bipp, Tanja, and Ad Kleingeld. "Goal-Setting in Practice: The Effects of Personality and Perceptions of the Goal-Setting Process on Job Satisfaction and Goal Commitment." Personnel Review, vol. 40, no. 3, 2011, pp. 306–323., doi:10.1108/00483481111118630.

lxxv.   Locke, Edwin A., and Gary P. Latham. "New Directions in Goal-Setting Theory." Current Directions in Psychological Science, vol. 15, no. 5, 2006, pp. 265–268., doi:10.1111/j.1467-8721.2006.00449.x.

lxxvi.   Dalton, Amy N., and Stephen A. Spiller. "Too Much of a Good Thing: The Benefits of Implementation Intentions Depend on the Number of Goals." Journal of Consumer Research, vol. 39, no. 3, 2012, pp. 600–614., doi:10.1086/664500.

lxxvii.   Tosi, Henry L., et al. "A Theory of Goal Setting and Task Performance." The Academy of Management Review, vol. 16, no. 2, 1991, p. 480., doi:10.2307/258875.

lxxviii.   Frazier, M. Lance, et al. "Psychological Safety: A Meta-Analytic Review and Extension." Personnel Psychology, vol. 70, no. 1, 2016, pp. 113–165., doi:10.1111/peps.12183.

lxxix.   Ogbeiwi, Osahon. "Why Written Objectives Need to Be Really SMART." British Journal of Healthcare Management, vol. 23, no. 7, 2017, pp. 324–336., doi:10.12968/bjhc.2017.23.7.324.

lxxx.   Zenger, John H., and Joseph R. Folkman. The Extraordinary Leader. McGraw-Hill, 2009.

lxxxi.   Zenger, John H., and Joseph R. Folkman. The Extraordinary Leader. McGraw-Hill, 2009.

lxxxii. Selye, Hans. The Stress of Life. McGraw-Hill, 1984.

lxxxiii. Simmons, Bret L., and Debra L. Nelson. "Eustress at Work: The Relationship between Hope and Health in Hospital Nurses." Health Care Management Review, vol. 26, no. 4, 2001, pp. 7–18., doi:10.1097/00004010-200110000-00002.

lxxxiv. Freudenberger, Herbert J. "Staff Burn-Out." Journal of Social Issues, vol. 30, no. 1, 1974, pp. 159–165., doi:10.1111/j.1540-4560.1974.tb00706.x.

lxxxv.   American Psychiatric Association. Diagnostic and Statistical Manual of Mental Disorders. American Psychiatric Association, 2013.

lxxxvi. Selye, Hans. The Stress of Life. McGraw-Hill, 1984.

lxxxvii. Edmondson, Amy C. The Fearless Organization: Creating Psychological Safety in the Workplace for Learning, Innovation, and Growth. Wiley, 2018.

lxxxviii. Edmondson, Amy C. The Fearless Organization: Creating Psychological Safety in the Workplace for Learning, Innovation, and Growth. Wiley, 2018.

lxxxix. Edmondson, Amy C. The Fearless Organization: Creating Psychological Safety in the Workplace for Learning, Innovation, and Growth. Wiley, 2018.

xc.   Edmondson, Amy C. The Fearless Organization: Creating Psychological Safety in the Workplace for Learning, Innovation, and Growth. Wiley, 2018.

xci.   Peters, Achim, et al. "Uncertainty and Stress: Why It Causes Diseases and How It Is Mastered by the Brain." Progress in Neurobiology, vol. 156, 2017, pp. 164–188., doi:10.1016/j.pneurobio.2017.05.004.

xcii.   Ajzen, Icek. "The Theory of Planned Behavior." Organizational Behavior and Human Decision Processes, vol. 50, no. 2, 1991, pp. 179–211., doi:10.1016/0749-5978(91)90020-t.

xciii. Byron, Kristin, et al. "The Relationship between Stressors and Creativity: A Meta-Analysis Examining Competing Theoretical Models." Journal of Applied Psychology, vol. 95, no. 1, 2010, pp. 201–212., doi:10.1037/a0017868.

xciv. Fink, George. Stress: Concepts, Cognition, Emotion and Behavior. Academic Press, 2016.

xcv. Rock, David, and Linda J. Page. Coaching with the Brain in Mind: Foundations for Practice. Wiley, 2009.

xcvi. Fink, George. Stress: Concepts, Cognition, Emotion and Behavior. Academic Press, 2016.

xcvii. Fink, George. Stress: Concepts, Cognition, Emotion and Behavior. Academic Press, 2016.

xcviii. Fink, George. Stress: Concepts, Cognition, Emotion and Behavior. Academic Press, 2016.

xcix. Fink, George. Stress: Concepts, Cognition, Emotion and Behavior. Academic Press, 2016.

c. Fink, George. Stress: Concepts, Cognition, Emotion and Behavior. Academic Press, 2016.

ci. Fink, George. Stress: Concepts, Cognition, Emotion and Behavior. Academic Press, 2016.

cii. Bennett, Edward L., et al. "Chemical and Anatomical Plasticity of Brain: Changes in Brain through Experience." Science, vol. 146, no. 3644, 1964, pp. 610–619., doi:10.1126/science.146.3644.610.

cii. Bennett, Edward L., et al. "Chemical and Anatomical Plasticity of Brain: Changes in Brain through Experience." Science, vol. 146, no. 3644, 1964, pp. 610–619., doi:10.1126/science.146.3644.610.

civ. Rampon, Claire, et al. "Effects of Environmental Enrichment on Gene Expression in the Brain." Proceedings of the National Academy of Sciences, vol. 97, no. 23, 2000, pp. 12880–12884, doi:10.1073/pnas.97.23.12880.

cv. Rampon, Claire, et al. "Effects of Environmental Enrichment on Gene Expression in the Brain." Proceedings of the National Academy of Sciences, vol. 97, no. 23, 2000, pp. 12880–12884, doi:10.1073/pnas.97.23.12880.

cvi. Levi, Lennart. Stress and Distress in Response to Psychosocial Stimuli: Laboratory and Real-Life Studies on Sympatho-Adreno-Medullary and Related Reactions. Pergamon Press, 1972.

cvii. Levi, Lennart. Stress and Distress in Response to Psychosocial Stimuli: Laboratory and Real-Life Studies on Sympatho-Adreno-Medullary and Related Reactions. Pergamon Press, 1972.

cviii. Levi, Lennart. Stress and Distress in Response to Psychosocial Stimuli: Laboratory and Real-Life Studies on Sympatho-Adreno-Medullary and Related Reactions. Pergamon Press, 1972.

cix. Levi, Lennart. Stress and Distress in Response to Psychosocial Stimuli: Laboratory and Real-Life Studies on Sympatho-Adreno-Medullary and Related Reactions. Pergamon Press, 1972.

cx. Vygotskiĭ Lev S., and Michael Cole. Mind in Society: The Development of Higher Psychological Processes. Harvard University Press, 1978.

cxi. Vygotskiĭ Lev S., and Michael Cole. Mind in Society: The Development of Higher Psychological Processes. Harvard University Press, 1978.

cxii. Vygotskiĭ Lev S., and Michael Cole. Mind in Society: The Development of Higher Psychological Processes. Harvard University Press, 1978.

cxiii. Wood, David, et al. "The Role of Tutoring in Problem-Solving." Journal of Child Psychology and Psychiatry, vol. 17, no. 2, 1976, pp. 89–100., doi:10.1111/j.1469-7610.1976.tb00381.x.

cxiv. Wood, David, et al. "The Role of Tutoring in Problem-Solving." Journal of Child Psychology and Psychiatry, vol. 17, no. 2, 1976, pp. 89–100., doi:10.1111/j.1469-7610.1976.tb00381.x.

cxv. Wood, David, and David Middleton. "A Study of Assisted Problem-Solving." British Journal of Psychology, vol. 66, no. 2, 1975, pp. 181–191., doi:10.1111/j.2044-8295.1975.tb01454.x.

cxvi. Wood, David, and David Middleton. "A Study of Assisted Problem-Solving." British Journal of Psychology, vol. 66, no. 2, 1975, pp. 181–191., doi:10.1111/j.2044-8295.1975.tb01454.x.

cxvii.   Wood, David, et al. "The Role of Tutoring in Problem-Solving." Journal of Child Psychology and Psychiatry, vol. 17, no. 2, 1976, pp. 89–100., doi:10.1111/j.1469-7610.1976.tb00381.x.

cxviii.   Martin, Roger L., and Alan G. Lafley. Playing to Win: How Strategy Really Works. Harvard Business Review Press, 2013.

cxix.   Aristotelēs. Aristotle's Prior and Posterior Analytics. Edited by William D. Ross, Oxford University Press, 2000.

cxx.   Aristotelēs. Aristotle's Prior and Posterior Analytics. Edited by William D. Ross, Oxford University Press, 2000.

cxxi.   Aristotelēs. Aristotle's Prior and Posterior Analytics. Edited by William D. Ross, Oxford University Press, 2000.

cxxii.   Aristotelēs. Aristotle's Prior and Posterior Analytics. Edited by William D. Ross, Oxford University Press, 2000.

cxxiii.   Martin, Roger L., and Alan G. Lafley. Playing to Win: How Strategy Really Works. Harvard Business Review Press, 2013.

cxxiv.   Martin, Roger L., and Alan G. Lafley. Playing to Win: How Strategy Really Works. Harvard Business Review Press, 2013.

cxxv.   Martin, Roger L., and Alan G. Lafley. Playing to Win: How Strategy Really Works. Harvard Business Review Press, 2013.

cxxvi.   Martin, Roger L., and Alan G. Lafley. Playing to Win: How Strategy Really Works. Harvard Business Review Press, 2013.

cxxvii.   Martin, Roger L., and Alan G. Lafley. Playing to Win: How Strategy Really Works. Harvard Business Review Press, 2013.

cxxviii. Martin, Roger L., and Alan G. Lafley. Playing to Win: How Strategy Really Works. Harvard Business Review Press, 2013.

cxxix.   Martin, Roger L., and Alan G. Lafley. Playing to Win: How Strategy Really Works. Harvard Business Review Press, 2013.

cxxx. Martin, Roger L., and Alan G. Lafley. Playing to Win: How Strategy Really Works. Harvard Business Review Press, 2013.

cxxxi. Fink, George. Stress: Concepts, Cognition, Emotion and Behavior. Academic Press, 2016.

cxxxii. Schön Donald. The Reflective Practitioner: How Professionals Think In Action. Basic Books, 1983.

cxxxiii. Kauffeld, Simone, and Nale Lehmann-Willenbrock. "Meetings Matter Effects of Team Meetings on Team and Organizational Success." Small Group Research, vol. 43, no. 2, 2011, pp. 130–158., doi:10.1177/1046496411429599.

cxxxiv. Kauffeld, Simone, and Nale Lehmann-Willenbrock. "Meetings Matter Effects of Team Meetings on Team and Organizational Success." Small Group Research, vol. 43, no. 2, 2011, pp. 130–158., doi:10.1177/1046496411429599.

cxxxv. Luong, Alexandra, and Steven G. Rogelberg. "Meetings and More Meetings: The Relationship between Meeting Load and the Daily Well-Being of Employees." Group Dynamics: Theory, Research, and Practice, vol. 9, no. 1, 2005, pp. 58–67., doi:10.1037/1089-2699.9.1.58.

cxxxvi. Liden, Robert C., et al. "Social Loafing: A Field Investigation." Journal of Management, vol. 30, no. 2, 2004, pp. 285–304., doi:10.1016/j.jm.2003.02.002.

cxxxvii. Karau, Steven J., and Jason W. Hart. "Group Cohesiveness and Social Loafing: Effects of a Social Interaction Manipulation on Individual Motivation within Groups." Group Dynamics: Theory, Research, and practice, vol. 2, no. 3, 1998, pp. 185–191., doi:10.1037/1089-2699.2.3.185.

cxxxviii. Cozolino, Louis J. The Neuroscience of Human Relationships: Attachment and the Developing Social Brain. W.W. Norton & Company, 2014.

cxxxix. John, Karen A., et al. "Why Differences Make a Difference: A Field Study of Diversity, Conflict, and Performance in Workgroups." Administrative Science Quarterly, vol. 44, no. 4, 1999, p. 741., doi:10.2307/2667054.

cxl. Allen, Joseph A., et al. "Let's Get This Meeting Started: Meeting Lateness and Actual Meeting Outcomes." Journal of Organizational Behavior, vol. 39, no. 8, 2018, pp. 1008–1021., doi:10.1002/job.2276.

cxli.   Rogelberg, Steven G., et al. "Lateness to Meetings: Examination of an Unexplored Temporal Phenomenon." European Journal of Work and Organizational Psychology, vol. 23, no. 3, 2013, pp. 323–341., doi:10.1080/1359432x.2012.745988.

cxlii. Newport, Cal. Deep Work: Rules for Focused Success in a Distracted World. Piatkus, 2016.

cxliii. Lotto, Beau. Deviate: The Creative Power of Transforming Your Perception. Weidenfeld & Nicolson, 2018.

cxliv. Crawford, Dylan W., et al. "Intelligence Demands Flexibility: Individual Differences in Attentional Disengagement Strongly Predict the General Cognitive Ability." Learning and Motivation, vol. 71, 2020, p. 101657., doi:10.1016/j.lmot.2020.101657.

cxlv.   Zenger, John H., and Joseph R. Folkman. The Extraordinary Leader. McGraw-Hill, 2009.

cxlvi. Lotto, Beau. Deviate: The Creative Power of Transforming Your Perception. Weidenfeld & Nicolson, 2018.

cxlvii. Lotto, Beau. Deviate: The Creative Power of Transforming Your Perception. Weidenfeld & Nicolson, 2018.

cxlviii. Sandi, Carmen. "Stress and Cognition." Wiley Interdisciplinary Reviews: Cognitive Science, vol. 4, no. 3, 2013, pp. 245–261., doi:10.1002/wcs.1222.

cxlix. Fink, George. Stress: Concepts, Cognition, Emotion and Behavior. Academic Press, 2016.

cl.   Wass, Rob, et al. "Scaffolding Critical Thinking in the Zone of Proximal Development." Higher Education Research & Development, vol. 30, no. 3, 2011, pp. 317–328., doi:10.1080/07294360.2010.489237.

cli.   Wass, Rob, et al. "Scaffolding Critical Thinking in the Zone of Proximal Development." Higher Education Research & Development, vol. 30, no. 3, 2011, pp. 317–328., doi:10.1080/07294360.2010.489237.

clii.   Nussbaum, Michael. "How Introverts versus Extroverts Approach Small-Group Argumentative Discussions." The Elementary School Journal, vol. 102, no. 3, Jan. 2002, pp. 183–197., doi:10.1086/499699.

cliii. Friedman, Alan F. "Extraversion and Introversion." Journal of Personality Assessment, vol. 46, no. 2, 1982, pp. 185–187., doi:10.1207/s15327752jpa4602_17.

cliv.   Corr, Philip J., and Gerald Matthews. The Cambridge Handbook of Personality Psychology. Cambridge University Press, 2009.

clv.   Corr, Philip J., and Gerald Matthews. The Cambridge Handbook of Personality Psychology. Cambridge University Press, 2009.

clvi.   Corr, Philip J., and Gerald Matthews. The Cambridge Handbook of Personality Psychology. Cambridge University Press, 2009.

clvii. Wass, Rob, et al. "Scaffolding Critical Thinking in the Zone of Proximal Development." Higher Education Research & Development, vol. 30, no. 3, 2011, pp. 317–328., doi:10.1080/07294360.2010.489237.

clviii. Gazzaniga, Michael S., et al. Cognitive Neuroscience: The Biology of the Mind. Norton, 2014.

clix.   Sandi, Carmen. "Stress and Cognition." Wiley Interdisciplinary Reviews: Cognitive Science, vol. 4, no. 3, 2013, pp. 245–261., doi:10.1002/wcs.1222.

clx.   Purves, Dale, and Beau Lotto. Why We See What We Do: An Empirical Theory of Vision. Sinauer Associates, 2003.

clxi.   Saks, Alan M. "Antecedents and Consequences of Employee Engagement." Journal of Managerial Psychology, vol. 21, no. 7, 2006, pp. 600–619., doi:10.1108/02683940610690169.

clxii. Sandi, Carmen. "Stress and Cognition." Wiley Interdisciplinary Reviews: Cognitive Science, vol. 4, no. 3, 2013, pp. 245–261., doi:10.1002/wcs.1222.

clxiii. Zenger, John H., and Joseph R. Folkman. The Extraordinary Leader. McGraw-Hill, 2009.

clxiv. Kong, Fang, et al. "How and When Group Cohesion Influences Employee Voice." Journal of Managerial Psychology, vol. 35, no. 3, 2020, pp. 142–154., doi:10.1108/jmp-04-2018-0161.

clxv.   Sinclair, Marta, and Neal M. Ashkanasy. "Intuition: Myth or a Decision-Making Tool?" Management Learning, vol. 36, no. 3, 2005, pp. 353–370., doi:10.1177/1350507605055351.

clxvi. Kolb, David A. Experimental Learning: Experience as the Source of Learning and Development. Prentice-Hall, 1984.

clxvii.   Kolb, David A. Experimental Learning: Experience as the Source of Learning and Development. Prentice-Hall, 1984.

clxviii. Lerner, Jennifer S., et al. "Emotion and Decision Making." Annual Review of Psychology, vol. 66, no. 1, 2015, pp. 799–823., doi:10.1146/annurev-psych-010213-115043.

clxix.   Lerner, Jennifer S., et al. "Emotion and Decision Making." Annual Review of Psychology, vol. 66, no. 1, 2015, pp. 799–823., doi:10.1146/annurev-psych-010213-115043.

clxx.   Lerner, Jennifer S., et al. "Emotion and Decision Making." Annual Review of Psychology, vol. 66, no. 1, 2015, pp. 799–823., doi:10.1146/annurev-psych-010213-115043.

clxxi.   Lerner, Jennifer S., et al. "Emotion and Decision Making." Annual Review of Psychology, vol. 66, no. 1, 2015, pp. 799–823., doi:10.1146/annurev-psych-010213-115043.

clxxii.   Lerner, Jennifer S., et al. "Emotion and Decision Making." Annual Review of Psychology, vol. 66, no. 1, 2015, pp. 799–823., doi:10.1146/annurev-psych-010213-115043.

clxxiii. Lerner, Jennifer S., et al. "Emotion and Decision Making." Annual Review of Psychology, vol. 66, no. 1, 2015, pp. 799–823., doi:10.1146/annurev-psych-010213-115043.

clxxiv.   Lerner, Jennifer S., et al. "Emotion and Decision Making." Annual Review of Psychology, vol. 66, no. 1, 2015, pp. 799–823., doi:10.1146/annurev-psych-010213-115043.

clxxv.   Klein, Gray. The Power of Intuition How to Use Your Gut Feelings to Make Better Decisions at Work. Doubleday, 2004.

clxxvi.   Lerner, Jennifer S., et al. "Emotion and Decision Making." Annual Review of Psychology, vol. 66, no. 1, 2015, pp. 799–823., doi:10.1146/annurev-psych-010213-115043.

clxxvii. Klein, Gray. The Power of Intuition How to Use Your Gut Feelings to Make Better Decisions at Work. Doubleday, 2004.

clxxviii. Klein, Gray. The Power of Intuition How to Use Your Gut Feelings to Make Better Decisions at Work. Doubleday, 2004.

clxxix.   Barrett, Lisa Feldman, et al. "Knowing What You're Feeling and Knowing What to Do About It: Mapping the Relation Between Emotion Differentiation and Emotion Regulation." Cognition & Emotion, vol. 15, no. 6, 2001, pp. 713–724., doi:10.1080/02699930143000239.

clxxx.   Tonetto, Leandro M., and Pia Tamminen. "Understanding the Role of Intuition in Decision-Making When Designing for Experiences: Contributions from Cognitive Psychology." Theoretical Issues in Ergonomics Science, vol. 16, no. 6, 2015, pp. 631–642., doi:10.108 0/1463922x.2015.1089019.

clxxxi.   Duval, Shelley, and Robert A. Wicklund. A Theory of Objective Self-Awareness. Academic Press, 1972.

clxxxii.   Duval, Shelley, and Robert A. Wicklund. A Theory of Objective Self-Awareness. Academic Press, 1972.

clxxxiii.   Eurich, Tasha. Insight How to Succeed by Seeing Yourself Clearly. Pan Books, 2018.

clxxxiv.   Quint, Eric, et al. Design Leadership Ignited: Elevating Design at Scale. Stanford Business Books, an Imprint of Stanford University Press, 2022.

clxxxv.   Quint, Eric, et al. Design Leadership Ignited: Elevating Design at Scale. Stanford Business Books, an Imprint of Stanford University Press, 2022.

clxxxvi.   Quint, Eric, et al. Design Leadership Ignited: Elevating Design at Scale. Stanford Business Books, an Imprint of Stanford University Press, 2022.

clxxxvii.   Quint, Eric, et al. Design Leadership Ignited: Elevating Design at Scale. Stanford Business Books, an Imprint of Stanford University Press, 2022.

clxxxviii. Quint, Eric, et al. Design Leadership Ignited: Elevating Design at Scale. Stanford Business Books, an Imprint of Stanford University Press, 2022.

clxxxix.   Quint, Eric, et al. Design Leadership Ignited: Elevating Design at Scale. Stanford Business Books, an Imprint of Stanford University Press, 2022.

Index

© Andrea Picchi 2022
A. Picchi, *Design Management*, https://doi.org/10.1007/978-1-4842-6954-1

Printed in the United States
by Baker & Taylor Publisher Services